Spoil Island

Toposophia
Sustainability, Dwelling, Design

Toposophia is a book series dedicated to the interdisciplinary and transdisciplinary study of place. Authors in the series attempt to engage a geographical turn in their research, emphasizing the spatial component, as well as the philosophical turn, raising questions both reflectively and critically.

Titles in the Series

Reading the Islamic City: Discursive Practices and Legal Judgment, by Akel Isma'il Kahera
Mysticism and Architecture: Wittgenstein and the Meanings of the Palais Stonborough, by Roger Paden
When France Was King of Cartography: The Patronage and Production of Maps in Early Modern France, by Christine Marie Petto
Environmental Dilemmas: Ethical Decision Making, by Robert Mugerauer and Lynne Manzo
The Timespace of Human Activity: On Performance, Society, and History as Indeterminate Teleological Events, by Theodore R. Schatzki
Metamorphoses of the Zoo: Animal Encounter after Noah, edited by Ralph R. Acampora
Reading the Islamic City Discursive Practices and Legal Judgment, by Akel Isma'il Kahera

Spoil Island

Reading the Makeshift Archipelago

Charlie Hailey

LEXINGTON BOOKS
Lanham • Boulder • New York • Toronto • Plymouth, UK

Published by Lexington Books
A wholly owned subsidiary of The Rowman & Littlefield Publishing Group, Inc.
4501 Forbes Boulevard, Suite 200, Lanham, Maryland 20706
www.rowman.com

10 Thornbury Road, Plymouth PL6 7PP, United Kingdom

British Library Cataloguing in Publication Information Available

Library of Congress Cataloging-in-Publication Data

Library of Congress Cataloging-in-Publication Data Is Available

ISBN 978-0-7391-7306-0 (cloth : alk. paper)
ISBN 978-0-7391-8534-6 (pbk. : alk. paper)
ISBN 978-0-7391-7307-7 (electronic)

∞™ The paper used in this publication meets the minimum requirements of American National Standard for Information Sciences Permanence of Paper for Printed Library Materials, ANSI/NISO Z39.48-1992.

Printed in the United States of America

In memoriam
Diana Bitz

Contents

List of Illustrations

Illustrations have been derived from nautical charts (National Oceanic and Atmospheric Administration), topographic maps and aerial photographs (United States Geologic Survey).

Preface

October 16, 2010

> What are the roots that clutch, what branches grow
> Out of this stony rubbish? [1]

Bunnies live on spoil islands. My son and daughter have returned to our camp breathlessly to share their discovery. It is an unexpected find within a landscape that is at once ordinary and strange. Only an hour has passed since our four-mile kayak trip across oyster beds, channel cuts, and drifting manatees brought us out to this island, and we are still adjusting to the flatness and to the sounds of the wind and water. We had seen birds, sea turtles, stingrays, and dolphin in the familiar waters of Withlacoochee Bay. On the island, we weren't surprised to find ants, no-see-ums, and more birds, but we did not expect rabbits. These islands have come to rest near the southern end of what is called the Big Bend, a still intermittently developed part of northwest Florida, eighty miles up the coast from Tampa. I joke that this island chain is the Big Bend's version of the Florida Keys, but the Keys sweep comfortably along the Gulf Stream to form a bay linking them to the state's main peninsula. The spoil islands jut out unnaturally—you might even say recklessly—into the Gulf's open waters. Our island camp feels like the middle of nowhere. That night, a half-moon illuminates what already looks like a lunar landscape. After it sets, the Milky Way—a rarity in Florida's incandescent lands—will structure the dark sky from horizon to horizon, shadowed only by the silhouette of passing clouds, just before dawn breaks, and the shrimp boats return to port.

It's unusual to be on dry land this far out in the Gulf of Mexico. Toward the mainland, the bay's flats are taut with suspended sediment and a turning tide. Shallow waters barely hide what might be land. When we look out to sea, channel markers stitch water and sky in a vertical suture. The tide is out now and this horizon line has also sewn in the mud banks revealed by the bay's "skinny water" far off the coast.[2] High tide on the island has deposited familiar clusters of shells and tannin-stained sand. Just above this line, cordgrass and Florida privet have rooted between fossilized sea urchins and other graying stones that support a collage of scrub—cabbage palms, mangroves, prickly pear cacti, and twisted driftwood. Further down the beach's slope, smaller white rocks and shells rest with a confetti of thinly bladed sea grass, and then at about mid-tide, grass and seaweed are rolled into elongated bundles. Tightly woven by tidal movement, these rolls look just like the geotextile "logs" that are laid across street drain inlets to filter pollutants during roadwork projects. On the island, receding tides have machined these tubes so that I question their provenance and wonder if either the Department of Transportation or the artist Andy Goldsworthy—known for land art that integrates ephemeral human intervention within natural settings—has been here. Below these matted rolls, larger sun-bleached stones have come to rest, revealing the same limestone that makes up the lowest, algae-browned rocks, glistening and slippery at low tide's quiet ebb and flow.

On our passage along the chain of spoil islands, we are able to read the topographic record of their formation. Islands closer to the shore, and consequently in shallower water, are higher because dredging was deeper. Vegetation on the islands diminishes along with contour, and the last is marl and sand. From above, you would see the clarity of the tapering archipelago's plan, with islands larger in area closer to land, and the last seaward mark visible as a spectral bank or reef just below the water's surface. Angled a few degrees south of due west, the datum that controls this island chain's incipient logic is the channel that also hints at this area's unlikely history as an industrial landscape. The channel is the terminus of the Cross Florida Barge Canal, partially dredged between 1964 and 1971. The consistency of its controlling depth and width yields greater volume of excavated material in the shallower waters closer to shore. History is made legible in topography.

This history begins with proposals made by Spanish explorers to shorten travel distances across Florida's peninsula between the Atlantic Ocean and the Gulf of Mexico. The United States Army Corps of Engineers followed with its first plans in the 1840s. With the next century's growing commerce and increasing concerns for national security and navigational safety, the Corps put forward Route 13-B, from among twenty-eight canal paths in 1932. Three years later, emergency funds were made available for the project, which would eventually create these spoil islands and link this remote area to national, even international, concerns. While a Florida senator proclaimed that the canal would transform the Gulf of Mexico into the Mediterranean Sea of the western world, Franklin Delano Roosevelt detonated the blast that was the project's inaugural and ceremonial ground-breaking.[3] The President saw opportunities for work, and the Senator imagined a stateside Venice. The proposal's scale spoke of something else. Funds ran out in the first year, but one can only imagine how the planned sea-level canal, with its thirty-four-foot controlling depth and five hundred foot width, would have transformed this coastal landscape along with the rest of the state's topography. In the Gulf, a seventeen-mile-long jetty would be where the five-mile chain of spoil islands is now. On the mainland peninsula, southern Florida's residents worried that their region would become an island, disconnected from the rest of the nation. Thirty years later, construction began with another explosion, this time triggered by Lyndon Johnson. Now scaled down to a barge canal with a twelve-foot controlling depth, the project continued for six years, and workers nearly completed this western terminus before Richard Nixon stopped work on the canal in 1971. The project was eventually deauthorized in 1991.

Our island is one of the nine—at low tide, twelve—dredge spoil deposits created by the canal project. Their presence makes visible the channel that was the project's extension into the Gulf, while their displaced geology records other histories. Not just the calcified fossils of shellfish and coral that formed Florida's ancient seabed, but also the operations that mined these deposits. Among the spoil island's stones are traces of the phosphate that initiated this area's industrial past. In 1890, only ten years after the resource's discovery, Dunnellon Phosphate Company transformed the coastal zone into a large-scale mining operation. After the turn of the century and before trade slowed after World War I, Port Inglis Terminal, on Chambers Island at the mouth of the Withlacoochee River, would send phosphate to Hamburg and London. Even though we did not see the barge and tow that still pass the islands every few months on their short journey up the canal for mined phosphate, our nautical chart attests to these otherwise obscure histories of commercial trade. The channel next to our island camp is labeled "Tug Gap," and shipwrecks dot the conveyor belt of "spoil areas." On our outbound paddle, we might have also seen the ruins of Chamber Island's once bustling port, altered by storms and now hidden by vegetation like this coast's other industries that ranged from Civil War salt works to Faber's red cedar millworks twenty miles north in Cedar Key.

In the wake of other, more fleeting infrastructure, the spoil islands have remained. As both witness and testament, they are places for seeing past and future, even if their marginality makes for an unlikely location to perceive mainland histories. The islands have come to rest at a nexus of vulnerability and stability in this transitional zone. They linger between human intervention and natural system, between water and land, and among conservation, recreation, and ongoing industry. At night, our outer position on the island permits a clear view of the Crystal River power station. Should we be camping so close? Its flickering lights tie us back to the mainland's energy systems, the hyperboloid shape of its cooling towers traces the curve of seashells, and its outsized scale enhances the delicate nature of what it does not obliterate. If the power station is juxtaposition, the spoil islands are camouflage. And although both projects participate in the industrialized, or "productive," landscape and both projects originate from the same time period—the barge canal broke ground in February 1964 and the first generating station began production in October 1966, their legacies diverge. Today, the Crystal River Energy Complex consistently ranks among the top ten plants in the nation for power generation, and its well-maintained jetties direct warm water from its generators out to sea, just past the Three Nautical Mile Line to about the same distance as the last spoil bank of the barge canal.

To look at a map of this area is to read an unnerving index of earth marks: jetties, spoil areas, canals and rivers all measure the transition from land to sea. Some are scars, with the river as their vein, and others are more like scabs, unmoored from the dredged incisions that have bled limestone, fossils, and sand. The barge canal borrows from the Withlacoochee River's natural flow to form an artificial delta, and the natural tendency for oyster bars to form at river confluences with this coast means that similar banks form at the canal's mouth. Unsettling to navigators of commercial barges, these earth marks are the weft to the dredged canals' warp and even pose hazards to the shallow-draft vessels of kayakers and recreational boaters. Although we left shore a few hours after low tide, we still had to negotiate the oyster bars at the mouth of the Withlacoochee River. A similar navigational impediment at the canal's outlet—set up by human intervention and then carried out by natural forces—requires that barges pause for high tide alongside the spoil islands, which have also been adopted by their natural setting. The power station's jetties follow another model. They refuse adoption, and their sheer size and continuous landform disrupts natural currents, spawns erosion, and disperses sediment throughout the bay. Though modest in comparison, these jetties give a sense of what the Corps of Engineers' first canal scheme might have yielded. The originally-proposed seventeen-mile extensions would have superseded not only the antiquated three-mile territorial line but also the twelve-mile demarcation of territorial seas. While the broken spoil islands submit to wind and wave, the jetties resist these forces; and where the spoil islands are public domain, the jetties are spikes of private land, off-limits to camping. No bunnies on those lands, but in winter months manatees flock to the power station's warmed effluent.

Other marks are reminders of the engineered landscape that persists. Walking on the spoil island's northern shore, I discovered a U.S. Geological Survey (USGS) benchmark. Erosion by littoral drift had exposed more than five feet of the green PVC pipe that supports its brass medallion. It is as if this pipe attempts not only to locate the island's position in cartographic space but also to pin it to the logic of its construction, even if its waste materials defy convention. Realities of the spoil island's transitional make-up—its industrialized origin and naturalizing context—swirl around this definitive locator, now shifting and certainly inexact by survey standards. Geographer Vaughan Cornish described the ocean currents that have exposed the PVC pipe and molded these spoil islands as the "eddy-curve," a complex form of drift, with a "blunt head and a fine tail," which is in essence the "effect of a current moving against a fixed obstacle."[4] The resulting ichthyomorphism gives the islands a windward head and a leeward tail. Perhaps suffering from a castaway's madness, I imagine that the rounded medallion of the survey post could be the eye of this fish-like form.

Their distinctive shape more like tadpoles, these islands swim out to sea, treading in waters that they alter by their very presence. Introduced into natural systems of ebb, flow, and current, the spoil islands interact with their context through a unique and artificially created reciprocity. The sea acts on these islands, and the islands have transformed the sea. When he reached the end of his thousand-mile walk at Cedar Key, John Muir remarked on the "burnished" waters of the Gulf. His arrival preceded the spoil islands by a century, but his description captures this combination of artifice and nature.[5] It is possible that Muir might have arrived at one of those moments when swells flatten and the Gulf's waters thicken like the viscous oil. The coast's thickly humid air might have also polished this aqueous surface in Muir's imagination. You often find a similar phenomenon on the leeward side of islands, particularly artificial islands that interrupt—more resolutely than their naturally formed cohorts—natural currents and wind-flow. Burnishing acts by way of a tool, even something as crude as the hand that polishes a surface; and in these burnished seas, the spoil islands are the implements that transform their context. Muir's term also holds another meaning for the Gulf landscape. To burnish is to make brown, and river flow and dredging have colored the sediment-laden, tannin-soaked waters of this bay.

When we explore the island, its ground shifts and seems to moan with the weight of each footfall. This island might be a sleeping whale like the one Sinbad awakened with his campfire, ending his first voyage when it plunged below water. The rocky terrain moves but it does remain in place. Sometimes it sounds hollow, and I am reminded of the unsettling resonance that comes with walking on the egg-smooth rocks along Scotland's North Sea coast. Muir was there too. Born on this coast in Dunbar, Scotland, the naturalist felt a sense of homecoming when he reached the Gulf coast of Florida after he had been landlocked in the American Midwest for almost two decades. Probably it was the smell of the sea's briny air, rather than visual or aural recognition, that returned Muir to thoughts of home. On first glance, the two environments of Florida and Scotland could not be more different. But perhaps Muir did read something in Cedar Key's coastal scrub that reminded him of another Scottish landscape. A mixture of invasive exotics and native scrub, the bristling uplands of the spoil island suggested for me the uncultivated setting of those same heaths. Could it be that this impossible terrain, in its combination of the familiar and the foreign, registers a new landscape rife with associations and potential lessons? Ones that Muir might have recognized and appreciated? Muir had already described this area's islands as "bouquets," suggesting that their form might have been arranged by human intervention and that their context might constitute a surprisingly human-made garden.

In its past uses, heath identified waste land, more precisely the wasted land of untilled country.[6] Heath also designated the plants that could grow in such neglected places. By extension, it is the garden that has returned to a wild state out of its disuse. Applied to the spoil island context, this latter meaning provides one answer to T. S. Eliot's query about what "branches might grow out of this stony rubbish." According to its etymology, those who lived in the heath, were heathens, displaced as the "other" in what is perceived as a waste land of belief or primitive custom. Nearly fifty years after their creation, these islands out in the Gulf are no longer merely spoil created from dredging. They are also material that has been spoiled. They are abandoned sites left to become either what we imagine—heaths in Florida—or what we fear—places overrun by exotic plants. As heathen landscapes, spoil islands are the transitional landscape turned marginal.

We soon discover that our island has, to a degree, been cultivated. Among the larger spoil deposits, our island is one-third mile long and six hundred feet wide in its broadest area. We found our unlikely heath in this widest section on an upland, seven feet at its greatest elevation, and we began to recognize traces left in the landscape by previous visitors to the island. Campfire rings, outdoor rooms cleared from the scrub, compacted sand areas, torn bits of tarpaulin, and—most surprisingly—faint paths, more like garden tracks, here and there lined with stones, which on closer inspection are the rounded sea urchin fossils, some of them charred gray and black by time or by fire. This would be the heath's wild garden, piecemeal but communal, colonizing the spoil island's dredged waste. It seems as if there is a firepit for each visitor to the island. Like us, the island's periodic residents are more than sporadic campers. They have, in a sense, served as its caretakers. There is garbage, but it is isolated to particular campsites—left perhaps by more iconoclastic visitors to this heathen landscape.[7]

We can camp here because the islands are public land managed as part of the Marjorie Harris Carr Cross Florida Greenway. Its namesake led mid-century opposition to the canal project that now provides the land for public recreation. The spoil islands reflect in miniature the greenway's larger transformation of the land from industrial to recreational use. While inland sections of the barge canal project—its dams, locks, and canals—have often languished in political and environmental debate, the spoil islands have flourished in relative isolation as bird sanctuaries, primitive camps, and channel references. More broadly, and perhaps counterintuitively, the islands are opportunities to reverse the failure of an industrialized landscape and redefine what "productive landscape" means. Some are conserved as bird sanctuaries, others provide backcountry camping, and most offer other forms of resource extraction—recreational fishing. During our first night camping on the island, bright lights woke us in the chilled early morning air. Coming from a flat-bottom boat less than twenty feet from our tent, the lights shone just under the water's surface to illuminate the bottom where flounder disguised themselves in the rocks and sediment. The well-defined shores of spoil islands are well-suited to this type of sport fishing, known as "gigging" for flounder. In this way, the islands are not unlike artificial reefs that have been installed as fish habitats and are similar to the power station's warmed effluent that offers extended fishing seasons during winter's cooling months.

When Armanis F. Knotts fished this area in 1919, the lawyer-politician was so impressed by its natural resources that he purchased four thousand acres of land along the Withlacoochee River. Up north, he had helped plan the city of Gary, Indiana, known today for its history of industrialization. In late 1923, he established a sportsman's resort town, its name soon changed from Knotts to Yankeetown to acknowledge its residents from the north. For Knotts, sporting meant not only hunting and fishing but also land speculation and boosterism, and he energetically lobbied for Yankeetown to become the western terminus of the proposed cross-Florida canal. A portion of the canal eventually came, and to a certain degree, it is Knotts' legacy. But his idea that recreation and industry are not mutually exclusive is much more complex, and it is doubtful that he could have envisioned the subtle synthesis of naturalized and industrialized ground that the spoil islands represent. Now marketed as Florida's "Nature Coast," this region may feel wild and untouched, but its landscapes tell the story of a new nature. The spoil islands in particular tell of an ongoing process that shuttles between a past as technical by-products of canal construction and a future as natural resources.

Yankeetown provides another staging of the spoil islands' many narratives. In 1961, Elvis came to Yankeetown to film *Follow That Dream*. The spoil islands were not yet built, but residents were asked to fabricate another artificial island by transforming the low wetlands of Pumpkin Island into a tropical beach, with white sand, palm trees and cabanas. In the film, the island is connected to the mainland by a new bridge where Elvis's character Toby Kwimper sets up a roadside homestead. The film is based on Richard Powell's 1959 novel *Pioneer, Go Home!*, which tells the story of a northern family who squats on the right of way of this recently constructed bridge after their car breaks down. Powell drew from a real story further south in Florida where vacationers camped on excavated spoil from bridge construction that connected Ft. Myers to Pine Island.[8] The film and novel both draw from the ambiguities that arise when development creates indeterminate land. The spoil bank provides Elvis with an idyllic, mostly carefree zone to sing and romance. It also offers the framework for a small degree of narrative conflict to move the story and heighten Elvis's coolness at the margins of authority and his laid-back heroism in the face of mild adversity. The spoil area provided the film's dreamer and the novel's pioneer with temporary homes on a strange and not always welcoming frontier. Though squatting on spoil islands is now restricted, they still make room for dreamers and pioneers. The put-in beach for our kayaks at the start of our trip was on Highway 40—also named "Follow That Dream Road," just west of Elvis's bridge. There is a totemic head carved into one of the two cabbage palm trunks that made perfectly spaced climbing footholds for my son and daughter. These are not Elvis' coconut palms, but I like to think that they formed a part of that film set.

These islands have not been named. On the east coast and along the Gulf further to the south, the Corps of Engineers assigns numbers to spoil islands, but here they remain in anonymity. Relatives rather than absolutes identify their location. We camped on the third from the end at high tide, which would become sixth from the end once the tide was out. Between boat launch and island campsite, we passed Chambers Island, Deadman Key, and Captain Joe Island. But these are naturally formed islands, and they have been named, while the spoil islands—distinguishable only in aerial photographs by the white blaze of their chalky shores—remain unnamed. Because they are public and because they are set out into the Gulf's frontier, these spoil islands might have a thousand names, or at least a dozen, one for each stone-ringed fire circle found on our island.

These islands should not exist, and on topographic maps, they don't. In USGS surveys, they fall at the corner of three separate quadrangle maps, and their inclusion would have required the USGS to issue a separate quad for this land that is topographic but perhaps not considered topography in a landward sense. As Herman Melville wrote of the island Kokovoko: "It is not drawn on any map; true places never are."[9] Melville's island "far away to the West and South" extends his narrative's geography to its furthest extent, links marginal character with remote place, and underscores the limits of map-making. Like Kokovoko, spoil islands eschew precise cartographic documentation. They are unnamed and unseen only because they are not widely perceived.[10] But they are not blank spaces waiting to be filled. They are there to be understood on their own terms. Surreal, free, and maddeningly difficult, spoil islands offer a range of experience like nowhere else, and they are places where the unknown incites adventure—what Joseph Conrad called "fabulous geography."[11] We find these islands in this gap, that topographic blind spot, between what we haven't seen and what we still might find. For the better and for the worse.

Spoil islands are waste. As by-products of dredging, their residual nature causes problems. They are settings for exotic plants, impediments to navigation, and indeterminate public spaces. As offshoots of infrastructural systems and transportation networks, they also tell us about ourselves and our relation to this environment—part of the possible "truths" that Melville cites. Melville's fellow traveler Charles Olson picks up on the type of space that affords such opportunity. He draws from an American frontier and an inner self on the "half sea half land" of the American plains: "Some men ride on such space, others have to fasten themselves like a tent stake to survive."[12] Land and sea, spoil and island tether a wildness that combines relinquished control with self-will. To locate our spoil island campsite is not to set coordinates to 28° 58' 18.6"N by 82° 47' 39.5"W. It is to read the islands as an unprecedented archipelago where human and natural forces collide and where each is vulnerable. For us mainlanders, these are places not easily found but worth the trip. Birds know where they are, and so evidently does the lonely rabbit dancing in the island's broken karst.

A waste turned over to wildness. A complex wildness and a syncretism of progress and hubris within nature's humbling systems. Amidst this new wildness, spoil islands ask us to rethink conventional distinctions of nature and culture. Wildness, even if pressed to a near margin, can be proximate. Remember the flopsy bunnies who found the rubbish heap outside Mr. McGregor's garden so "soporific" in Beatrix Potter's story? This wildness is not part of a man-nature opposition, but is irrepressible and emergent, even in the most human-made of landscapes. Camping on these islands, we imagine ourselves like *peregrini*, the sixth-century monks described by Robert Macfarlane, occupying the "remote littorals" of Britain and Ireland.[13] We also relish our solitude, however rugged, obscure, and indefinable our place might be. At these moments, it is possible, but not necessary, to forget that this land is human-made. These islands grow, recede, and extend. They weren't meant to be here, but the sea has made them its own. And it remains that these islands have placed us—and the bunnies—where we would normally not be.

NOTES

1. T. S. Eliot, "The Waste Land," *The Complete Poems and Plays* (New York: Harcourt, Brace, and World, Inc., 1980), 37.
2. "Skinny" describes water that is exceedingly shallow and is often neither navigable nor passable.
3. See Steven Noll and David Tegeder, *Ditch of Dreams* (Gainesville, FL: University Press of Florida, 2009), 3, 81.

4. Vaughan Cornish, *Waves of Sand and Snow and the Eddies Which Make Them* (London: T.F. Unwin, 1914), 93. See also Vaughan Cornish, *Ocean Waves and Kindred Geophysical Phenomena* (London: Cambridge University Press, 1934).

5. John Muir, *A Thousand-Mile Walk to the Gulf* (New York: Houghton Mifflin Co., 1916), 124.

6. See the entry for "heath" in the *Oxford English Dictionary*, 2nd ed. (New York: Oxford University Press, 1989).

7. This form of caretaking only works when campers minimize permanent transformations, pack out what they pack in, and reuse previous fire rings and camping areas. Encouraging these practices, the Greenway's ranger asked us to camp in an area that had already been cleared.

8. Richard Powell, *Pioneer, Go Home!* (New York: Charles Scribner's Sons, 1959). United Artists released the film *Follow That Dream* in April 1962. The story draws from squatters who set up temporary dwellings and fishing shacks along the landward edges of Matlacha Bridge, which connects Pine Island to the mainland areas of Cape Coral and Ft. Myers. It came to be known as the "fishingest bridge in the world." By 1943, the squatters' actions came to the attention of the Internal Improvement Fund's Board of Trustees, who oversaw the sale and lease of state lands. The board's June 29th minutes show that the board leased the land as a defensive measure against the unauthorized construction: "Application was presented . . . on behalf of Charles Dean, offering fifty dollars for a one-year lease on approximately 800 feet of road frontage on the fill going from the mainland to Pine Island on State Road No. 183. Information was that applicant desired to lease in order to prevent squatters from putting up fishing shacks." See June 29, 1943, *Minutes of the Trustees of the Internal Improvement Fund* 24 (Tallahassee, FL: 1945): 229.

9. Herman Melville, *Moby-Dick, or, The Whale*, eds. Harrison Hayford and Hershel Parker (Evanston: Northwestern University Press, 2001), 55. The name of the fictional South Pacific island has varied between Kokovoko in American editions and Rokovoko in British editions of the book.

10. Robert Macfarlane, *The Wild Places* (London: Penguin, 2008), 225. Macfarlane draws from Paul Nash's idea of the "unseen landscape," which results in a kind of invisibility. In a similar vein, Macfarlane points out that wildness is irrepressible, emerging even in the most human-made of landscapes.

11. Joseph Conrad discusses "fabulous geography" in his essay "Geography and Some Explorers," *Last Essays*, ed. Richard Curle (London: J. M. Dent & Sons, 1926), 10–17.

12. Charles Olson, *Call Me Ishmael* (San Francisco: City Lights Books, Inc., 1947), 12.

13. MacFarlane, 24.

Withlacoochee

Acknowledgments

I am grateful to my family for inspiring and supporting this project. Explorations with Melanie, Aidan, and Phoebe seeded its early ideas, and it was a joy to discover these extraordinary places with them. My colleagues in the International Association for the Study of Environment, Space and Place have also contributed to the project, and I particularly thank Susie Allen, Gary Backhaus, David Bello, John Murungi, Troy Paddock, Christine Petto, Kip Redick, and Richard Wilson. And Peter Nekola added insights as we kayaked New York City's roiling waters. During its early stages, Siddhartha Lokanandi made vital comments about the work. I also thank Christo for discussing his Miami project and the Samuel Proctor Oral History Program at the University of Florida for helping with the interview. Colleagues in Macedonia—Armir Ferati and Kujtim Elezi—were important interlocutors for ideas about urbanism and ecology. In Florida, Mark Crosley, Robert Day, Marc Virgilio, and Dereck Winning helped with background information on the state's spoil islands. Series editors Robert Mugeraver and Gary Backhaus offered critical insights for the project, and editors Jana Hodges-Kluck, Jay Song, and Karen Ackermann were instrumental in bringing the book to completion. Research for the project could not have been completed without the crucial assistance of Kimberly Nofal, who also helped create the book's illustrations.

Makeshift

TOPOGRAPHIC BLIND SPOT

Is there an allure of spoiled places? Spoil islands are overlooked places that combine dirt with paradise, waste-land with "brave new world," and wildness with human intervention. Seemingly mundane by-products of dredging, these islands are the unavoidable residue of technological process. At the same time, they are readily adapted for other, often unintended, functions and demonstrate the potential value and contested re-valuation of landscapes of waste. To examine the marginalized topographies of spoil islands is to understand emergent concerns of twenty-first-century place-making, public space, and natural and artificial infrastructure. This book sets a course southward from New York City to Key West to read this uninvestigated archipelago.

Systems of dredged channels, ports, and cuts have yielded a manufactured landscape of diverse human activity that ranges from homesteading to conservation, from transient squatting to recreation, from memorialization to artistic endeavor, and from global transport to local land-use politics. But the technical origins of this landscape are much less varied. This dredged system, in its early stages in the United States, was an exercise in efficiency, one that illuminates attitudes about the interaction of technology and nature in a nation's transport network and its growing economy. With a focus on navigation, the piling up of dredged material was a means to an end, a way to ensure flows of trade as well as national defense. Early on, removal of natural obstructions superseded concerns about environment, but the question of where to put the unwanted material always remained.

Not all dredged islands are incidental consequences. Procedures for constructing ports and maintaining navigable waters have informed the deliberate creation of islands. Made possible by advances in the technology and the scale of dredging operations, purpose-built dredge islands manufacture land—not where sea recedes but where earth artificially advances, not conventional land reclamation but land invention. This highly-intentioned, human-made landscape includes offshore platforms, micro-nations, city-state land expansion, transportation infrastructure, feudal *crannogs* in the British Isles, lock islands (*aits*) along the Thames River, and foreign outposts of trade and quarantine. Spoil islands fit into the broader cultural histories and contemporary frameworks of these artificial islands, but they are also early test cases of how artificial islands are made and what they might be.

Spoil islands put you in places you normally wouldn't—and probably shouldn't—be. Their insular status presents burden and opportunity and mixes ambiguity with ambition. Until recently, these islands constituted topographic blind spots, forming a part of the visual field but frequently overlooked in their double marginality as island and waste. They combine isolation with accessibility. They are in the public eye, but they also linger unseen. You might look past them as you drive across a causeway, ignore them as insignificant epiphenomena and secondary outcomes of another process, or squint and try to rub them from your vision like an irritating mote, small in size and muddy in provenance. This unique isolation now attracts attention from a varied group of users and planners. Campers, conservationists, engineers, and developers are all drawn to these residual places. What they find is nature already spoiled and artificial land already naturalized. Spoil islands teach us how to make an island; they also show us what living on the mainland can be. Such islands reveal mainland truths.

Spoil islands are disinherited landscapes. From a mainland perspective, these islands have been separated by space and substance. Their insularity prevents ready-made connection or easy communication, and their basis in dredged materials makes for unstable foundations and capricious performance. And yet the resemblance to naturally formed islands is uncanny, complicating the dispossession of this landscape. Spoil islands inherit the soils, sediments, and stones once hidden beneath the water's surface. Sometimes stratigraphy is reversed, and deeper levels occur on top while surface materials are buried. Although this grotesquerie is kept at a safe distance from boat traffic, the islands remain reasonably close—usually about six or seven hundred feet—to their progenitor waterways and channels. This double displacement—locational and material—is a critical aspect of spoil islands' ecology and is a basis for their extended phenomenology of insularity and disposal. Islands formed from waste, and waste experienced as island.

ISLAND

Their displacement makes spoil islands particularly difficult to define in terms of conventional island categories. Geographers define two main island types. Continental islands can be associated with a much larger land mass, and oceanic islands reside separately in the earth's seas, though sometimes associated with large rifts hidden beneath the water. Spoil islands elude this classification but borrow from each. They do parallel the depositional and transitory formations of continental barrier islands. And like the remote locations of many oceanic islands, spoil islands appear as isolated, unpredictably placed, reminders that land is possible in open water.

The general term "artificial island" also falls short as an umbrella term that might include spoil islands because it connotes the intentionality of human agency and artifice that most spoil islands do not have. Spoil islands are best understood as accidental islands, unplanned in how they are placed but fortuitous in what they tell us. As displaced earth, they remind us that the sea rests on the globe and that the ground exists as a substrate even where it cannot be seen. Their double-displacement means that islands separate and create. They are distinct nodes of possibility. The production of spoil islands parallels the imagination of islands: "It is no longer the island that is created from the bowels of the earth through the liquid depths, it is humans who create the world anew from the island and on the waters."[1] With this framework, in which geography and imagination collude, I will define island, and then waste.

Islands are contracted worlds. Their displacement affords isolation, whether by choice as a space of retreat or by power as a site of imprisonment. They are microcosms of the diverse ways we occupy and transform environments. Islands host narratives of the paradise and the castaway, the utopia and the shipwreck. They also harbor exiles. St. John famously found his banishment to the island of Patmos as an opportunity to reflect on and write about the future, recorded in the Book of Revelation. Other literary islands have served to show us what we are or might become. Shakespeare's island in *The Tempest* eludes precise location as it also borrows from the historic accounts of world explorers. In this imagined island, Miranda sees a "brave new world" where innocence and nascent civilization might unite. Yet Prospero's reply "new to thee" counters the novelty of the place and grounds this idealized world in timeless burdens of isolation and conflict, characteristics that will haunt him on "this bare island" by the play's end.

D. H. Lawrence also links "island-ness" to self-definition. His "lover of islands" leaves the larger island of his birth to seek another that fits his individual persona: "An island, if it is big enough, is no better than a continent. It has to be really quite small, before it *feels like an island*; and this story will show how tiny it has to be, before you can presume to fill it with your personality."[2] What constitutes the definitive area of an island is a fluid debate, but the Scottish *crannogs*, artificial lands built as lake-bound refuges, constitute a version of the islet that Lawrence's character seeks. One definition sets the threshold between island and islet at twenty thousand square feet, the limit of what one extended family can construct and the lower limit of what constitutes the larger island.[3] According to this formulation, spoil islands are typically large islets or very small islands. Though some spoil areas extend much further, they remain emergent features with limited land area.

Not just the size but also the instability and multiplicity of spoil islands might satisfy D. H. Lawrence's character. His process of island-seeking as self-discovery implies that there is one island for each person. A famous psychoanalytic exercise asks patients to draw their own island within a piece of paper's open ocean. This activity picks up on Carl Jung's idea that each of us is an island of consciousness amidst an inherited oceanic expanse of the collective unconscious. Each of us forms his or her own island. That makes for quite an extensive archipelago, geographically untenable for an island-loving populace, but at least made fathomable with the array of spoil islands found along U.S. coasts, particularly Florida's east coast and its Indian River. Lawrence takes this individualization of the island further and suggests that you don't just find the island but the island also begets you. Islands incubate, and by extension—through an architecture of the nest—mold, their residents: "an island is a nest which holds one egg, and one only. This egg is the islander himself."[4] The stability of the nest also holds the structure's expected temporality and transience. Nests break down and are rebuilt, just as islands—particularly those made by humans—might rise and subside. Here, insularity connotes both physical disorientation and psychological disquiet, at a distance from the mainland and prey to oceanic instability: "But once you isolate yourself on a little island in the sea of space, and the moment begins to heave and expand in great circles, the solid earth is gone."[5] Islands might provide land, but their position can be disorienting and their dislocation can be unbearable.

Even if we do not all reside on islands, many of us identify with them. At a distance, we might project on them our dreams and desires. Closer, they might inspire us to imagine other worlds. And if we inhabit them, however briefly, those experiences might provide a look back at the mainland, constructive reflections made possible by this temporary displacement. Spoil islands are typically small and close to land. Lawrence and his characters came to revile the former quality but might have been assured by the latter. Delivered from the Ile de Port Cros near the end of his life, Lawrence's declaration "I don't like little islands" parallels the conclusions of his story's island-lover.[6] His island was too small—nearly forty thousand times smaller than his home island of Great Britain—and too far out—even if only six miles from the French coast.

Islands temper the idealism they might initially inspire. Another island utopia, outlined in Lawrence's final story, becomes a wasteland context for the failed ideal society of Rananim.[7] In their contracted spaces and often "disheveled" state, spoil islands reflect many of Lawrence's reservations about idealized places, but their proximity to the mainland means that these qualities are both accessible and inescapable. For some, islands are freedom; for others they are lawless. For some, islands make up a picturesque landscape; for others they interrupt the view. Still others don't consider them at all. From afar, spoil islands occupy a public domain, with a viewership of thousands or even millions. Up close, the islands offer a range of sights—undeveloped green space, non-native growth, washed-up garbage—and an array of experiences—watching birds, having a picnic, camping, or long-term squatting.

Islands are also like stones. "So many stones, all different," writes the artist Bruno Munari. Stones displace volume, and they can be cast into the water. They also inspire imagination. In *From Afar It Is an Island*, Munari posits a scale shift between stone and island to redefine each: "She runs toward us. She has discovered an island. Look! She shows it to us, all rust-colored and bare. So close, it is just a stone, like all the others. But if we hold it as if it were far, far away in the sea, it is an island."[8] Like stones, islands are vehicles of child-like discovery. In this way, both are worlds in miniature. In each stone, each individual object veined with deposits and sculpted by natural forces, Munari finds an index of possibility: "But here is a real island that looks like a stone. But this is the first stone which we thought was an island! Three inches long, the size of a hand, it cannot be an island! Don't be too sure!"[9] Spoil islands, essentially piles of stone or at least excavated piles of rubble, are microcosms of what we make and how it defines us.

SPOIL

Spoil islands are by-products of dredging. This waste material is the unavoidable, and necessary, consequence of the technical and mechanical process that deepens waterways and allows for unhindered navigation. Dredging produces a particular kind of material displacement, the waste of dregs. Pulled from the bottom, dregs are sediments that remain out of a liquid's solution. They imply that the process of dredging—to which the term "dregs" is linked etymologically—has excavated and stirred up something that is undesirable or at least no longer wanted. Certainly today this latter aspect is a particular concern with the toxicity of many sediments found in busy waterways and ports. But dredged waste also includes a broader concept of residue and surplus that, through displacement, might have the potential to become something else—in this case, an island.

Dredged waste lingers. It comes to exist between systems. This waste is excess, out-of-place matter that engenders ambivalence. Nobody wants it. Considered incidental to the goals of the dredging process, this material hovers between classification as residue—something that came about necessarily from a useful process—and as garbage—something that is patently no longer useful. Displacement of dredged material also introduces disorder within the complex order of tides, currents, and wave action. Between land and water, it was previously submerged sediment that is now placed within a liquid matrix. Not yet an island, it is piled-up material that has been discarded. When it does not intrude on its originator's purpose—in this case, clear navigation—it remains residual to the system. When it is created, it is nameless. Dredged material first occupies areas generically, such as the "places of deposit" noted in early navigational maps.

The designation "spoil" further captures this material's particular characteristics as residual and in-between waste. From the nineteenth century, spoil has identified the remnants of dredging, mining, and manufacturing processes. These excavated, burnt, or rejected materials were piled up to form banks, mounds, and islands. For the anthropologist Mary Douglas, dirt occupies a similarly liminal zone. Like dirt, spoil initially has the potential to upset the order of the systems around it. As dredged material, its displacement might disrupt navigation, disturb hydrology, or introduce toxins. According to Douglas, such materials have seemingly lost their utility and value and come to represent impurity. Threats to order, these rejected bits and pieces are recognizably "out of place."[10] What they lack in utility, they gain with a residual identity that links them back to the process of dredging. The Corps of Engineers sometimes uses "dredge spoil island" to denote clearly the technological process along with the material's procedural origins. This full acknowledgment of the material's source also seeks to avoid spoil's negative connotations, like those suggested by Douglas' dirt, and to side-step undesirable environmental results—spoiling natural systems. Euphemistically applied, other terms move even further from spoil. "Ocean dredge" adds a natural prefix, and "dredge island" suggests clear technical purpose. Regardless of their classification, the spoils of the dredging process remain in that liminal zone between waste and garbage.

At the same time, the form of the waste material—whether dirt or spoil—remains in question. In the process of dredging and deposition, spoil resides between the formed and the formless. Deployed by Georges Bataille in the introduction to his "Critical Dictionary," *informe* translates literally to "formless" but in effect describes the process of the loss of form. Like the "gob of spittle" Bataille uses as one analog for *informe*, spoil is inassimilable waste with a proclivity toward formlessness.[11] In this sense, it cannot be neatly returned to its previous place or form and yields to a chaotic context. With *informe*, Bataille is also challenging the certainties of formal categories. Can forms be based on unstable, or even destabilizing, materials? A mixture of sand, sediment, and fluid, spoil is an unlikely material for island formation, susceptible to the natural forces that will reshape it. The material properties of spoil islands are innately formless, they are at the whim of the sea, and a spoil island's waste cannot be neatly returned to its previous, proper place. Malleable by technologies and nature, spoil islands occur between formlessness and form.

In the dredging process, spoil is also plundered earth. In ancient Roman times, *spolia* were valuable goods taken by the victor from the vanquished enemy. Spoil's etymological origins tell the story of this redistribution of resources, and to dredge is to raid the seabed of its sediments, residents, and habitats.[12] Dredging digs, scrapes, and scars the littoral environment. It opposes natural forces—currents and tides—that fill and shift channels; and a successful dredging project was a victory over nature, a narrative explicitly conveyed in early projects around Manhattan. Triumph translated nature into something useful as shallow waterways were cleared of obstructions and harbors were deepened. Such gains were also linked to the less tangible, but equally potent, assurances of economic trade or national defense. Although no longer usually framed as a battle against nature, dredging continues in an effort to counteract natural forces. Dredging seizes excavated earth by technologically driven forces to counteract natural entropy, and the undersea environment—besieged by this process—is not unlike a battlefield.

It was a similar mindset, prevalent in the nineteenth century, that understood wetlands as wastelands. Such waterlogged lands impeded development, harbored disease, and fostered a general wildness. By this logic, submerged land was wasted land and should be returned to the ordered stability of terra firma. Equipment and procedures similar to the tools of ocean dredging helped build dikes, cut canals, and shore up newly made lands against the natural systems that made places like the Everglades complex, misunderstood, and consequently maligned environments. Spoil, as fill material, was a means to control nature. This process irrevocably damaged a pristine natural environment. At the time, these first human-made transformations to the Everglades might have constituted a change from unspoiled to spoiled. But opposing these terms today is a naïve and no longer tenable attempt to differentiate "untouched" environments from those unaffected by human influence. Environments cannot escape the effects of human manipulation, and the spectrum now moves between least and most spoiled. Spoil islands offer a unique environment in this range. As islands, their adaptation to natural systems tempers their artificial origins, which are entirely and quite literally based on spoil.

When dredging produces spoil, there is a unique interaction of waste and value. In one respect, spoil connotes a loss of value. In spite of associations with victory, spoil's meaning carries with it the decay and ruin caused by destruction. Devaluation can arise with the stripped material's change in context. Historically, spoil is also the skin shed and sloughed off by a snake. But shedding implies rebirth and hints at the hidden values that spoil represents. Friedrich Nietzsche alluded to waste's potential value when he wrote that "waste, decay, elimination need not be condemned: they are necessary consequences of life, of the growth of life."[13] In this respect, waste is a signal that there is something of value.[14] Dredging yields navigable waters, and spoil is its "necessary consequence." Spoil areas point back to the process of dredging deemed valuable by the useful channels, waterways, and ports that result.

But when spoil becomes island, the concept of value extends to the waste material itself. It is not surprising that ancient *spolia* were also building materials salvaged and repurposed by military victors for their construction projects. The rubble of a defeated enemy's architecture—its columns, capitals, shafts, sculptures, marble fragments—could become useful once again as walls, floors, and defensive bulwarks. Even when nature was no longer considered an enemy, transformation of nature's waterways and harbors yielded unwitting spoils of an unwinnable war. These spoil islands demonstrate the value of waste, and to study waste is to understand value. This spoil has an intrinsic value and suggests that waste is a potential source for revitalization.[15] Displaced from its natural environment, spoil is an indicator of the nature of our relation to this world. When it accumulates as an island, spoil avoids absorption in the anonymity of the rubbish heap and instead comes to rest as a singular, identifiable place.

RESIDUAL AND RUDERAL PLACES

> All Horehound delights in ruderous places.[16]

Spoil islands are residual places. The double-displacement of insular waste and residual island yields a particular kind of place, not frequently considered by practitioners of architecture, landscape architecture, and planning but critical to the concerns of each in defining new ecological frameworks. Leftover, often unwanted, and always indeterminate, spoil islands eschew standardized and legal definitions. As a result, it is difficult to classify such places. In one respect, this difficulty presents what Edward Casey has called the "threat of atopia" and provokes "an irrational desire to have and to know as much determinate presence as possible." Such conditions and desires privilege certainties of "being" over the subtle ties and often indefinite qualities of "place."[17] Spoil islands are certainly displaced matter, but they do not resist the fullness of experience. As "matter out of place," they still rely on context for definition and classification. They are wastelands without emptiness, residual sites that allow for "implacement."[18] Practically speaking, these island sites allow for direct occupation of landscapes—typically inland waters—that are normally difficult to experience without a boat.

The double-displacement of these places does not result in desolation but instead provides a surfeit of opportunity. Some of this potential comes from paradox: the proximity and accessibility of the island's isolation and the palatable, even paradisiacal, transformation of waste. But, as remnants, these places offer other possibilities. They are places we would not normally occupy in two respects: location and disposition. As residual waste, they should not be where they are. And as waste sites, they do not quite look the way they should. These conditions resonate with marginal landscapes where you might have played as a child, those zones not far away from home, but just out of sight, in areas your parents may not have realized were there, places that you could colonize and set up a parallel life. As residual places, spoil islands invoke a similar mix of the real and the imagined.

Weeds probably preceded you in this playful colonization. Weeds started a process that you continue. Often defined as plants that are in the wrong place, weeds suggest that a particular landscape is uncultivated. Residual places are not only leftovers from the process of cultivation, whether it is the conversion or maintenance of land for agricultural use. They are also contexts where devalued vegetation grows. Context makes the weed. Classification of weed as weed has as much to do with its location as its characteristics.[19] These sites share characteristics with other marginalized, human-influenced topographies: ditches, berms, borrow-pits, borrow-ponds, containment berms, drainage ditches, and impoundments. It is their diversity as much as their marginality that incites an array of designations to convey rejection, dismissal, or repulsion. Terms like "disturbed area" or "substantially altered zone" attempt to classify these sites but project more psychology than topography and hint at the dismissive bias similarly associated with weeds. Official classifications often place spoil islands together with these sites under the heading of "ruderal landscapes."[20]

Ruderal describes vegetation that occurs in waste areas and places significantly changed by human intervention. Some definitions of the term limit these areas to a condition where "natural" growth has been interrupted. Such designations privilege agriculture over other land uses and presume a clear boundary between what is human-made and natural and what is weed and non-weed. Originally, the definition of ruderal did not entail these oppositions but instead connoted waste sites in close proximity to the human-altered landscape—near roadsides, along walls, and in piles of garbage. So-called ruderous places "abounded" in rubbish but nonetheless supported botanical growth. Further back in its etymological lineage, ruderal was in even closer contact with this human context and with places of unlikely growth.[21] In its plural form *rudera*, was rubble, the fragments and ruins of a building. And the singular *rudus* was a broken stone. "Ruderal" told of instances of ecological adaptation close to, or even in the wake of, human-made landscapes.

Ruderal plants consequently found use in the no longer useful and occupied the fissures of accident. Not only did these residual sites support what would seem to be ecological impossibilities, but the realities of their plant growth also pointed toward adaptability, resiliency, and innovation—qualities that parallel a child's playful reoccupation of marginalized landscapes. Ruderal landscapes attract the pioneer, whether human or plant. Some horticulturists will attest to this boldness and tell you that ruderal plants are the first to arrive.[22] Such areas also offer the exhilaration of discovery. When apothecary Jean de Renou writes of the Horehound's "delight in ruderous places," he captures this sense of possibility and discovery. The Horehound plant is well-suited to ruderal sites where it demonstrates that something playful—its woolly leaves and flower whorls—and something indeed useful—the herb has been used medicinally since ancient times—might come out of waste sites. Throughout this book, the ruderal landscapes of spoil islands will afford similar discoveries, surprising in their closeness to human intervention and compelling in the ecologies they model.

MAKESHIFT LANDSCAPE

As residual and ruderal places, spoil islands constitute a makeshift landscape. They are roguish yet vulnerable, mechanically produced yet haphazardly arranged, stand-ins for islands yet hardly distinguishable from their naturally-formed siblings. These sites fall within a general frame of cultural histories of overlooked places. Their study parallels Colin Ward's delineation of a makeshift Arcadia within the twentieth-century English landscape. Along with Dennis Hardy, Ward explored the relation of property, freedom, and control in self-built and self-governed settlements created in England's countryside, particularly along its coasts and riversides. Inhabitants of these areas knew them as "sweet Arcady" because of their provision of fresh air and inexpensive holiday lodging, while government authorities tagged them as "plotlands," with a perception that they desecrated the environment. Forming a patchwork of land holdings, plotlands were leftover pieces of land mostly deemed too small for agricultural use and too fragmented for real estate development.[23]

Characteristics of the plotland settlements parallel the makeshift landscapes of spoil islands. Their marginality came out of economic necessity.

The plots were not needed for conventionally defined uses, such as agriculture, industry or development; and plotlanders could afford these sites on disused agricultural lands often deemed "three-horse land" because of its heavy clay and in vulnerable coastal areas like Jaywick Sands and Canvey Island. This latter site began as four thousand acres of marshland mostly below sea level at high tide. Over time, marsh became island, its subsequent plotlands not attracting the ire from environmentalists that similar settlement patterns inspired on the South Coast of England. Although providing apparent distance through an idealized nature and a determinedly rustic setting, many plotlands were not far from urban centers, some as close to London as the banks of the Thames River. Others were more far-flung, though still geographically and economically within reach—thirty miles travel to the east or fifty to the southern coasts.

In this process, disinherited landscapes became appropriated land. Ward notes that a burgeoning suburban population decried the "haphazard distribution" of these sites. In the 1950s and 1960s, many complained that "hut dwellers both get the view and spoil it," but their placement was effectively a function of the "benign neglect" of these areas by planners and local authorities.[24] Like spoil islands, plotlands proved to be testing grounds for the balance of public control and individual initiative.[25] Plotlanders met the challenges of these marginal sites even though they lacked the funds and skills to become unanticipated curators of the remnant properties. Similar practices of squatting and homesteading are found on spoil islands, where their classification as public land further complicates settlement practices of the makeshift landscape. And like spoil islands, plotlands were experiments in defining environmental management.

Spoil islands extend the meaning and function of the makeshift landscape with their underpinnings in island geomorphology and waste. Ward's makeshift landscape focuses on alterations to a pre-existing environment, largely ex-urban, pastoral, and conventional in its topographic and geologic formation. The makeshift quality of the spoil island landscape includes not only what is built on them—there have been experiments in housing, squatting, and development—but also how the islands themselves, as substrate landscapes, are constructed. Geomorphology studies the topographic features and forms of the earth, and spoil islands are geomorphic landscapes initiated by humans and then modified by the natural forces that define coastal landforms. Consequently, these sites have already been built when the first vacationers, developers, conservationists, or squatters arrive. Once occupied, they are as much cultural landscape as topographic feature. To study spoil islands, the geomorphologist must also be the phenomenologist by watching geomorphic natural phenomena while acknowledging human agency and experience.

Spoil islands mix environmental with social histories. Natural analogs inform human construction, which in turn transforms its natural context. Language provides one way of tracking this genealogy. Cays, known as keys in the Caribbean, are islands of coralline sand and gravel that have been deposited on a coral reef. In the language of coastal development, "keye" became "quay" to describe human-made projections into the sea, particularly those for safely landing ships in contrast to the shoaled cays that posed navigational hazards and wrecked so many vessels. Spoil islands do not have the same homologies and etymologies, but they do share with "key" its depositional origins and its shifting semantics.

The dual marginality of spoil islands—on the coast between land and water and as waste material to the system that produced them—constitutes a makeshift landscape where humans seek to maintain control of nature's influence. Given this marginality, "makeshift" describes a landscape exposed to natural and human-made transformation. Spoil islands mix geography with metaphor as much as earth with water and natural force with artifice. Spoil islands have natural analogs in geomorphic phenomena of spits, tombolos, and tumuli. But they are also "artificial" analogs to the architecture of platforms and docks, and they bring to mind architectural earth marks such as berms, ditches, and impoundments. On spoil islands, coastal phenomena of wind and water—flow, current, and drift—overlap with spatial experiences of restlessness, retreat, and recreation. The geomorphic setting of spoil islands provides a glimpse of what occupying a makeshift landscape—already augmented by man and transformed by nature—might be like.

Makeshift combinations of spoil and island provide frameworks for revaluing waste and for continuing to examine landscapes of a new ecology. Their island-ness made them credible landscape features, and their origins in waste opened them to informal use. For many years leading up to policies such as the Clean Water Act in the 1970s, spoil islands—as incidental waste areas—were ignored by the state but occupied, appropriated and embraced by the general populace. Homesteaders, squatters, naturalists, boaters, even pirates valued these places before government mandates formalized them as sites of conservation and recreation. Spoil islands are in the public domain, and these uses resonate with definitions of waste that denote not only wild and desolate areas but also uncultivated land that lies in common but has little or no use value for agriculture. As these islands—dredged, deposited, and then largely forgotten—became a part of the naturally-made archipelago, their prolonged existence also made them a part of what Michael Thompson has called historical waste: "waste that has become residue by already being deposited in its proper place."[26] Over time, these waste sites were accepted as appropriate components of the landscape.

The story of this appropriation and assimilation outlines waste's potential for revitalization. Like islands, waste contracts the world around us. Considering waste brings what was previously dismissed, made invisible, backgrounded, or forgotten back into our near field of perception. Similarly, when we talk about the environment, it is no longer that separate zone "out there." The distance from this other place is reduced, and we establish contact with our more immediate context. And neither nature nor waste is separated from what we have made. As insular condition and waste, spoil islands also participate in the environmental legacies hypothesized by Peter Sloterdijk when he noted that one legacy of ecology will be "to transform the phenomenon of refuse into a 'high' theme."[27] Matter out of place can become suitably placed matter and this makeshift landscape provides new readings of human intervention and nature. Many of this book's island subjects began as unintended works. They are sometimes exercises in how not to make an island, but the tools and techniques of dredging nonetheless frame the technical side of island-making and expand how we understand this makeshift landscape.

HOW TO MAKE AN ISLAND

It is said that Alexander the Great paused during his campaigns along the Indian Ocean to observe—and, we might assume, to admire—how mangroves made land. At the time, these island-makers reversed expectations of where trees should grow and demonstrated how land might rise out of the sea. In addition to mangroves, another naturally-occurring landform inspired and provided foundations for Alexander's engineering triumph on the Phoenician coast, where residents of the island city Tyre offered singular resistance. To access the fortified city, the Macedonian army augmented an emergent—though still submerged—tombolo, a wave-built ridge of sand connecting island to mainland. Depositing the first bin of earth at the head of his army, Alexander himself began the construction process after prophetically challenging the Tyrians' opposition: "I perceive that reposing on the security of your situation, because you inhabit an island, you despise these land forces: but I will shortly show you that your place is part of the continent." Roman historian Quintus Curtius invoked this mix of land and sea to hyperbolize the army's ensuing work: "Rising from the bottom of the sea, the mole had already swelled to the magnitude of a mountain."[28] To fabricate this new landmass, Alexander exploited the tombolo's natural formation with a combination of dredged materials and stones salvaged from Tyre's abandoned mainland fortifications.

Alexander's project links spoil—from earth and from battle—to making land. Among the thousands of human-made islands along the eastern Atlantic coast of the United States, some followed in the tradition of Alexander's military objectives, others consolidated spoil for purpose-built islands, but most came about less intentionally, as by-products of a dredging process that was focused on the construction and maintenance of coastal waterways. These latter spoil islands can be traced to a mandate and a problem. In his 1808 report, Albert Gallatin, Thomas Jefferson's Treasury Secretary, called for a waterway—what would later become the Atlantic Intracoastal Waterway—as a component of proposed infrastructure to satisfy the growing transport needs of a new country. Supported by Benjamin Latrobe's poetic descriptions of a compliant coastal topography, Gallatin's recommendation alluded only obliquely to the dredging process his directive would entail. The Atlantic's characteristically shallow coastal waters required substantial deepening. A century and a half later, Lyndon Johnson summarized the situation and the implications of Gallatin's mandate:

> God was good to this country. He endowed it with resources unsurpassed in their variety and their abundance. But in His wisdom the Creator left something for men to do for themselves. . . . He left it to us to control these carriers of commerce. He gave us great estuaries and natural locales for harbors. But He left us to dredge them out so we could use them with modern ships. He gave us shallow waters along most of our coastline, which form natural routes for protected coastal waterways. But he left us to carve out the channels to make them usable.[29]

Over the intervening years between Gallatin's proposal and Johnson's observation, making the shallow waterways "usable" also meant inadvertently making islands. Rationalization of the waterway preceded articulations of how it would be accomplished and the effects it would have on the surrounding environment.

Before Gallatin's commercial interests could supersede concerns of war, the Atlantic's shallow waters continued to host more purposeful island-making for coastal defense. Early in 1829, Lieutenant Henry Brewerton of the Corps of Engineers placed an advertisement in U.S. newspapers for thirty thousand tons of stone. Not exactly spoils of war, but this project followed in the traditions of Alexander and other ancient land-makers, whom the nation's inchoate army corps had studied for engineering precedents. With the advertised material, Brewerton's project strategically augmented a natural feature, a sand shoal in Charleston, South Carolina's harbor. On December 5th of the previous year, the Corps had approved plans for a new, human-made island to shore up national defense as a part of the "third system" of fortifications following the War of 1812. Fort Sumter would slowly emerge from the harbor's waters as a pentagonal masonry garrison with a dedicated wharf. This combination of engineered earthwork and national defense reflects the Corps' early military mission, but this time period also marked the early stages of the division's more sustained involvement in civil work. In 1824, Congress appropriated funds to remove obstacles from the Ohio and Mississippi Rivers and solidified the Corps' role in maintaining the nation's aquatic transport network. Natural accumulations of sand and sediment that in one place might serve as foundations for wartime defense would deter domestic and international commerce in another. It is this charge—in the beginning, a relatively modest directive to remove sandbars, snags, and wrecks—that would define the Corps' modern role and develop into a complex geotechnics of dredging and depositing.

Charleston's harbor also hosted an early innovation in dredging. Although the harbor's strategic naval location was established and its interior region's agricultural production was increasing, a sand bar located near its harbor's mouth was inhibiting the city's growth as a major seaport. In 1852, Alexander Bache, great-grandson of Benjamin Franklin and superintendent of the U.S. Coast Survey, studied the harbor's dynamic tidal cycles and noted how ebb flow met the newly incoming tide and dropped its suspended sand along the bar. Bache compared this condition of suspended sediment and shifting harbor bed to "being alive," and his mid-century report advised that a cut made through the bar would benefit from the natural forces of "scour" that kept other areas of the harbor at depths of more than thirty feet and effectively stabilized the bar's pre-existing escape channels.[30] This project's goals exceeded the rake-and-drag dredging techniques offered by steam-driven tugs of the time period; and, with new Rivers and Harbors Acts on the horizon, federal engineers saw the project as an "excellent experiment" and advertised for innovative equipment.[31] The first proposal, a self-propelled barge with clamshell diggers on either side of the bow, seemed promising—it was the nation's first hopper dredge—but failed to hold its position and to manipulate the diggers in the tidal fluctuations outside the harbor.

A year later, Nathaniel Lebby, a local inventor whose steam-driven pumps were used on rice plantations, proposed what would be the world's first hydraulic suction dredge. By late spring 1857, Lebby's centrifugal pump, mounted on the *General Moultrie*, was drawing out 350 cubic yards per day from the vexing sand bar that separated the inner harbor from Maffit's seaward channel. By June, the vessel's "spoil holds" had discharged 145 thousand cubic yards of material at Cummings Point along the harbor's southern edge.[32] Captain George Washington Cullum of the Corps of Engineers reported that the nineteen-inch-diameter iron suction hose pulled up not only sand and shell but also bricks, rocks, and cannon balls; and city leaders lauded Lebby's homegrown invention for the new channel, almost fourteen feet deep at mean low water.[33] Despite this success, Lebby did not patent the machine, and only Captain Cullen's short descriptions survive. Dredging continued the following year, but city officials could not agree on purchasing the vessel for regular maintenance. With the Civil War, dredging operations ended. The 365-ton *Moultrie* was converted to a troop carrier, slipped undetected through the Union's blockade, and delivered 163 contraband bales of government cotton to Nassau, where it was sold as a dredger for the Bahamian harbor.[34]

Over the next hundred years, dredging followed a course similar to the *Moultrie*'s irregular voyage. Even if the Corps' mission was clear, the technologies for carrying it out remained piecemeal. Early dredging and snag-removal operations used hand-powered digging equipment, winched shovels, and amphibious plows. After early dredging experiments with French engineer Le Demour's centrifugal pump in the 1730s, James Watts' steam engine made possible Henri-Émile Bazin's suction dredge, which was displayed at the Paris Exposition of 1867 and was used in the excavation of the Suez Canal. Preceding Bazin's innovation, Lebby's "excellent experiment" with the hydraulic hopper dredge illustrates the relative isolation and fragmented histories of dredgers' tools in North America that continued through the middle of the twentieth century. In 1970, John Huston noted the indefinite chronologies and messy technological legacies of dredging:

> The dredging pioneers recorded very little of their experiences. They saw the development of the dredge happen through random trial and error rather than by plan. Interested in the practical aspects of making money, rather than academics, they put into writing little of what they did and started the circumstances which constituted the situation today—there is a dearth of useable, available, organized dredging literature. . . . Less has been written intelligently about the dredge than any other piece of excavation equipment in existence. . . . There were no pressures on dredgemen to write a book or publish a paper. They just dig mud and tend to keep their innovations and improvements to themselves.[35]

In practice, sidecaster dredges broadcast spoil out over the water in coastal inlets, and hopper dredges or mechanical buckets consolidated much of the material for deposition. In most cases, cost-effectiveness made off-channel discharge common within intracoastal waterways, and dredgers allowed natural currents to carry away or consolidate disturbed sediment.[36] In the latter half of the twentieth century, modern dredging updated and enlarged these tools with techniques of suction and water injection; and on the east coast of the United States, the Corps of Engineers and its contractors have typically used cutterhead dredging vessels that rely on fittings at the suction inlet to dislodge and pull material through hydraulic pipelines.

The informal process of dredging produced an equally informal system of spoil islands. Spoil became island by expediency and haphazard circumstance. With early dredging focused on channel construction, engineering documents rarely noted areas of spoil placement. Although an attitude of ambivalence to these sites persisted until the middle of the twentieth century in many locations along the eastern U.S. coast, dredging operations in the Delaware River presented early attempts to manage spoil and prefigured the transition from spoil as bothersome but innocuous by-product to spoil as environmental risk. Mariners navigating the Delaware River had documented the river's array of obstructions and shoals since the seventeenth century.[37] After a series of ineffective dredging innovations, a Delaware state board, convening in the 1870s, recommended the deposit of spoil on inland terrain "beyond the reach of tides."[38] Holding basins dug within diked enclosures, which could be emptied by landward rail systems, proved expensive, and riparian landowners, who feared decline in real estate values, generally rejected deals to place spoil on their land. As dredging increased with more commercial traffic and deeper draft vessels, the problem of spoil deposit continued, and by 1896 more than ten million cubic yards of material were dumped at nine different locations in the river.[39]

A survey authorized by the Rivers and Harbors Act of 1896 proposed a six million dollar project along fifty-six miles of river to remove thirty-five million cubic yards of material. Six locations were identified simply as "places of deposit." In the framework of navigability, this terminology candidly shows waste that has been generated but at the same time demonstrates that it has not impeded shipping channels. The use of "place" also connotes that nature has already suggested placement at these locations—the shoals, spits, and tombolos created by currents and tides.[40] Deposit sites ranged from a "deep hole below Ben Davis Point Shoal" to a proposed area that would come to be known as "artificial island," a term acknowledging what the deposits formed but retaining the land's genericism.[41] With bulkheading, this site later became the principal deposit basin for the Lower Delaware. In the 1920s, maps exchanged "place of deposit" for the less innocent but more accurate label "dumping ground"; and in 1943, charts showed a fully-fledged "Artificial Island" and denoted the area between its eastern edge and the New Jersey shore as "being filled." In 1977, the island's fill would bear the full weight of a nuclear power plant.[42]

In 1970, the same year that Huston diagnosed the trials and errors of dredging, the rationalization of waterways yielded to the rationalization of waste. Approved in 1969 and enacted a year later, the National Environmental Policy Act (NEPA), along with a set of other acts and legal constraints outlined criteria for material disposal in ocean and inland waters.[43] NEPA included dredging among the activities that required an Environmental Impact Statement (EIS) and had an immediate influence on the cost and logistics of waterway maintenance and, by extension, island construction. Up to 1970, a major limiting factor had been the economics of dredging itself, with spoil deposit determined by cost-effectiveness. After 1970, environmental impact, with a particular emphasis on spoil deposition itself, became the determining factor.[44] Questions of what to do with dredged waste had long been a part of administrating waterways, and the Rivers and Harbors Act of 1899, approved by Congress on March 3rd, is often mentioned as the nation's oldest environmental law. Ostensibly, this Act protected navigation by addressing the alteration and obstruction of navigable waters, but this goal paralleled the preservation of the waters themselves and prefigured subsequent legislation like the Clean Water Act in the 1970s. The 1899 Act also expanded the role of the Corps of Engineers, charging it with determination of what is refuse and where it can be placed in addition to oversight of navigability.[45] Throughout the century, culminating with NEPA, modifications to environmental policy fundamentally changed how spoil islands were made and resulted in a clear delineation of unconfined and confined spoil. Also, in the 1970s, recognition of the environmental potential of dredge spoil islands began to link activities of dredging to conservation. These Acts started a process of stewardship that has culminated in an archipelago of islands unrivaled in their location, now strategically placed at the inflection point of waste and conservation.[46]

After 1970, environmental imperatives made confined spoil the preferred option for island-making. Craney Island was the Corps of Engineers' first confined disposal facility (CDF).[47] No longer an island, the original land formed as a river-mouth bar where Craney Creek and the Elizabeth River meet—and where early settlers mistakenly identified its eponymous cranes—at the southern part of Chesapeake Bay's Hampton Roads, one of the world's largest natural harbors. Strategically located for navigation, commerce, and defense, the original island was the scene of pivotal military conflicts. British soldiers held Revolutionary War prisoners on the island until it was captured by Colonial forces in 1775, it was the site of resistance to British forces in 1812, and its shoals hosted the famous Civil War battle between the ironclad vessels *Monitor* and *Merrimack*.[48] After the War Department purchased the island in 1817, its fortifications were quickly outdated, and the island hosted a series of functions—a molasses refinery in the late 1800s, a quarantine facility and hospital in 1918, a fuel depot in the 1920s, and spoil deposit beginning in 1942.

The island received nearly four million cubic yards of spoil material between October 1956 and May 1957—the inaugural dredging cycle for its newly diked containment areas. In 1971, as the Corps' districts felt NEPA's effects, the island received material from unconfined sites, and the total deposition was more than seventeen million cubic yards. At this early stage as a CDF, the island was essentially a massive pen for material, contained but not always completely organized. As the island approached its projected capacity of one hundred million cubic yards, improvements in 1984 sought to manage the spoil with a more mechanized system of telescoping weirs and three distinct sub-containment cells that allowed alternate cycles of depositing, draining, and drying.[49] That same year, coordination began to provide avian habitats in selected areas of the artificial island.[50]

Today, the island makes its own landscape. Officially designated as Craney Island Dredged Material Management Area (CIDMMA), the island encompasses more than 2,500 acres, and each of its three cells is the equivalent area of Central Park. At the intersection of technology and terrain, it is a manufactured landscape that has its own weight, its own sedimentation, its own microclimates, and its own cycles of seasons and dredging deposit. The dredge material dries and leaves a parched, cracked landscape that is very much like a desiccated desert plain, its two-mile long southern cell stretching to a distant horizon. The dredged material is fifty percent water by weight, and its process of pan evaporation offers clues for even greater efficiency in spoil management. Strip drains along the fill's vertical edges accelerate consolidation, and the island sinks as it also rises, subsiding more than two feet over a period of three months. But even with ocean disposal to help disperse excess dredge material, the island's four-square-mile area reaches heights of more than thirty feet.

Three miles southeast of Craney Island up the Elizabeth River, a buoy labeled Mile Marker 0 initiates the Atlantic Intracoastal Waterway. It is an unremarkable beginning compared to Craney's immense mound still visible to the north. The latter monument testifies to the process that makes the former possible. It also evidences a portion of the four hundred million cubic yards of material that must be dredged from the United States' waterways and ports each year. This slurry would cover all of Manhattan Island in almost eighteen feet of mud, sand, rock, and detritus. If it were deposited further south, the island of Key West could be shored up by more than seventy feet. In annual dredging operations, some of this material goes out to sea, some to isolated landfills, and the rest to a continuing process of making islands.

READING THE ARCHIPELAGO

Each spoil island is a hypothesis. Each island proposes a new location, lays down a conditional foundation, and tests a set of materials. In spite of its displacement, each island supposes place. Most of the islands in this book are offshoots of another process and are not intentional. They equate to so many unwitting test cases of what an island, and a habitable landscape, might be. These largely unauthored working hypotheses could pose a dilemma to coherency, but instead they provide an opportunity to explore a working method, albeit one at full scale, that is not unlike the design process. Each island is a sketch study, a straw man, or a diagram that holds ideas and concepts. In the creative design process, some of these would be rejected, others altered, or a few carried forward. In the world of spoil islands, the island proposals—venerated and maligned, well-placed and misplaced—live on within an invented landscape. Then the sea goes to work sculpting, reducing, even erasing the work.

Deposition yields disposition. What the dredging process sets in motion, the wind and tides subsequently transform, and tendencies or themes emerge. Spoil islands play at order and disorder. They demonstrate how disorder might give way to a new kind of ordering system. Their fluid landform and their indefinite status as engineering by-products—what Robert Smithson called the "wreck of precision"—provide a context where use precedes meaning and where new settlement patterns might emerge. The distribution of spoil islands reflects this dynamic. When they occur as a set, the size and spacing of spoil islands are often irregular, but their overall layout follows a system of deposition. Along a dredged channel, the islands record the pace of cut and fill operation. Sometimes, a depositional rhythm can be found, such as the half-mile spacing between islands in the Indian River, although that pace is just as easily broken by changes in contour and the force of currents. Distances between cut channels and deposited islands are usually less than seven hundred feet, offering separation between boating and potential obstruction. Traveling up or down these channels, boaters might notice an alternation between port and starboard, three islands this side, then two more that side. To order spoil islands is to understand their complex origins in nature and in human enterprise.

Spoil islands also demonstrate how waste becomes resource. As the islands gain meaning, some wish to conserve them, others play on them, and a few find ways to dwell on their sometimes hostile lands. While these uses are adaptive, qualitative and often elusive, the translation from waste to resource can be quantified to an extent, or at least "read" in the landscape. Vertical depth and material properties determine the size of deposits and the horizontal distance between islands. Port cuts also reflect the exchange of cut and fill. At strategic points of entry or channel connections, land has been shifted. Subtracted mass becomes additive island, and many times the island's figure closely matches its now empty point of origin, like an oversized puzzle piece. The islands become resources as material and by location.

Spoil islands relate formlessness to form. Currents shape each new island: eels, eggs, cashews, leaves, twigs come to mind. Aerial views show half-hidden sandbars tethered to raised areas of rock colonized by vegetation. Sometimes the deposition process—helped along by subsequent tidal action—has deformed the new land so that it drifts nearby like a ghost, left to haunt landmasses in its waywardness and incorrigibility. These islands are frail in their mutability but potent in their unpredictability. In either case, this new land is made surreal by its exposure of the invisible and its displacement of the stable. Such formlessness is not anti-form, but a possibility to rethink aesthetic ordering. Christo's project for *Surrounded Islands* found in Biscayne Bay's spoil islands the sites for temporary monuments to the environment that did not freeze or stultify the natural processes he sought to engage. He made visible the variable forms of wind, tide, and light; he augmented the islands as found objects; and his project's technical detailing discovered new geometries in the amorphous landforms. In the process, the islands showed how unstable foundations and capricious performance make up our experience of the environment.

To test these ideas and to "sound" their depths of meaning, this book reads the spoil island archipelago from multiple perspectives. The narrative course follows traditions of piloting guides for cruising the east coast's intracoastal waterways. These guides typically pair first-hand accounts with navigational charts. They operate from two viewpoints—the mariner and the cartographer, and this book adapts both perspectives. From above, the chain of spoil islands can be read as a map of relations and patterns and as a part of infrastructural and geomorphic systems. From the water, lived experience of the islands can inform their phenomenological presence and their social and cultural histories. A third perspective—what might be called a portolan mentality—operates as a navigational bridge between the aerial and experiential views. Portolan maps were based on a navigator's experience of the coastal edge. Visible information, gathered by sea rather than by land, was translated into a map of peripheries, leaving the interior mainland as a blank—sometimes unknowable—area. Navigation made the map, and the edges came to define the mainland. Spoil islands work similarly. Their presence at the edge of waterways, cities, and infrastructure delineates a particular kind of map, one that offers insights into economic, political, and environmental workings that those on the mainland have not—and perhaps cannot—always see.

Reading the archipelago uncovers other systems. Geographically, the east coast's dredge spoil islands form a linear chain that runs from New York to Key West, mostly tracking the coastal waterway. Along the way, clustered archipelagoes bring spoil islands into contact with natural and purpose-built islands. In the early chapters, four of New York's confetti of more than thirty islands are studied: one that was destroyed, two that were purpose-built, and one that was piled up and ignored for many years. Thirteen hundred miles south, Biscayne Bay's linear chain of spoil islands cross through a ready-made Venice—the nation's first purpose-built residential islands and the world's busiest port for passenger ships. Linking these urban centers, the intracoastal waterway that spawned many of this book's spoil islands forms a narrow, unprecedented ribbon of public space along which nearly one third of the U.S. population lives. New York's cluster archipelago is home to eight million, and about the same number inhabits the 350-mile line of Florida's east coast.[51] Colloquially referred to as the "Ditch" and variously labeled AICW, AIWW, and ICW, this waterway technically begins in Norfolk and ends in Key West, but its fragmented system, technical histories, and extended course along the entire east coast provide a background for this book's islands.[52]

Birds and boaters use this corridor. The overlap of migration routes with patterns of travel and tourism means that the spoil islands provide stopovers and nesting sites along the East Coast's flyways and also host sailors, bathers, vacationers, squatters, and pioneers, often on a similarly seasonal basis. This combination of flyway and lifeway is not new, and it points toward shared histories. In Manhattan, the Dutch first extended their new island home with garbage; and in places like Florida's central Atlantic coast, Indians created the first spoil islands with discarded shells that formed refuse mounds. Along the Indian River's one hundred mile course, more than two hundred spoil islands remind us of the equally extensive shell midden archipelago that existed a thousand years ago.

The history of spoil islands does not follow a linear chronology. Just as ancient shell middens might be compared to modern spoils, dredging is a cyclic process. It often requires repeated maintenance, and it is at the mercy of the cycles of the tides. And dredging is a kind of industrial archeology in which the dredged islands reveal geologic time. If New York City builds islands on its substrate of granite, then Biscayne Bay must rely solely on sand and marl and Key West constructs its islands like a reef, speeding up the slow, porous accumulation of delicate polyps with sand and broken coral. In Miami, this sand is enough to hold a real estate sign—as Will Rogers quipped. In New York, two of its human-made islets almost held billboards. In Key West, one spoil island boasts some of the nation's most prized real estate, while the other remains undeveloped. And the East Coast's shallow waters harbor many other unexpected juxtapositions.

Throughout this archipelago, the spoil islands have been variably contested, preserved, and forgotten spaces. Paired terms at the start of each chapter highlight these paradoxical uses, contradictory experiences, and administrative debates. These terms set up the geographic, philosophic, and environmental frameworks for each section. They are not deployed in strict opposition but in the spirit of the rich dialogue and negotiation that spoil islands necessitate. The dialectic of site and nonsite proposed by Robert Smithson, whose own work parallels the functions and meanings of spoil islands, serves as an overarching frame for these paired terms. For Smithson, the terms were not antithetical but relational. The site was both the expanded field and its precise territories. The nonsite incorporated the "raw materials" of the site. Smithson might bring back excavated—and now displaced—samples from a remote site to a new context, such as a museum floor or studio space. The nonsite "pointed to" the site and "contracted" its area. For Smithson, the relation of site and nonsite was also a "dialectic of mind and matter."[53] As sites, spoil islands can be occupied as places for picnicking or camping. As nonsites, they are displaced materials that become places of reckoning. Smithson's description of his project *Spiral Jetty* demonstrates how that project mixed site and nonsite just as spoil islands provide a setting to negotiate often contradictory conditions: "My dialectics of site and nonsite whirled into an indeterminate state, where solid and liquid lost themselves in each other."[54]

Each chapter investigates sections of this archipelago on the route from New York City to Key West. Along the way, the east coast's bays, ports, channels, and waterways hold a diverse array of dredged islands that register emergent relations between culture, waste and nature. Studying their historical background, process of formation, status, multivalent uses, and their relation to the broader typology of artificial islands supports the objective to understand the contemporary meaning of spoil islands. Having begun in the preface with the outlier islands off Florida's Gulf coast, the book's narrative sequence follows a chain of dredged islands from the New York Harbor, southward along the Intracoastal Waterway to Florida's east coast, down the Indian River, past Peanut Island, across Biscayne Bay, to a pair of islands just west of Key West at the southern limit of the continental United States. Between each chapter, shorter accounts of singular islands focus on key concepts, provide transitions between larger case studies, and approximate station points, rest stops, and navigational beacons. These intermezzo texts require two detours from the east coast's waterways—the Kissimmee River in south central Florida and Fred Howard Beach in the Gulf of Mexico. Both islands relate to the broader network of waterways and consequently point toward conceptual extensions beyond the book's geographical focus.

Smithson's project for a barge initiates the circumnavigation of New York City's harbor and its human-made islands. In chapter 1, the artificial islands of New York challenge the concept that the modern city can abandon the urban condition for a new nature. These islands' blurred distinction between the human-made and the natural requires that an ecological perspective also include public space, social structures, and infrastructure, just as it must also address the genealogies of urban waste. After a look at the removal of natural islands, which were seen as obstacles to the infrastructure of navigation, the construction of Hoffman and Swinburne Islands frames how purpose-built islands provide places of institutional isolation, afford flexible sites for the expansion of urban space, and register social hierarchy in their geographical placement. The chapter then follows the changing perceptions of the Harbor's constructed islands, which have been at the center of modifications of the city's natural and infrastructural contours, moving from isolation to recreation and development and finally to conservation. Remaining in the Harbor, the transitional section analyzes a famous photograph of Robert Moses poised above the East River's "rip-rap" where he might survey some of the islands that for him became tools of urban planning and where we might reflect on the nature of infrastructure. Chapter 2 then closely reads a small, often overlooked, spoil island in the East River. More recently known as U Thant Island, the rocky outcropping originally named for August Belmont demonstrates how infrastructure, where nature comes to reside, might provide spiritual focus. For United Nations employees and Sri Chinmoy's followers who view it from the United Nations headquarters building, the East River's landscape approximates a garden, and the island's spoil material becomes a substrate for meditation and peace.

A brief study of how mounded spoil becomes both monument and memorial makes the transition to the middens that introduce the next chapter. Chapter 3 explores the dynamics of public and private on spoil islands. In the Indian River, the dredged channel was a sporting paradise for fishermen and yacht owners—many of whom avidly followed the progress of dredging in 1933 editions of the *New York Times*' sports pages. The waterway's commercial value was ceded to its recreational potential. The Indian River's shallows also yielded hundreds of unclaimed, public-domain islands where local residents soon began to build weekend cabins across from their mainland properties. With increased awareness of the islands, management shifted from private caretakers who operated with little state oversight to environmental agencies who sought "optimal use and control" of the islands. Disregarded islands became central features of coastal management policy identifying the islands' value for deposition, conservation, recreation, education, even the aesthetics of landscape. In their residual presence as spoil material, these islands have also endured as areas for common wildness. After a stop on Nettles Island's dredged material supporting an exceedingly dense settlement pattern, chapter 4 focuses on Peanut Island where utilitarian functions coexist with cultural and recreational programs to transform dredged sites into multi-use and multivalent places. The island includes John F. Kennedy's vacation bomb shelter, the port-of-call for the replica HMS *Bounty*, a Coast Guard station and maritime museum, a preserved coastal hammock, and a depository for dredged material. This mix of use and culture makes the case for the spoil island archipelago as a cultural landscape of national significance.

Between chapters 4 and 5, a spoil island fish camp on the Kissimmee River shows how spoil islands, in spite of their marginal placement, foster this communal space and answer debates about land tenure and state sovereignty. Chapters 5 and 6 then focus on Biscayne Bay as a surreal landscape of developers' aspirations, informal recreation, and environmental art. Inclining toward the utopian, even at times the piratical, visions for the bay's artificial islands defy economies of scale, planning conventions, and societal standards. Tourists arrive at the world's largest cruise terminal—the purpose-built Dodge Island, while a few miles north criminals on probation squat on a spoil areas below the Julia Tuttle expressway, their driver's licenses issued to addresses on Picnic Island #4. The bay provides an unprecedented zone for colliding rationales of purpose-built islands, sometimes irrationally conceived, with the unintended function and settlement patterns of dredge spoil islands, formed as leftovers from the technical exigencies of the bay's intricate dredging program. Sixty years before Christo surrounded the spoil islands with frangipani pink, developers envisioned a "modern Venice" and fought a much easier battle for even more radical changes to the bay's disposition—purchasing bay bottom, applying for dredge permits, setting wood piles to claim territory, and pumping dredge material to form the Venetian islands. In 1983, Christo made visible this collision of the realities and surreal natures of the bay during his two-week installation titled *Surrounded Islands*. Impossible to execute without the indefinite status of the eleven spoil islands it wrapped with nearly seven million square feet of pink woven polypropylene fabric, the project demonstrated possibilities—even if flawed—for occupying landscapes on a large scale without permanently altering their physical presence. Its island afterimage points toward a natural contract that draws social contracts into the worlds of art and experience.

The transitional section "Sand" looks at spoil's use in island forms of beach augmentation, and then chapter 7 demonstrates how spoil islands resist political and social order and at the same time confirm an unlikely logic of manufactured land. At a nation's terminus, Key West defers its westernmost status to two spoil islands, one developed for high-end residence and the other resisting annexation in spite of its rich histories of utility. To those looking west from Mallory docks, the two islands present a stereographic view that oscillates between order and disorder. On the left, Sunset Key, previously called Tank Island, is densely developed with houses offering a formal plan of development but patently disrupting the natural sunset. On the right, Wisteria Island accommodates transient sailors and squatters and waits for decisions about its fate, while its invasive Australian pines (it is also known as Christmas Tree Island) wave riotously in eastern trade winds. Between order and disorder, these spoil islands offer a place of reckoning, not unlike the 2005 construction of Robert Smithson's barge project, which makes the transition to the book's concluding chapter. This final chapter addresses the provocation of the allure and revaluation of spoiled places and proposes that spoil islands are symptomatic of twenty-first-century discussions of nature, waste, environment, and public space. Working from Smithson's archipelago of "hypothetical islands" and his *Island of Broken Glass* project, the chapter summarizes the idea of spoil islands as fluid landscapes that are consistently altered by human intervention and currents and tides—being built up as much as they are eroded and shifted and recalling explorer Louis-Antoine de Bougainville's query: "Is this extraordinary land rising, or is it in ruins?"[55]

ISOLARIO

Illustrations for this book come out of the tradition of *isolario* books. These sixteenth-century texts mapped the known world's islands. For Italian cartographer Benedetto Bordone, in particular, his islands were as much story as geography. Bordone hoped his *isolario* would serve as a navigational guide for readers as well as sailors. His rendered woodcuts present stylized, scalloped shorelines that approximate each island's shape and hint at a largely unexplored global archipelago. Drawn by skilled hands, the curved arcs of coastline are truly human-made, born from manual craft rather than natural process. While not subscribing to artifice, this book's illustrations acknowledge the sometimes indefinite nature of mapping and narrating its subject. But this blurring of boundaries is also an opportunity, and the illustrations seek to capture the multivalence of residual waste and the complexity of land-to-water and water-to-land processes—what Robert Smithson called the "accident of precision."

Topographic maps, nautical charts, and aerial photographs have served as base imagery for the illustrations. Within traditions of ichnographic map-making, these resources generate documents that contrast with Bordone's pictorial map-making and, for a moment, counter the unexpected nature of spoil islands. But their cartographic translation also retains the geographic—sometimes even psychogeographic—narrations of the *isolario.* To account for the figure-ground play of dredging operations, land above the high tide line has been left white. With spoil islands, solid material becomes voided ground within watery expanse. The islands share the figural characteristics of the public spaces rendered as white zones in Nolli's map of Rome. Dredged material becomes an outdoor room—not carved out of an urban fabric as in Nolli's map, but deposited within a fluid framework.

The illustrations occur at the end of each chapter as afterimages that ask readers to reflect on the preceding island stories in relation to the map's layered spaces. I have included two maps at the end of this introductory chapter to set up the range of scales and intentionality that will be found in subsequent illustrations. Craney Island offers the grand vision of the engineered island, while the Delaware River's "places of deposit" acknowledge the informal nature of the islands that make up the majority of this book's cases. Set against the backdrop of mainland shorelines and urban edges (as in Miami or New York), the islands are figures, open-air elements that have the constructed aspect of a park. Together, they suggest an unprecedented ribbon of inherently public space cast along a nation's margins. This book of islands—now spoil islands—provides a form of illustrated travel literature for the undiscovered, post-industrial, and ruderal archipelago. It tracks this makeshift landscape to understand how residue and waste inform human activity at new intersections of cultural, technological, and natural imperatives.

NOTES

1. Gilles Deleuze, *Desert Islands and Other Texts* (Los Angeles: Semiotext(e), 2004), 10. Here, I am working from Gilles Deleuze's analysis of island types in his essay "Desert Islands." Deleuze discusses continental and oceanic islands, and spoil islands borrow from the characteristics of each.

2. D. H. Lawrence, "The Man Who Loved Islands," *Dial*, vol.83, no.1 (July 1927): 1–25. The story has also been anthologized in *Selected Short Stories*, ed. Michael Lockwood (Cambridge: Cambridge University Press, 1997).

3. Mark W. Holley, *The Artificial Islets / Crannogs of the Central Hebrides* (Oxford: British Archeological Reports, 2000), 2.

4. Lawrence, 1.

5. Lawrence, 3.

6. D. H. Lawrence, cited in Edward Nehls, *D. H. Lawrence: A Composite Biography* (Madison: University of Wisconsin Press, 1959), 265.

7. In *The Plumed Serpent*, written late in his life, Lawrence's ideal island has become a wasteland, and the "disheveled island" was "all dry rock and thorny bushes, and scorpions," signaling apocalyptic endings. See *The Plumed Serpent*, ed. L. D. Clark (Cambridge: Cambridge University Press, 1987), 258, 316.

8. Bruno Munari, *From Afar It Is an Island* (Cleveland, OH: World Publishing Company, 1971), 2.

9. Munari, 29.

10. Mary Douglas, *Purity and Danger* (New York: Routledge Classics, 2002), 50.

11. Georges Bataille, ed., "Critical Dictionary," *Encyclopaedia Acéphalica: Comprising the Critical Dictionary and Related Texts*, eds. Robert Lebel and Isabelle Waldberg, trans. Iain White (London: Atlas Press, 1995), 51–52.

12. See the entry for "spoil" in the *Oxford English Dictionary*, 2nd ed. (New York: Oxford University Press, 1989).

13. Friedrich Nietzsche, *The Will To Power*, trans. Walter Kaufman and R. J. Hollingdale (New York: Vintage, 1968), 25.

14. See Sophie Gee, *Making Waste: Leftovers and the Eighteenth-Century Imagination* (Princeton: Princeton University Press, 2010). See also Gary Hawkins and Stephen Muecke, eds., *Culture and Waste: The Creation and Destruction of Value* (Lanham, MD: Rowman and Littlefield, 2003), in particular Suzanne Raitt's chapter titled "Psychic Waste."

15. For discussions of the revitalizing potential of waste, see Douglas, *Purity and Danger* (1966/2002) and Michael Thompson, *Rubbish Theory: The Creation and Destruction of Value* (Oxford: Oxford University Press, 1979).

16. Jean de Renou, *A Medicinal Dispensatory*, trans. Richard Tomlinson (London, 1657), 318, cited in the *Oxford English Dictionary* (1989).

17. Edward Casey, *Getting Back into Place* (Bloomington: Indiana University Press, 1993), xi.

18. Casey, 3.

19. See Richard Mabey, *Weeds: In Defense of Nature's Most Unloved Plants* (New York: Harper Collins, 2010), 3, 77.

20. For an example of these classifications, see Office of Coastal and Aquatic Managed Areas, *Indian River Lagoon State Buffer Preserve Land Management Plan* (Tallahassee, FL: Florida Department of Environmental Protection, 2004), 1. Ruderal areas comprise approximately seventy-five percent of the buffer preserve's total acreage. The three major types of disturbed areas in the preserve's parcels are spoil islands, citrus groves, and impoundment berms for mosquito control.

21. See the entries for "ruderal" and "ruderous" in the *Oxford English Dictionary* (1989).

22. There is some ambiguity about whether pioneer plants are only exotics because some designations oppose pioneer to native, but others use the term for plants such as lichens that attach to rocks.

23. See Colin Ward and Dennis Hardy, *Arcadia for All: The Legacy of a Makeshift Landscape* (Nottingham: Five Leaves Publications, 2003).

24. Ward and Hardy, 3–4.

25. Ward and Hardy, vii.

26. Thompson, 99.

27. Peter Sloterdijk, *Critique of Cynical Reason*, trans. Michael Eldred (Minneapolis: University of Minnesota Press, 1987), 151. See also the chapter "Acculturation of Waste" in *Waste-Site Stories*, eds. Brian Neville and Johanne Villeneuve (Albany: State University of New York Press, 2002).

28. Quintus Curtius Rufus, *The History of the Life and Reign of Alexander the Great*, trans. P. Pratt (London: Samuel Bagster, 1809), 354. The made land was two hundred feet wide and one half mile long. The word *molos* is Greek for "mass."

29. Lyndon B. Johnson, "Remarks at the Ground-Breaking Ceremony for the Florida Cross-State Barge Canal," February 27, 1964, online by Gerhard Peters and John T. Woolley, *The American Presidency Project*, http://www/presidency.ucsb.edu/ws/?pid=26085, accessed January 11, 2012.

30. John B. Bonds, "Opening the Bar: First Dredging at Charleston, 1853–1859," *The South Carolina Historical Magazine*, vol.98, no.3 (July 1997): 236.

31. Bonds, 238.

32. Bonds, 244. Cummings Point anchors the northern end of Morris Island, which has continued to be a spoil deposit site throughout the twentieth century. A private landowner on part of the island contested the state's ownership of a seven-hundred-acre spoil area on the island in 1975 and again in 2011. South Carolina's Supreme Court ruled in favor of the State Ports Authority's ownership in 2012, and the Corps of Engineers will maintain its easement until 2017.

33. Bonds, 245.

34. Bonds, 250.

35. John Huston, *Hydraulic Dredging* (Cambridge, MD: Cornell Maritime Press, 1970), 1–2. In 1977, the same year as the Clean Water Act, J. B. Herbich also pointed out that "until recently" methods of dredged-material disposal were based "almost entirely on economics, with little consideration of the physical properties of the area." See Y. K. Vyes and J. B. Herbich, "Erosion of Dredge-Material Islands Due to Waves and Currents," *Proceedings of the 2nd Annual Symposium on Dredging Technology*, vol.1 (Cranfield, UK: Center for Dredging Studies and BHRA Fluid Engineering, 1977), n.p.

36. Benjamin T. Blankinship, "Problems and Challenges in the Dredging Program of USACE," *Proceedings of WODCON World Dredging Conference* (1975), 26.

37. The Delaware River continued to pose navigational challenges throughout the eighteenth and nineteenth centuries. In 1804, Oliver Evans drove his amphibious digger—reputedly North America's first wheeled vehicle—in the river to clear impediments and deepen the channel. In 1829, Congress provided funds for the "procurement" of a dredging machine to deepen the river. The ladder dredge was soon supplanted by dipper dredges. See Frank Snyder and Brian Guss, *The District: A History of the Philadelphia District of the U.S. Army Corps of Engineers, 1866-1971* (Philadelphia: U.S. Army Engineer District, 1974), 65–66.

38. See Chief of Engineers, United States Army, *Annual Reports of the Chief of Engineers* (Washington, D.C.: Government Printing Office, 1879), 463–464.

39. Snyder and Guss, 68. See the illustration at the end of this chapter.

40. Twentieth-century dredging manuals have sometimes used the Delaware River as a case study of deposition strategies.

41. Snyder and Guss, 69–70.

42. The island also bears anomalies of "local magnetic disturbances" and surveying errors that place its northern tip in Delaware rather than New Jersey. In the 1970s, researchers estimated that sites for spoil deposit would reach capacity by 1990 and recommended the development of new strategies for disposal. Proposals included a dredging plant designed

exclusively for the river along with equipment to transport dredged material up to fifty miles away. See W. L. Murphy and T. W. Zeigler, *Practices and Problems in the Confinement of Dredged Material in Corps of Engineers Projects*, DMRP Technical Report D-74-2 (Vicksburg, MS: U.S. Army Engineer Waterways Experiment Station, May 1972), 13–15.

43. In addition to NEPA, a set of other policies regulated dredging, spoil deposition, and island construction. The Fish and Wildlife Coordination Act (FWCA) from 1934 already provided a degree of oversight, and 1972 amendments in Section 404 increased the reach of the Federal Water Pollution Control Act (FWPCA), which was originally enacted in 1948. Also in 1972, the Marine Protection, Research and Sanctuaries Act (MPRSA), known as the Ocean Dumping Act, controlled the disposal of waste in open waters of the ocean.

44. Problems with unconfined dredge include increased water turbidity, disruptions to biotic communities, and shoaling along the channel that can interfere with drainage, natural flows and currents, and navigation.

45. Among these Acts, Section 9 prohibits the construction of dams, dikes, causeways, and bridges without Congressional consent, and Section 10 extends this restriction to projects—including excavation and fill activities—that will alter the waters' course. Section 13 of the Rivers and Harbors Act prohibits the discharge of refuse into navigable waters without a permit. Known as the Refuse Act, this section expanded the Corps' role administering and overseeing the maintenance and construction of ports and waterways. It also allowed the Secretary of War, with the Chief of Engineers' judgment that navigation would not be impeded, to "deposit any material above mentioned in navigable waters . . . provided application is made prior to depositing such material." The Clean Water Act of 1977 built upon this earlier, turn-of-the-century legislation.

46. In the 1960s, a more global approach to dredging research had begun to take shape. The industry's first technical periodical *World Dredging and Marine Construction* appeared in 1965, and a year later the Organization of World Dredging Association (WODA) was formed. New environmental constraints also led to the Waterways Experiment Station (WES) in Vicksburg, Mississippi and the formation of the Dredged Material Research Program (DMRP) in 1973. A specific task for WES was "to assess the technological and economic aspects of developing disposal areas as landfill sites and recreational areas." WES also looked at possibilities for habitat development. Early conflicts between environmental oversight and the Corps' use of island deposits are evident in one author's comparison of the Sabine Lake spoil island along the Texas coast to a "dredged material island" in the Cape Fear River of North Carolina. The author notes that Cape Fear's island hosted royal terns in the first year after it was first deposited. The island soon became a refuge and was off limits in 1975 for further deposition. At Sabine Lake, the spoil island includes a habitat for roseate spoonbills, but it was still being used for deposit. The author notes ironically that continued dredging and island augmentation are the only things that will keep the island in place for the bird colony. (Blankinship, 32, 34)

47. Confined disposal organizes spoil in diked intertidal or upland facilities. It is one of three categories for material placement currently used by the Corps. Open water disposal places material in rivers, lakes, estuaries, or oceans without containment. Beneficial use puts the material toward a productive purpose, such as beach nourishment or wetland creation. Island formation comes about from all three types of disposal, but most of the dredge deposition that now builds islands is confined. In the early 1930s, USACE districts in Sacramento and Philadelphia used confined disposal sites, but projects to confine dredge material began intermittently and at a small scale. See W. L. Murphy and T. W. Zeigler, *Practices and Problems in the Confinement of Dredged Material in Corps of Engineers Projects*, DMRP Technical Report D-74-2 (Vicksburg, MS: U.S. Army Engineer Waterways Experiment Station, May 1972), 3.

48. The island was a strategic point of defense and guarded the Elizabeth River entrance to Norfolk with seven guns. On June 22, 1813, the USS *Constellation* repelled a British offensive, which had planned to march on the island during low tide. Craney also witnessed the Battle of Hampton Roads, famously known as the first battle of the ironclad vessels during the Civil War. The USS *Monitor* fought to an indeterminate finish with the CSS *Virginia*, which had been salvaged from the USS *Merrimack*. Early on May 11, 1862, with Union forces occupying Norfolk, the *Virginia* was run ashore on Craney's southeast shoals. The vessel was set on fire, and the explosion of its powder magazine sent debris across the island and into the river's

channel. The wreck was designated a navigational hazard after the schooner *Priscilla* wrecked on its remains, which were variably salvaged for metal casting and souvenirs until USACE's 1942 dredging operations deepened the fuel terminal dockside from eighteen to forty-two feet. Archeologists noted the dredging scours that marked the wreck's site.

49. The Craney Island Management Plan (CIMP) was created in 1981. See Michael R. Palermo, F. Douglas Shields, and Donald F. Hayes, *Development of a Management Plan for Craney Island Disposal Area* (Vicksburg, MA: U.S. Army Engineer Waterways Experiment Station, 1981). See also Michael R. Palermo and Thomas E. Schaefer, *Craney Island Disposal Area: Site Operations and Monitoring Report, 1980–1987*, Misc. Paper EL-90-10 (Vicksburg, MS: U.S. Army Engineer Waterways Experiment Station, July 1990), Appendix A.

50. In 1984, the College of William and Mary and the Virginia Department of Game and Inland Fisheries formalized the island's bird protection project that has increased populations of piping plovers and least terns; and in 1988, the Bird Habitat Management Effort was initiated with least terns decoyed to five sites. In 1989, the first pair of piping plovers nested on the island, spurring a memorandum of understanding between William and Mary and the Corps.

51. The 2010 Census recorded a population of 112,642,503 (36% of the country's total population) for the East Coast states, extending from Maine to Florida. See "Resident Population Data," accessed March 17, 2012, http://2010.census.gov/2010census/data/apportionment-pop-text.php. For a look at global implications of coastal growth, see Don Hinrichsen, "The Coastal Population Explosion," accessed March 5, 2012, http://oceanservice.noaa.gov/websites/retiredsites/natdia_pdf/3hinrichsen.pdf.

52. Technically, ICW, for "Intracoastal Waterway," identifies the three-thousand mile route between Norfolk, Virginia and Brownsville, Texas, and includes both the AICW and the Gulf Intracoastal Waterway (GICW). Because of its common usage, ICW will be used throughout this book to refer generally to the intracoastal waterway. AIWW and AICW both refer to "Atlantic Intracoastal Waterway" as technical terms commonly used by the Corps of Engineers and coastal navigators. IWW is often used to designate the waterway between Jacksonville and Miami in Florida. Navigational practices along its route link the coastal pilot to far places. Since the ICW runs parallel to the coast, and the return to port is relative to each boat's itinerary, the mnemonic "red right return" is extended to include "red right return Texas," where the waterway reaches its southwest terminus. The Atlantic mariner thus imagines the furthest reach of the ICW to orient the boat to navigational markers.

53. Robert Smithson, *Robert Smithson: The Collected Writings* (Berkeley: University of California Press, 1996), 197-199.

54. Smithson, 8.

55. Louis-Antoine de Bougainville, *Voyage autour du monde* (1771), eds. Michel Bideaux and Sonia Faessel (Paris: Presses de l'Université de Paris-Sorbonne, 2001), cited in Alistair Sponsel, "Lords of the Ring," *Cabinet* 38 (Summer 2010): 55.

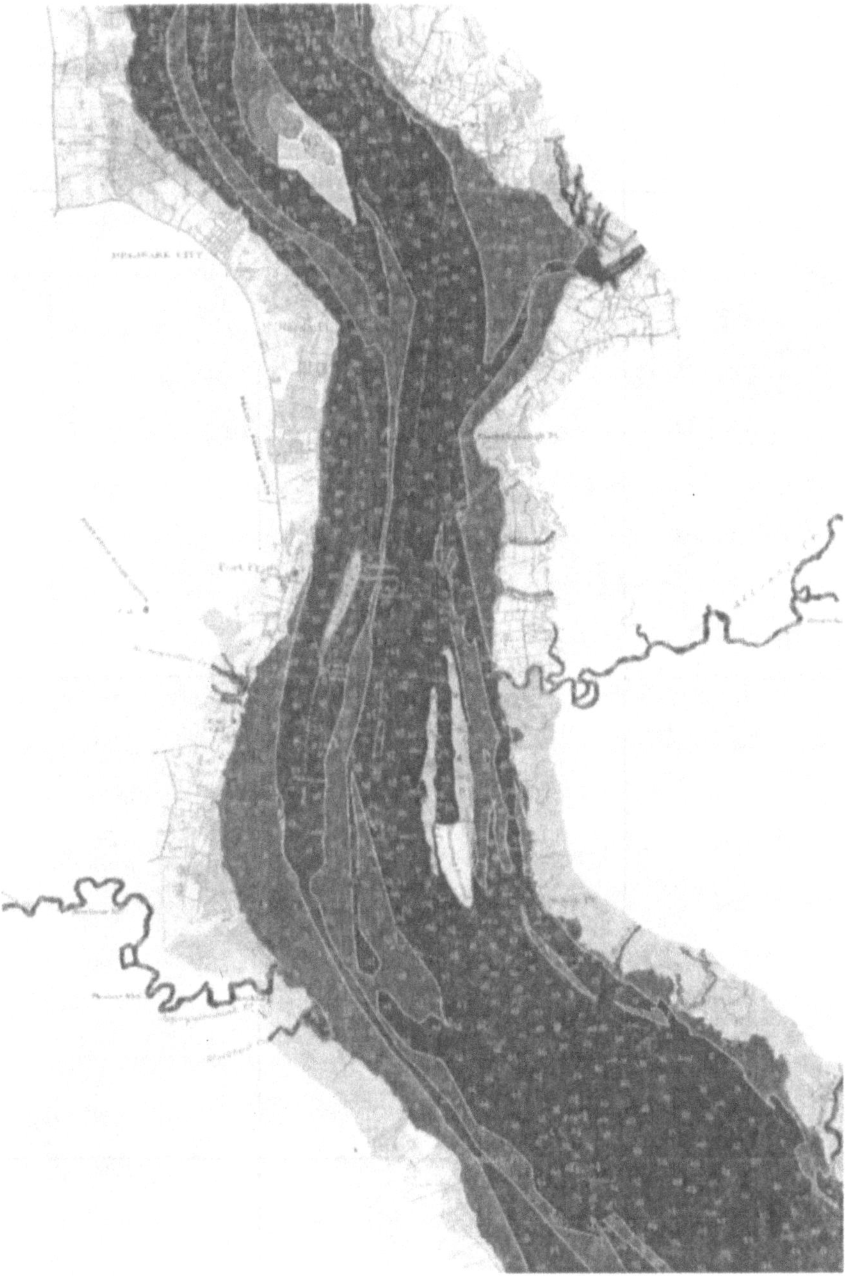

Churchland
Port
Norfolk
Parkview

Barge

Early in 1970 Robert Smithson made a simple sketch of a tugboat pulling a barge. The tug is reminiscent of a plastic bathtub toy, the barge is a squared planter, the New York skyline is so many wooden blocks, and the sketch is really the diagram of an idea. In this document titled "Floating Island to Travel around Manhattan Island," Smithson proposed to fill a barge with a weeping willow, bushes, a rock, a path, and trees "common to the region." This collection of landscape elements asks us to reconsider island, park, and the idea of nature itself. When he made this sketch, Smithson's applications for island projects near Vancouver had recently been denied, and in one respect the New York project mobilized the island site to circumvent further permitting problems.

The project's circumnavigation also serves to remind us of Manhattan's insularity. The barge's unstable ground offers an opportunity to rethink what that larger island's ground might be, and the floating island is a fragment of Central Park that moves Manhattan's center to its periphery. Smithson believed that the park's landscape architect Frederick Law Olmsted had found the "dialectics of landscape" in his original vision for the park, blending the beautiful and the sublime. Smithson's project excavates not only New York earth but also the underpinnings of Central Park as a "man-made wasteland."[1] As a nonsite to Manhattan's islandness and Central Park's "grotesque gardens," the barge holds Smithson's own vision for parks as "carriers of the unexpected and of contradiction on all levels of human activity, be it societal, political, or natural."[2] Displacing air, water, and earth, the vessel reminds us that we also make nature.

Smithson's proposed project refuted the idealism of the enclosed garden. It liberated a fragment of Central Park from its confined location. The barge vessel itself contains—but does not enclose—the floating island as it circumscribes Manhattan, which is the urban island that securely wraps its internal park. Smithson opened up the barge-park to the elements. Dragged behind its earnest tug, the barge exposes a fragment of parkland to New York harbor's roiling currents, its sea spray, coastal winds, and the public's gaze. And to heighten this sense of vulnerability, the barge carries dirt instead of garbage. Smithson hoped to exchange one infrastructure for another. Green infrastructure takes the place of brown and gray infrastructure, and parkland becomes wasted space. It is literally wasted because it cannot be occupied—no footprints mark the sketched path. It is also wasted because it appears to have been thrown away, discarded in the daily exodus of urban rubbish on its way out to the sea's dumping grounds.

Smithson's waste-park was destined not for open waters but for trips "around Manhattan Island." The artist was proposing an infrastructure that linked idealism to pragmatism, paradise to degradation, and landscape to garbage. He also asked the question what if dirt—or waste—is displaced and comes to rest at various points around Manhattan. Here? Or here? Or maybe over here? New York already had an extended history of land made from dredged and discarded material. Within this context, his barge was there temporarily to remind us that we make waste just as we make nature. And both become infrastructure. Traversing Smithson's dialectic of site and non-site, the barge also negotiated the interplay of nature and infrastructure—nature becoming infrastructure, and human-made systems becoming substrates for natural systems. The barge was a tool for understanding this interaction: "Dialectics of this type are a way of seeing things in a manifold of relations, not as isolated objects. Nature for the dialectician is *indifferent* to any formal ideal."[3]

NOTES

1. Robert Smithson, *Robert Smithson: The Collected Writings* (Berkeley: University of California Press, 1996), 158.
2. Smithson, 160.
3. Smithson, 160.

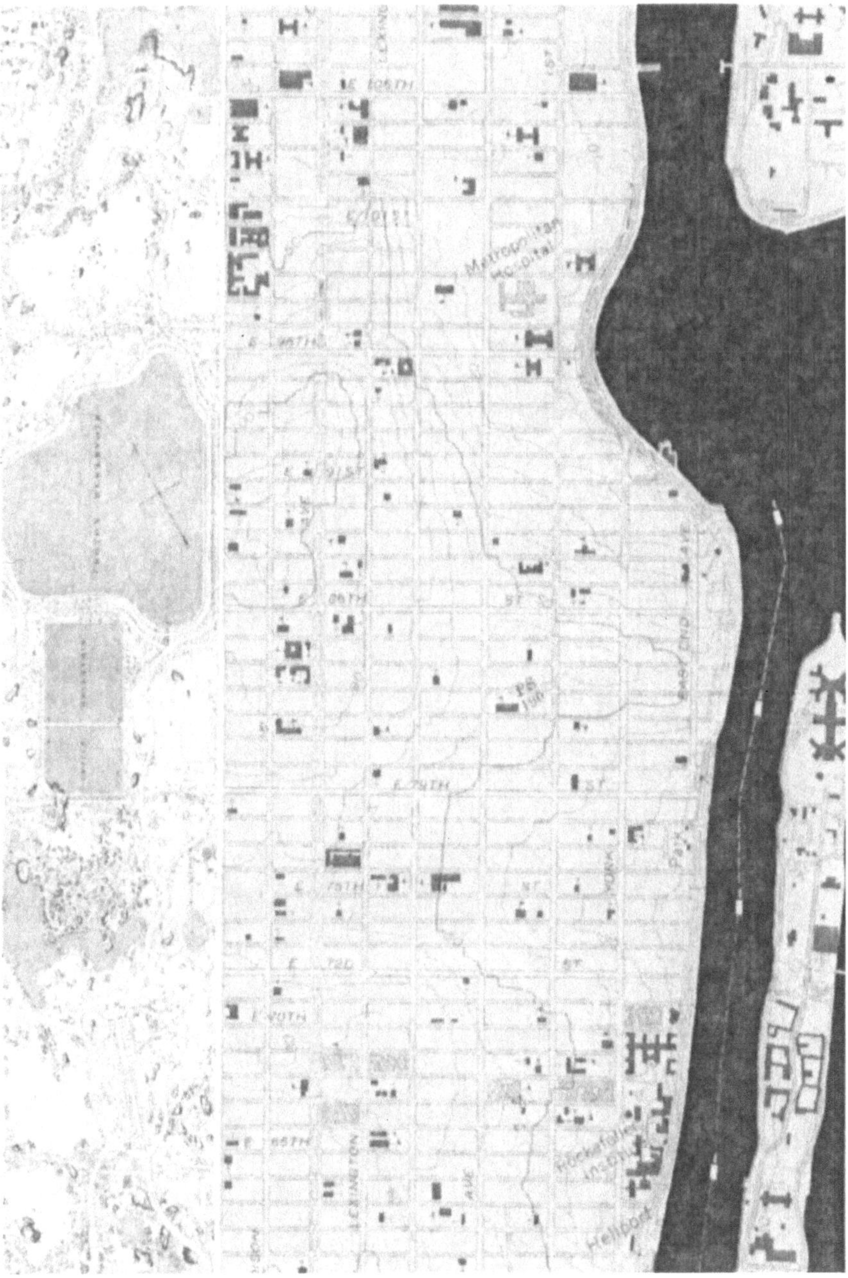
Metropolitan Hospital
LEXINGTON

Chapter One

Natural and Infrastructural

Building New York Harbor's Islands of Waste

As this interesting bit of property is entirely under water, the Commissioners charged with the Quarantine business will at once proceed to make an island.[1]

The long labor and thoughtful study of Gen. John Newton and his corps of assistants culminated in a momentary but magnificent display of upheaved waters, and another triumph of human skill over the resistance of nature was recorded.[2]

BURDEN AND OPPORTUNITY

On its circumnavigation of Manhattan, Smithson's barge passed through an archipelago that has represented both burden and opportunity for the city's planners, engineers, and residents. Its islands have offered protection, and they have also impeded navigation. They have been built up and taken down. Although often viewed as marginal places, the city's "other" islands—like the East Coast's spoil islands—have at times been ignored but other times have attracted attention, at the center of spectacles, accidents, and debates. In the mid and late nineteenth century, some existing islands were navigational obstacles, while other newly formed islands were prospects for separating social functions and extending buildable land. Through all of this, the islands—whether naturally formed, augmented, or human-made—demonstrate how nature serves infrastructure and how infrastructure serves nature. This dialogue between nature and infrastructure points out that neither term can stand on its own. The twining of the two elicits a revised framework, one that

Smithson alluded to when he exchanged spoil for soil and garbage for vegetation.

In the fluctuating Manhattan archipelago, Smithson found a zone of contention and transition that paralleled the artistic process and by extension informed the broader projects of urbanism and living. The islands' blurred distinction between the human-made and the natural requires that an ecological perspective also include issues of social power, capital, and public space, just as it must also address the genealogies of urban waste. None of New York City's islands have avoided human intervention. They challenge the concept that the modern city can abandon its urban condition for a new, or reinvented, nature.[3] Over time, islands of natural origin have been augmented or obliterated, and new islands and new land have displaced water to extend the urban context. Most of New York City's artificial islands are purpose-built, but one island in particular came about from construction waste, and its origins and legacy touch on the histories and implications of the rest of the city's archipelago. Named for two people, an engineer and a Buddhist diplomat, this island lingers—residual waste that it is—between islands constructed and islands destroyed. In so doing, Belmont Island—named also for UN Secretary General U Thant—approximates the dialogic space between nature and infrastructure, between mind and matter.

RUGGED BROWN SPOTS

> The crowds of people and boats were ranged in something like a circle around a space about a mile in diameter, in the centre of which lay Great Mill and Little Mill Rocks. People looked at these rugged brown spots with feelings that were deeper than respect. . . .[4]

Along with its changeable currents as a tidal strait, the East River's reefs and rocks have proven hazardous. Belmont Island rests on one of the natural rock outcroppings that plagued navigation of New York's waters. These reefs rise up in the middle of the East River to create two narrow channels between Manhattan and Long Island. A northbound navigator would first pass Man-o'-War rocks, on which Belmont rests, and then continue along the reefs of Roosevelt Island, originally called Blackwell's Island. Further north, Blackwell's reefs give way to the remnants of Flood Rock and a perilous turn known as Hell Gate, flanked—or, more accurately, guarded—by two other legendary rocks, Great Mill and Little Mill. These islands were joined, and Flood Rock was destroyed, but such obstacles have historically reduced the East River's ship traffic, limited vessel size, and impeded connections between New York's harbor and Long Island Sound.

According to estimates in the 1850s, more than one thousand ships ran aground each year, and one in fifty wrecked. French engineer Benjamin

Maillefert was hired in 1850 to address these navigational concerns, and his first project cut four feet from Pot Rock, yielding an eighteen-foot navigational clearance after the detonation of 284 charges.[5] But other hazards remained, affecting recreational and commercial boat traffic alike. On August 12, 1880, the excursion barge *Republic* was stranded on Flood Rock after the vessel's rotten hawser rope broke away from its tug. The barge's twelve hundred passengers, all members of the Corinthian Freemasons Lodges, landed on the rock by way of a wooden plank extended from the stranded vessel.[6] In 1909, another obstacle known as Little Hog Back Ledge was still wrecking ships, causing on average one million dollars in damage annually.

New Yorkers feared and admired these rocks, while navigators cursed and respected them. The names of the obstructions varied from the poetic to the severe: Frying Pan, Ways Reef, Heel Tap, Shelldrake Rock, Bread and Cheese, Gridiron.[7] Other naming conventions borrowed from racial difference and fear of otherness to characterize the full range of perception toward these outcroppings.[8] The rocky islands destroyed and also saved. The Freemasons could attest to this, but the islands' removal was tantamount to a growing city's infrastructure, and nature was seen as a surmountable obstacle. If tunnels could be cleared below the East River, rocks could certainly be removed above. The same year engineers began surveying the East River for a suitable tunnel location for rail connections between the boroughs of Queens and Manhattan, other engineers were preparing to obliterate Flood Rock from the northern part of the East River's channel. Five years after the stranding of the *Republic*, its unexpected island refuge would be gone.

At precisely 11:13:50.2 a.m. on October 10, 1885, Mary Newton, the eleven-year-old daughter of General John Newton, pushed the button that detonated what was at that time the "greatest single charge of explosives ever used." The impulse was sent from the command center and small observation stand for officials across the water in Astoria, one thousand feet from the explosion under Flood Rock. Newton, at the age of two, had triggered a similar though smaller explosion at Hallet's Point nine years ago, and she stood unfazed as reporters took notes to describe the scene: "There was a single instant of breathless silence. Then there came a sudden jar of the solid rock underneath the feet of the little group of spectators, and with a choked crash and muffled roar, the great volume of seething and foaming water sprang into the air." Soon after, a "great tidal wave rolled in upon the beach."[9] Reports varied, but rock lifted at least 50 feet above the river and columns of water rose from 150 to 200 feet in the air.

The event conquered one part of nature, but reports relied on comparisons with other natural elements to measure its impact. The following day's *New York Times* article described the spectacle reverentially and with a hint of religious zeal, noting that fishermen not just harvested but "ladled" dead fish out of the "upheaved waters." Natural attributes were also transformed as "yellow waters of the river were riven" amidst a "ragged smoking dun-brown mass," and the "whole run of the tides had been altered." Another account called the explosion the "greatest artificial earthquake in history," and reflected that "as a spectacle, it was simply an Icelandic or Yellowstone geyser on an extended scale."[10] Lasting from thirty to forty-five seconds, the explosive charge was felt in Princeton, New Jersey, where windows rattled at 11:14:47 a.m., and it was measured as far away as Harvard University, 197 miles distant. Another commentator William Hosea Ballou noted that this amount of explosive material would have leveled—he says "obliterated"—Manhattan Island if it had been placed on the surface. But ten million cubic yards of salt water tamped down the subaqueous and subterranean explosion. Ballou ruminates whether the Flood Rock blast might help settle the question of what causes earthquakes. Unfortunately, it does not; and Ballou asks rhetorically and a bit wistfully: "What mighty . . . ingredients combined to explode Java?"[11] Not the rack-a-rock charges and Number 1 dynamite used on Flood Rock.

As many as fifty thousand spectators in buildings, on boats, and along the shore had watched three hundred thousand pounds of explosives reduce the nine-acre island to rubble. Dredges and barges would spend the next decade removing the debris and placing it between Great Mill and Little Mill. The leftover material had to go somewhere, and the newly formed island of Mill Rock is a reminder of the spectacle's scale and nature's resistance. After its destruction, twenty-two feet of water was left over the top of Flood Rock, and, although its removal did improve the chances of safe passage, the Hell Gate channel would remain a problem.

After the explosion, New Yorkers explored the ruins of Flood Rock. It remained as a surreal landscape that mixed natural and human-made rubble. One report referred to the site as a "temporary island." The natural island was now an ephemeral place that combined a shifting, broken terrain with the makeshift wooden platforms and scaffold that had been erected as the work's staging area over the outcropping. This October 12th account summarily described the island as a "complete wreck" and inventoried at length the destruction: towers fallen east and west, a "bottomless"—and no doubt sublime—pool where the shaft entrance had been, and remnants of shattered brick buildings, their pulverized clay now plastering the exposed rocks with a red coating, as if the rocks had been painted.[12] Destruction of the rock itself was also effusively described as fragments pierced by heavy bolts and broken rock "thrown in heaps like sand."

This newly formed wasteland became a tourist destination. More than fifty boats spent the following day ferrying thousands of visitors to and from the island's remains. Ad hoc entrepreneurs envisioned commercial success in the wasteland. One New Yorker sold makeshift souvenirs, peddling signal wiring as "original Flood Rock wire" for ten cents a foot; and an industrious boy used pieces of the wooden scaffold to make a raft, which he towed ashore to sell. Other "relic hunters" looked for mementoes in the upturned geology, finding unusual rocks and shells and lumps of red clay. At one point, more than two hundred visitors stood on what had become unstable ground. Mine workers reported that the island had sunk more than two and a half feet throughout the first day after the blast. High tide now washed over the island's new upland, which had previously risen more than nine feet above tidal floods. Harbor pilots ferrying visitors also noticed that alterations to the rock had transformed currents and eddies and warranted a completely "new study of the stream."[13]

A RESPECTABLE LITTLE ISLAND

While miners were preparing explosives for the Flood Rock demolition, engineers two miles down the tidal strait were surveying for a subway tunnel to link Manhattan and Queens. Engineers from the newly formed East River Railroad Tunnel Company soon chose a path that aligned with the east-west axis of Manhattan's 42nd Street. The line crossed under Man-o'-War reef south of Blackwell's Island, and plans called for the location of a mining shaft through the rocky outcropping. The East River was known for its rolling underwater topography. Its humps of granite gneiss were typically called "reefs," indicating that they posed dangers similar to coralline outcroppings further south in the tropics. Though impediments to navigation, the reefs were in this case natural expedients for island construction. And even though the site also posed challenges to digging, the centrally located shaft would provide intake for air, access for excavators, and the meeting point from each side of the operations. The reef's natural formation was a means to an end that would eventually create Belmont Island.

In 1887, the New York and Long Island Railroad Company took over planning of the project, but it was another four years before enough capital could be raised. The well-known piano-maker William Steinway became one of the project's main investors in July 1891, and excavators finally broke ground a year later. At the time, only the Brooklyn Bridge crossed the East River, and the tunnel would help link Steinway's estate interests in the 5th Ward of Long Island City to Manhattan's urban core. Construction continued from June 1892 until an explosion killed five workers and injured twenty after only seven months. Unlike Flood Rock's spectacle, this explosion brought negative publicity to the dismantling of New York's stubborn geology. The event combined with early indications of the 1893 financial panic and persistent water infiltration problems in the tunnels to stop the project. On February 2, 1893, the tunnel entrances were boarded up. Later that year, natural forces mirrored economic weather—and a possible fate for islands—

when a hurricane destroyed Hog Island on August 24, 1893. Belmont Island had not yet been built, and Man-o'-War reef remained a navigational hazard, the East River's oily waves lapping over granite gneiss exposed at low tide.

Steinway died in 1896, and the tunnels that kept his name would lay dormant for six years until another investor could hold the financial reins. Before he bred the racehorse Man o' War, August Belmont revived the project that would ultimately bring the first passenger tunnel under the East River and across Man-o'-War reef. Tunnel excavation began again from each side—on July 14, 1905 in Shaft Four on the Queens side, and on September 1st in the Manhattan side's Shaft One. The project's pair of tunnels would proceed rapidly in spite of accusations of land-grabbing, permit revocations, safety concerns, and numerous accidents. The East River posed particular excavating dangers because of its "geologic diversity" mixing solid gneiss rock with pockets of mud and sand and unexpected dolomite limestone deposits.[14] There were also allegations that Belmont's franchise for the New York and Long Island Railroad Company was defective.[15] When it was determined that they were digging with an invalid grant at the start of the Manhattan side's second tunnel, Belmont's crew shifted the location of Shaft Two, partially built on city property, to an adjacent site entirely owned by Belmont. Work began again at noon on September 28; and even though newspapers would report authoritatively and expectantly on the tunnels' progress, the project would continue to be surrounded by mystery.

The emergent island in the middle of the East River partially unveiled the project's secrecy. As yet unnamed, Belmont Island was a visual cue to the subterranean activities and their infrastructural requirements. After the state legislature granted permission to dig in the waterway's easement, workers constructed an air-tight crib and coffer-dam to protect Shaft Three from tidal fluctuations over the submerged reef.[16] In the first week of October 1905, excavated spoil began to transform a "mass of rocks below the surface at low water" into what news reports called a "respectable little island."[17] Though favorably reviewed, the island did not meet size projections. Before the start of construction, engineers had estimated that tunnel excavation would yield a much larger island than its functional area of less than three hundred square feet. These calculations had perhaps ignored the navigational disadvantages of infringing on the already narrow channels in the East River. In the end, the island was large enough to host the small electric power plant installed by engineers to run the compressors that kept water out of the tunnels and powered high pressure drills and other machinery.[18]

The mechanized island made visible the hidden geology of a city's substrate while it also hinted at much larger operations below. The project's main contractor Degnon Construction Company employed twelve hundred workers in the two tunnels, each of which measured fifteen-and-a-half feet wide and extended more than a mile between the two boroughs.[19] Even though the overall scope differed, newspapers compared the work and its demand on workers to the concurrent Panama Canal project. Excavation of the tunnels proceeded rapidly, so fast that on November 5, 1906, reports proclaimed the project had set a new digging record—"little short of a miracle"—and surpassed ongoing projects like the Pennsylvania tunnels under the Hudson River and the McAdoo tunnel under the North River in Harlem. New York is a city of islands, but at this time it was also known as a city of tunnels. Journalist Arthur B. Reeve noted: "With sixteen tubes now built or building under the rivers which make New York City an island and eight others projected, Manhattan may be described as a body of land surrounded by tunnels."[20] In this fanciful but true description by Reeve, who was a mystery writer, tunnels substitute for water; and islands are defined, not by their natural insularity within water, but by the infrastructure that connects them. Far below the surface, the project's "monster moles," newly improved tunneling machinery designed by English engineers Sir Benjamin Baker and Henry Greathead, worked steadily on the twin passages.[21] Tunneling moved at a rate of about two feet per day, and by the middle of May 1907 the excavating "sandhogs" from each side were "conversing" under the East River, lacking only twelve feet of granite gneiss before joining the north tube's east and west sections.[22]

On August 14, 1907, Chief Engineer John Clark walked with fifty "adventuresome" invitees—including Columbia University professors and engineering students—from Manhattan through the iron-lined north tube to Queens and returned home in the yet-to-be completed south tube.[23] And on September 24th, the first electric car to pass under the East River reached speeds of forty-five miles per hour. The trips became a tourist attraction, but they did not yet link with other transportation networks. Their connection point in Queens on Long Island was still "as lonely a spot as there is in Long Island City," with marshland stretching four to five miles in circumference.[24] After the exhilaration of passing below the East River's tremulous waters wore off, the tunnels fell idle. Interborough Rapid Transit (IRT) used them for electrical transmission lines until 1913 when the city bought the tunnels to resolve a rapid transit regulatory dispute. Two years later—twenty-three years after the original start of construction—subway trains finally passed through the tunnels in the summer of 1915. On April 3, 1914, Shaft 3 on Belmont Island had been backfilled and sealed, officially ending the island's direct connection to the infrastructure project that produced it. The island now shudders—perceptible only to the birds nesting on it—with the passing of each subway car on the IRT Flushing Line #7 in the Steinway Tunnels below.

HOFFMAN AND SWINBURNE

The "respectable little island" fell into anonymity, silent to its origins, and still hazardous to navigation. To the south in the city's Lower Bay, two other artificial islands—carrying their own hazards—defined the southern margin of New York's extended archipelago. Hoffman and Swinburne were built as quarantine facilities and served the city's immigration offices from the early 1870s until the late 1920s. Their origin as purpose-built islands contrasts with Belmont's status as a by-product of excavation, but these islands nonetheless share connections to the early development of public works, and each island now provides a unique site to consider the changing relations of that infrastructure to other urban and natural systems. While Belmont Island was first linked to transportation, Hoffman and Swinburne became a part of another administrative archipelago and formed an isolated infrastructure of the institution.

Islands offered institutions clearly defined territories and, as necessary, isolation from planning restrictions or from a populace opposed to that institution's function. In New York, North Brother Island hosted tuberculosis patients, Randall's Island included potter's fields and homes for delinquent children, Wards Island was home to Civil War veterans and a homeopathic hospital, Rikers still imprisons about thirteen thousand inmates, and Blackwell's Island at one point held eleven institutions—for prisoners, mentally ill patients, and those afflicted with smallpox, paralysis, epilepsy, fevers, and other "incurables."[25] Construction of quarantine islands solved the city's problem of segregating incoming disease. Larger than Belmont and the harbor's quarantine ships but smaller than Ellis and Governors, Hoffman and Swinburne offered a malleable solution that could expand as the numbers of immigrants and transit passengers increased. The purpose-built islands were adaptable in size and function. Like Belmont, they are also easily forgotten. Hoffman is more than a mile and a half from Verrazano Narrows Bridge, and Swinburne hovers on the horizon the same distance from Staten Island. They are visible from the bridge and from the borough but occupy a middle ground, already crowded with vessels and navigational infrastructure. This relative distance offered the separation that city officials needed for a marginalized and maligned institution. And from Staten Island's South Beach, the Lower Bay's waters stretch out in what must have been a reassuring boundary to the borough's residents.

New York's quarantine facilities were first housed on Bedloe's Island in the mid-eighteenth century before they were moved to Governors Island and then relocated to Staten Island's Tompkinsville at the start of the nineteenth century. A local yellow fever outbreak in the borough at mid-century spawned protests from its residents who burned relocated quarantine structures in 1857 and 1858. Health commissioners immediately set up floating quarantine facilities on two ships, the *Florence Nightingale* and the *Illinois*, anchored in the Lower Bay. With political pressures mounting and the floating hospitals inadequate and vulnerable to rough seas, commissioners would "at once proceed to make an island."[26] This announcement in 1866—ten years before the first New York island would be destroyed at Hallet's Point—presents island construction as if it were the irrefutable option. It saw a logical solution to place permanent ground where temporary ships anchored, but the matter-of-fact approach elides the full implications of making new land in more than twenty feet of water at low-water mark with "three thousand miles of ocean beating against them."[27] This vulnerable—and isolated—position in the middle of the bay tempered the intransience Staten Islanders had feared in the institution. Compared to the floating quarantines, the proposed islands—what would become Hoffman and Swinburne—were more fixed, if still not completely permanent. But the islands affirmed the fleeting nature of quarantine that in other places might suggest a chronic residency of the diseased population. Lower New York Bay's West Bank offered a substrate for the deposition of material dredged from Orchard Shoals and silted harbor channels to form the islands. Between 1866 and 1868, dredging operations laid the foundations for the two islands, and work proceeded with heavy rip-rap to protect concrete walls and cribwork, which was filled with sand.[28]

Initially known collectively as Quarantine Islands, each insular site provided two and a half acres for passengers, patients, and staff. Completed in 1873, Upper Quarantine became known as Hoffman Island, named for the state's current governor John T. Hoffman. Geographically closer to Staten Island's shores and Manhattan, Hoffman Island was planned as a "quarantine of observation" and housed passengers who did not show symptoms but had been exposed to diseases such as typhus and smallpox.[29] Detention and observation lasted from a few days to two weeks, depending on the estimated incubation period of the disease. Further out and southward, Lower Quarantine was completed in 1870 and initially named Dix Island for the state's new governor John Dix and then Swinburne for John S. Swinburne, who had fought in the Civil War and directed the islands' development. Swinburne Island replaced the floating hospitals, which had been anchored at the port entrance, with permanent buildings—wards, fumigating rooms, crematorium, and morgue. On the island, travelers infected with contagious diseases—mainly yellow fever and cholera cases—were detained for indefinite periods of time. In the years following their completion, authorities defended the costs of the artificial islands by citing the reduced rate of cholera mortality and by asserting that the islands "at the gate of the metropolis" gave New York the "most extensive and complete quarantine establishment in the world."[30]

The position of the artificial islands comes about from their purpose to isolate citizens from incoming disease and to distinguish degrees of exposure and infection. The islands also make visible hierarchies of class. "Upper" and "lower" distinguished not just geographic location and virulence of infection, but also social structure. It was expected that cabin-class passengers, identified as travelers exposed to disease, would proceed quickly through the observation process of Upper Quarantine's Hoffman Island back to Manhattan. In contrast, lower income travelers and immigrants might languish for extended periods of time in Swinburne's facilities without medical care.[31] The two islands also received different amounts of funding. While Swinburne languished as a "disgrace to the state" and was inadequately maintained, health officials with state support expanded Hoffman Island to nine acres in 1896. Hoffman's population peaked at almost eight thousand detainees in 1901, and both islands were partially renovated between 1916 and 1923 before new laws reduced immigration numbers and made the quarantine facilities obsolete. Swinburne's station finally closed in 1928, at a time when Hoffman itself only had three detainees, and the quarantine institution officially returned to Staten Island soon after.

Purpose-built islands are particularly adaptable not only in terms of size but also function. And the inimitable isolation of Hoffman and Swinburne in the Lower Bay's open waters has fostered diverse uses, official and unofficial, sanctioned and unauthorized. Starting in the 1890s, the United States government gained jurisdiction, and during World War I, Hoffman hosted soldiers with venereal disease. For the first part of the 1930s, Hoffman also served as the quarantine location for imported parrots carrying *psittacosis*, known as "parrot fever," and in the late 1930s and 1940s, it hosted the U.S. Maritime Service training school for merchant marines. Attesting to the islands' strategic location at the entry to the Narrows and New York's Upper Bay, Swinburne also provided military function as an "electronic control center for underwater mines," laid to protect the city's harbor during World War II.[32]

The islands' positions afforded varying degrees of visibility. As quarantine facilities, they had been far enough away. As targets of a city's "morbid fascination"—not unlike the way in which Flood Rock's remains became a tourist attraction, they were just close enough, mostly out of physical reach but still sketches of eccentricity along an indistinct horizon. Before the opening of the Verrazano Narrows Bridge in the mid-1960s, Staten Islanders had the best view of Hoffman Island from South Beach or Fort Wadsworth. Swinburne was further offshore, but its alignment with Seaview Avenue, running through the Midland Beach neighborhood, made for a lightly tethered urban connection. Vigilant cruise ship passengers and Brooklyn-bound drivers, from their newly elevated vantage point on the bridge, have also glimpsed the islands, no longer catching sight of the yellow flags that once flew over the islands to signal quarantine but sometimes viewing evidence of the vandalism and unofficial use that has plagued the islands. Reports of nude sunbathing on Swinburne, pornographic film-making on Hoffman, and general hedonism—described in one 1929 case as "disturbing use"—have attracted mainland attention from time to time. In 1966, a fire started by stranded Wagner College students stopped traffic on the bridge and the Belt Parkway. A 1934 description of Swinburne Island echoes this marginality with a tinge of melancholy: "Peopled only by shades of those who were banished there by disease to die or recover, the isle has held the morbid fascination for occasional visitors curious to see this forsaken spot."[33]

In spite of their histories of disease and reputations of impropriety, the islands have afforded escape from the city and offered salubrious environments in the natural elements. Only islands, particularly those opportunistically placed in the middle of an open bay near the Atlantic Ocean, could offer such a rapid transition from infection to recreation. In the 1930s, mostly after Hoffman's quarantine station officially returned to Staten Island, the borough's ferry shuttles brought picnickers to the artificial island's clear light, fresh air, sea change, and forgotten hardships. Reflecting this populist trend, Staten Island Borough President Joseph Palma proposed an amusement site for Hoffman. Art historian and writer John Van Dyke had already foreseen this shift when he wrote in 1909 that islands like Hoffman and Swinburne were "waste places that could be spared readily" but that they also had the potential to be "the most livable and lovable portions of the town."[34]

Thirty years before Van Dyke's prediction, Dr. William M. Smith began his tenure as health officer for the islands. Smith vowed to sanitize—in his words to "scour"—the islands to eliminate "every trace of disease."[35] One of Smith's recommendations was to cover the islands in asphalt. The New York Academy of Medicine concurred and included the following among its suggested improvements: "The surface of the island should be cemented or asphalted, and . . . should be smooth and hard to allow thorough cleansing, and prevent the lodgment of filth or germs."[36] At the time, experts on disease saw asphalt as a readily-sanitized surface and as a barrier to block dampness and "to exclude emanations from the soil," and this vision worked from the perspective that the island was human-made for a singular purpose to control disease. At the same time, Lower Quarantine was called "salubrious little Swinburne Island" by the *New York Herald*, which had initially been critical of the facility. In the same report on Hoffman and Swinburne, articles advocated the recuperative—even if no longer viewed as curative—properties of fresh air, especially ocean air, because "three thousand miles of ocean beating against them" was also three thousand miles of sea air. These islands were the best and worst places—what might be a sea of asphalt was also the open sea of bracing salt air. The buildings would crumble, the paving break, and natural forces would adapt to the island's presence. Meanwhile, the public would alternately forget, ogle, and reconsider them as pleasant anomalies to the urban condition. The ruined infrastructure of one project became the foundation for other proposals.

ROBERT MOSES' WALDEN

The islands also entered the imagination of politicians, speculators, and planners. In addition to Borough President Palma's scheme for a dog track and amusement park on Hoffman, other officials proposed functions that either paralleled quarantine's characteristic isolation or benefited from the site's open-air location. Plans included a homeless shelter, a drug rehabilitation center, and a summer camp. New York City Mayor Fiorello LaGuardia planned to develop this latter program. In 1937, Hoffman Island hosted eight hundred urban youth for a massive picnic and recreational outing, but Parks Commissioner Robert Moses opposed the mayor's plan. For Moses, Hoffman was an "absolutely undesirable place" to organize a camp for disadvantaged children. He believed the surrounding water was not suitable for swimming, but he also contended that so many campers on the island would simply duplicate urban blight—what he called "tenement house conditions."[37] Guard rails were added to protect campers from the bay's polluted waters and the dangers of swimming, and LaGuardia's camp tenuously—and fleetingly—occupied Hoffman for the summer. By 1956 Moses no longer dismissed the islands, and they would become major components of his 250-acre proposal to transform wasteland—and wasted space—into parkland. The islands became new kinds of tools for planning, and Moses found in waste an infrastructure for public space.

In greater New York, Robert Moses discovered another Walden. Early in his career, he professed a particularly deep connection to Long Island's South Shore, having played Thoreau in this once marginal place. In a November 16, 1968 article for *Newsday*, he presents himself as something of a seaman. He dismisses those "inland people" who see Suffolk's south shore as "flat, tame, and tiresome." Posing as a self-described "aficionado" and seeming to recast himself as the great American transcendentalist, Moses lauds the area as timeless, beautiful and isolated: "The South Shore is the product of rivalry of the sun and the moon, the residue of the Ice Ages, tides, winds, and clouds. Here you can reflect on the eternal verities and keep the human spirit alive in a vast, heartless, mechanical world."[38] On the South Shore, Moses found room to think, to ponder tidal as well as seasonal change, and to understand how the residue of natural processes might form emotive landscapes. He also envisioned public access.

Moses was an avid swimmer. He paddled, sailed, crabbed, and fished the Atlantic's waters. Like much of the patchwork material in his autobiographically inclined tome *Public Works: A Dangerous Trade*, Moses' South Shore accounts establish a personal background for his decision-making process and universalize the significance of his personal experiences. The beaches he planned would open this compelling landscape to the public. Moses hoped to translate private experience into populist experiment, and Jones Beach was his masterpiece. Moses' plan transformed Jones Island in Long Island's Nassau County from a swampy lowland to a sandy upland with dredged material from the Atlantic. The beach opened on August 4, 1929, a date still celebrated as its "birthday." And today its six million annual visitors make it the east coast's busiest beach. Throughout the summer of 1929, Moses drove the completed but unopened causeway across Long Island Sound to swim at Jones Beach. Historian Robert Caro recognized the importance of this translation of solitude to spectacle and during his research visited the beach to imagine Moses' experiences.[39]

In these stories, Moses is also responding to critics of his projects, first as the president of the Long Island State Park Commission and later as New York City's Parks Commissioner. He is also laying out arguments and evidence that both nature and waste are infrastructure. In the brief manifesto that was his November 16th *Newsday* column, he replies to "witless critics" and contends that his beach projects—Jones, Oak, Captree, and Fire Island—did not "violate or attempt to improve on nature." With the directness for which Moses was known and with no irony—although this passage follows right after his South Shore elegy in *Public Works*—he says: "we simply moved nature from one sandy shore to another."[40] Writing in a colder than normal November, Moses bemoans the season's deserted beaches, in contrast to summer's much-discussed overcrowding. He even proposes an "ingenious use of glass to break the wind and concentrate the sun."[41] If he could move sand, he might also divert wind and harness solar warmth.

Not improving, but moving nature. Beaches, like islands, were also tools of planning and management for Moses. In one sense, he invokes beach augmentation procedures to make land. In another sense, he implies that a general public has gained access to nature, in this case the southern edge of Long Island. To bring this about, Moses' projects required other infrastructure. The smaller barrier island and its beaches were linked to the main island with three causeways, one named the "Robert Moses Causeway" connecting New Yorkers to "Robert Moses State Park." Aerial photographs show the white sand of beaches matched in area by the gray blaze of parking lots. At its height, Jacob Riis Park on Rockaway, offered nine thousand parking spaces to beach-goers, and today Jones Beach accommodates three times that number. Moses unabashedly illustrated his book *Public Works* with these bird's-eye views to index the components of his urban Walden. In one photograph, the Marine Parkway Bridge crosses Rockaway Inlet; the "Parking Field," completely filled with cars, occupies the middle ground; and the beach, less than half the parking lot's area, is impossibly filled with sunbathers and beach-walkers. The rigidity of the parking grid contrasts with the human confetti of the beach. The Atlantic Ocean itself, cropped by the oblique view, remains out of the frame. Unlike Thoreau, you do not go to nature. Although bumper-to-bumper weekend traffic on the Meadowbrook Parkway may come to feel like a ritual crossing, nature has been brought to you.

And in the frontispiece photograph for *Public Works*, Robert Moses, in distinctive profile, looks pensively from a helicopter window down to Jones Beach. Over his shoulder, the Italianate water tower anchors Wantagh Parkway's axial approach from a hazy horizon. At the center, its parking lots—for the moment, empty of cars—are two perfectly-formed, high-contrast inland seas, and its beaches—off-season and vacant—are human-made accessories to this mobilized landscape. Moses is again alone in his new nature. H. G. Wells called Jones Beach "one of the finest beaches in the United States, and almost the only one designed with forethought and good taste." Moses includes this quotation in *Public Works* along with the words of a British journalist who calls Wells' comment the only understatement he ever made: "This is the finest seashore playground ever given the public anywhere in the world."[42] Nature, as infrastructure, has been moved—in Moses' plan, "given"—to a public through two systems. The work of dredgers has augmented an island, and the supporting transportation networks of bridges, causeways, and parkways have mobilized a populace. On the first page of his seminal written work, we follow Moses' line of sight from helicopter to island, and he presents us with a vision of public space that can be made as easily as it is seen.

TWO WRECKS OF SMALL ISLANDS

Along the Lower Bay's main channels leading from the Atlantic to the inner harbors, Hoffman and Swinburne Islands form a critical part of the sea approach to New York City. The islands also occupy the extended littoral zone that Moses so loved. We might imagine this zone extending west from Long Island's South Shore, across the Lower Bay, along Staten Island's eastern shore, to the shifting spit of Sandy Hook. Moses' divergent proposals for Hoffman and Swinburne make it difficult to understand whether he dismissed them as blight—small but permanent by-products of earlier institutions—or if he reserved them for future projects. In the 1930s, he opposed Mayor LaGuardia's summer camp program for disadvantaged youth on Hoffman Island, but in the 1950s he considered the island a viable site for a similar program. In 1955, at Moses' suggestion, Charles Silver, president of New York City's Board of Education, inspected Hoffman Island as a temporary site to house delinquent youth until a permanent home could be built. The final decision not to use the island may have had as much to do with the protracted inspection voyage as the island's physical conditions. In the eight years following the improvement of Hoffman's buildings for World War II operations and use by the Coast Guard as a training center, the site had been abandoned and vandalized. On inspection day, the trip from the Coast Guard pier at Old Slip in lower Manhattan took nearly an hour amidst delays and rough seas. Although now favored by Moses, these islands were at the outer reaches of his New York archipelago.

In spite of this marginality, Moses came to see the islands as an opportunity and sought their official incorporation into the city's park system. Newly discovered planning tools, they would potentially solve two vexing problems: a ruined sea approach to the city if they were privatized and the continued difficulties of the city's waste disposal. In 1948, Hoffman and Swinburne Islands reverted to the Public Buildings Administration, and articles described this surplus property variously as "orphaned," "forgotten," "ghostly," and "waiting." When the General Services Administration finally put the islands up for auction on November 23, 1956, their public status came into question, and Moses persuaded the federal agency to delay the auction for thirty days. Thirty-one days later, on December 24, 1956, he announced that financier and philanthropist Bernard Baruch had agreed to fund the ten thousand dollar purchase of the islands. A jubilant Moses framed the dilemma and summarized its solution in his speech thanking Baruch for his gift:

> Will the first sight seen from an ocean liner entering New York Harbor be—not Liberty lifting her lamp beside the golden door—but two wrecks of small islands with decaying buildings supporting advertising billboards? Mr. Baruch says 'No' and the public will say 'Amen.' The city will be grateful to Mr. Baruch for his farsighted and public-spirited contribution to insure protection of our harbor and to prevent the sale of these islands to commercial interests.[43]

Ostensibly, this purchase protected the city's harbor entrance from commercial development. But compared to other New York real estate, these islands were tiny—big enough to hold a billboard but too small and remote to support large projects. By Moses' account, these "wrecks of small islands" threatened public experience of and within the city. If their decaying countenance were the problem, economies of scale might suggest simply removing them—treating them as waste to be used as fill somewhere else. Hoffman and Swinburne were not meant to be there in the first place, so removing an already human-made construct might have made sense. Instead, these "wrecks"—amounting to less than twelve acres—became the cornerstone of a 250-acre park project. In Moses' scheme, the value of these misplaced islands was no longer their insularity. They had become place-holders for waste disposal, and what held them in place was rhetoric of public space and the image of the city. Waste would save the islands from privatization.

The day after he announced Baruch's contribution for the islands' purchase, Moses provided details of his plan for the new park. In an editorial on December 25, 1956, he wrote that his plan for the 250-acre park would "forestall a possible use adverse to the public interest, and create a new needed and rewarding dumping ground for Sanitation Department landfill for some years to come."[44] Behind the previous day's rhetoric, Moses knew that the space between these two islands—about a mile of Lower Bay—would hold twenty-five million cubic yards of uncompacted garbage, equaling about ten million cubic yards compacted. Waste disposal as land reclamation was not new. It was a strategy that had been used at the earliest stages of Manhattan's settlement by Europeans. Dutch settlers had already expanded the island's near-shore waters with waste and other displaced materials, and eighteenth-century maps show additional rapid extensions of Lower Manhattan's blocks into the East River.[45]

Moses also knew the city's history of garbage problems. Among other constraints, a 1934 Supreme Court decision limited the city's offshore dumping practices, and the garbage had very few places to go. In 1947, the year before the islands became surplus property, New York City opened the nearby landfill at Fresh Kills as a short-term solution to its growing problem of waste disposal. In 1948, Moses also became chair of New York's newly-formed Slum Clearance Committee, ahead of federal Housing Act legislation that would enable re-development of confiscated urban land. Its construction having finally begun, the project for the United Nations headquarters had recently demonstrated the acute difficulties of finding—and making—space in the city. More politically charged than ever, land acquisition and eminent domain influenced Moses' opportunistic view of the islands—whether as parkland or wasteland. Away from the boroughs' densities and constraints, this new decentralized public park in the Lower Bay's open waters would be almost one third the size of Central Park.

But visions built out of garbage, particularly at this scale, proved difficult to sustain. The following year, while Moses sought support for his Hoffman and Swinburne scheme, garbage was filling a portion of Jamaica Bay. The ongoing project for Brooklyn's Marine Park illustrated what could be expected with Moses' island project. By August 1957, more than one thousand acres of marshland and baywater had been reclaimed with layers of garbage and other waste. Two decades earlier, in 1938, Moses had written to Mayor LaGuardia about this area's potential in a report that invoked his vision of Long Island's South Shore idyll: "Here in Jamaica Bay lies the opportunity for a place within the limits of the city where the strain of our city life can be relieved, where the nerves of tired workers may be soothed, where the old may rest and the young may play."[46] Many reports looked positively on this project that Moses had hoped would be his masterpiece but would remain unfinished at his career's end: "Jamaica Bay is tangible proof that the fifteen thousand acre recreation program planned by Parks Commissioner Robert Moses has been carried forward in cooperation with the Department of Sanitation, the Department of Public Works, and the Department of Health."[47]

But first-hand accounts during the construction process invoked another landscape, one that seemed an impossible mix of waste, recreation, and the picturesque. In one vignette, the Mau-Mau Bridge links various stages of the human-made wasteland: "A 500-foot wooden bridge is arching from former wasteland to a man-made island to connect what will be three golf courses grown from refuse."[48] The procedure for making the parkland was layered and seasonally defined. In the winter, areas of water were filled, with the idea that cooler temperatures and the aqueous matrix would keep odors to a minimum. Additional garbage in the summer often created twenty-five-foot "mountains" that would be leveled with machinery into fifteen- to twenty-foot thick, incrementally compacted layers covered by "carpets of sand." Once the required elevation was reached, tankers pumped "many millions of gallons of sludge" from sewer treatment plants onto the compacted layers to form a "synthetic topsoil."[49] In spite of this scene, media coverage steadfastly noted the project's growth "out of vision and garbage" and sought to refute the project's critics whose concerns were realized in Jamaica Bay's pollution, a problem that continues today with prohibitions on swimming. More than two-thirds of Marine Park was made with garbage fill, and the overall area of reclaimed land easily exceeded that of Central Park. Original plans had called for twice that area.

A REALLY GREATER NEW YORK

Unwrapped on Christmas Eve, Moses' gift of Hoffman and Swinburne to the city had technical precedents that were not dedicated to the use of garbage as fill. Techniques of dredge and fill, like those used to form Belmont Island's coffer-dam and to extend Governors Island, had also helped expand many of the city's smaller islands with spoil material. Beginning as Fort Gibson with three acres in 1813, Oyster Island grew throughout the nineteenth century to yield Ellis Island's twenty-seven acres by 1890. Proposals were also made to link islands to the mainland. In 1900, a project for what was called "dock island" would have linked Ellis Island to the New Jersey coast. Proposed by local business interests, this expansion of the harbor's terminal facilities as a freight clearing house foreshadowed the privatization of islands and submerged lands that concerned Moses later in the century.[50] Other projects joined islands to each other. Great Mill and Little Mill Rocks had been joined, and Moses himself planned the fusion of Randall's and Wards Islands as well as Hunter and Twin Islands. Carried out in the 1930s, this latter project followed his previous use of sand as a planning tool at Jones Beach and developed a mile-long crescent of sand to form Orchard Beach that connected the two islands to Rodman Neck and the eastern edge of Pelham Bay Park. The success of this project buoyed Moses' grander—and more geographically prominent—plans for joining Hoffman and Swinburne.

The diminutive islands that prompted Moses' project would be more than joined. In his proposal, they would be engulfed by the area to be filled. But such radical scaling also had precedents in sometimes outlandish schemes to augment New York's landmass. Undaunted by concerns of private enterprise, T. Kennard Thomson laid out his plan for "A Really Greater New York" in 1911. His proposal sought to expand the city with fifty square miles of reclaimed land, setting up a megalopolis for twenty-five million residents. An engineer by training, Thomson itemized the project's attributes in an expansive article for *Popular Science*: filling the East River and cutting a new north-south river east of Queens and Brooklyn, extending Manhattan Island southward so that short tunnels might link with Staten Island and Brooklyn, adding two anvil-shaped accretions to Staten Island's southeastern shore, and creating a landform designated as "New Island," shaped like an axe-head and connected to Sandy Hook through another tunnel.[51] Thomson proposed the coffer-dam and fill technique on a grand scale and cited the concurrent success of developers building islands in Biscayne Bay. Thomson drew from Thomas Edison, who had proposed a more modest plan that only filled the East River and who, according to Thomson, had endorsed his own more ambitious 1911 scheme.[52] Taking cues from Edison's eight-square-mile plan, a second proposal scaled the overall land area back to nine square miles, absorbed Governors Island, and filled the Upper Bay so that the Hudson and East Rivers met neatly at the Narrows.[53]

Moses would have also seen the breakwater project devised by Cornelius Hall, Staten Island Borough President from 1946 to 1953, to protect the borough's South Beach from erosion. This human-made version of Sandy Hook's five-mile-long barrier spit would have linked Staten Island with Hoffman and Swinburne. Hall's breakwater and Thomson's "City of New Manhattan" dwarf Moses' park proposal, but the schemes share a view of urban expansion that has few limits. The engineer Thomson saw land-making as a boundless activity unconstrained physically and socially. Radical for the time, he proposed the inclusion of women in the decision-making process of his urban plan. Moses, the "power broker," also saw limitless potential in making land, and his idea for Hoffman and Swinburne combined waste and recreation, just as it mixed pragmatism with idealism. With deal-making and island-making, a park could be a dumping ground and still be the "golden door" to the "world's greatest seaport."[54] Writing in 1970, as the islands were in the process of being incorporated into the Gateway National Recreation Area, Moses stood by his proposal for them "to be bulkheaded, filled, combined, and turned into a park" that would preserve New York's entry from "unsightly private exploitation" and would also "afford unusual recreational use."[55] Such optimism sought not just to improve the island "wrecks" but to create a "really greater" New York.

DOORMAT AND GATEWAY

Moses' vision for Hoffman and Swinburne persisted, and the islands continued to reflect changing attitudes about nature, recreation, and waste disposal. Plans for the islands focused on their location rather than their landscape. Their outward position afforded natural attributes and infrastructural expansion. It was not yet about the islands' own natural systems. Floating in the Lower New York Bay, they were either doormat or gateway, mundane threshold or dramatic access, a place to dust your feet or a grandiose portal suitably scaled to the aspirations of the city you were entering. Moses' framework for the combination of wasteland and parkland had created this range, and there was very little middle ground. The islands could languish as undersized "wrecks," or they could achieve the status appropriate for a "really greater New York." And the city's waste could still make this vision a reality.

By September 1960 at the end of Moses' tenure as parks commissioner, the source of fill for the island-building project momentarily shifted to material excavated from a proposed tunnel under the Narrows between Brooklyn and Staten Island.[56] But attention soon turned back to waste, and in July 1967 Hoffman and Swinburne were said to be the "logical place to begin" a landfill project in the Lower New York Bay. Maurice Feldman, Deputy Commissioner of Sanitation for Engineering, also noted that the diking of this part of the bay could eventually connect to Staten Island at Fort Wadsworth, and that the "best possible fill" was incinerator ash.[57] Resistance to ash and island-building has a long history around the borough's shores. In February of 1880, residents of Staten Island's North Shore opposed plans for an artificial island near Robbins Reef Light, fearing that it would function as the City's ash dump and its developers would "obtain a nice revenue for themselves at the cost" of residents' comfort.[58]

Undeterred by this past, the Sanitation department's ensuing feasibility study built on Moses' project and drew from Cornelius Hall's original breakwater island proposal, but it also recalled the ash dumps of Rikers Island and Flushing Meadows, where F. Scott Fitzgerald had seen: "a certain desolate area of land. It is a valley of ashes—a fantastic farm where ashes grow like wheat into ridges and hills and grotesque gardens, where ashes take the forms of houses and chimneys and rising smoke, and finally, with a transcendent effort, of ash-gray men, who move dimly and already crumbling through the powdery air."[59] Moses had seen this too, and he planned the area's clean-up in preparation for the 1939 World's Fair, putting out the fires, removing fifty million cubic yards of garbage from Flushing's Corona mountain, and ending the dumping of ash on Rikers Island. Before finally deciding to remove the waste, Moses noted that they "fondly hoped to cover the garbage with a thin layer of topsoil" and to plant trees "at a price which would not subject us to indictment."[60] Three decades later, mayor John Lindsay could finally preside over the 1967 rededication of Flushing Meadows-Corona Park. The event's official brochure closed with the following statement: "We have sought, in the words of Isaiah, to give the City 'beauty for ashes.'"[61] That same year, only a few weeks after the announcement of the latest proposal for Hoffman and Swinburne, Mayor Lindsay countered conservationists' opposition to a landfill in Pelham Bay, a residually marshy area of the Bronx, with the recreational benefits that would come out of the plan to give the Bronx another year of landfill capacity: "Many of our best recreation areas are on landfill, and park authorities feel the extended area will enable them to construct a better park."[62]

The politics of city waste became irrelevant and the "gateway" moniker became official for Hoffman and Swinburne when the U.S. Interior Department published its plan for the Gateway National Recreation Area on December 11, 1969. In this early phase, Hoffman and Swinburne constituted a separate unit, even though they made up only fourteen acres of a twenty-six thousand acre park.[63] In his summary of the project's goals, Interior Secretary Walter J. Hickel highlighted conservation as well as public access: "The logical molding of these existing and available publicly owned properties would provide a recreation resource of unequaled precedent in the United States. Gateway is the first step in a major conservation effort to bring parks to people—a program to provide a better life for all citizens."[64] Gateway would rely on public transit, rapid transit lines and ferries, rather than automobility and highway infrastructure. Bringing "parks to the people" contrasts with Moses' goals of bringing people to the parks. Like the other units, planners believed that the islands offered valuable resources because of their lack of development and their Lower Bay environment, but planned ferry access would falter and the islands would languish in the isolation that plagued them since their inception.[65]

Despite their inclusion in the Gateway project, debates about the islands' future lingered. In the year following its announcement, competing projects and territorial disputes highlighted the problems facing the Gateway project. Consolidated Edison Company was seeking to build nuclear power plants on Hoffman and Swinburne Islands. City officials appeared supportive of Con Ed's proposal, but planners at the Interior Department opposed the project, often through comments laced with irony: "If adequate safety standards could be determined . . . I could see, perhaps, a revolving restaurant atop a nuclear plant on Hoffman-Swinburne Islands."[66] The Interior Department official's sardonic proposal placed recreation on power infrastructure, which was itself a human-made substrate. This comment may be ironic, but it recalls the artificiality of the islands and their continued dependence on human agency. The hybrid program of power plant and eatery is not unlike pairing garbage with park, dumps with recreational activities, and dredge spoil with picnics.

With Congress' authorization on October 27, 1972, Gateway became one of the first urban parks in the United States' national park system. Con Ed went elsewhere, Hoffman and Swinburne remained, and the islands were officially conserved as a part of Gateway's Staten Island Unit. In 1979, Gateway's general plan called for a dynamic management strategy "based on changing environmental and social conditions."[67] The islands were not in the designated protection zone, and the Interior Department was still considering the possibility of joining the islands and connecting the new land mass by bridge to Staten Island's beaches: "Stage III at Staten Island will entail a study of the feasibility of constructing a major offshore island complex, connected by a bridge to the beach. These islands could be built up around the Hoffman and Swinburne Islands in an area where depths are extremely shallow, as little as 2 or 3 feet. After a relatively short walk, people would be able to picnic right next to the main ship channel. The islands could also be developed to make them destinations for private boat and ferry trips."[68]

The 1979 plan zoned Hoffman Island for "unstructured recreation." Swinburne's more nuanced zoning designation suggests not ambiguity of purpose but the indefinite conditions of the islands themselves. The zoning map calls for a mix of "unstructured recreation" and "use-by-reservation," but the explanatory text claims the latter, though tentatively: "Tiny Swinburne Island has been tentatively placed in this zone [use by reservation], based on subsequent study. Use-by-Reservation lands will be managed to perpetuate natural conditions, and uses will be limited to educational and low-intensity recreational activities."[69] The 1979 plan recognized the unique urban conditions of New York City—what it referred to as the "most compressed and heterogeneous urban landscape in the nation," exerting "powerful demographic, political, and psychological forces on the national recreational area."[70]

Within this context, Gateway's main purpose in the Staten Island unit, to which Hoffman and Swinburne were later added, was to expand recreational sites. The management plan noted that "few undisturbed natural areas remain" around Staten Island, and from some perspectives the islands still called to mind a waste-site past. Observing the islands from a helicopter in 1976, one journalist noted that they lay across the bay like "doormats."[71] But, in spite of their human-made provenance, Hoffman and Swinburne also offered what the larger island could not—self-contained, distinct places for the restoration of natural systems. The application of the management plan evolved over subsequent decades, and its original call for a dynamic mix of "low-intensity recreational activities" with environmental studies and with selective preservation would be ceded to a policy of more limited access.[72]

PROTECTED ZONE

Hoffman and Swinburne became infrastructure to support Gateway's goals to preserve wildlife. The recreation area's current plan has placed conservation on spoil—egrets and herons on Hoffman and the migrating harbor seals, wading birds, gulls, and cormorants on Swinburne. Restrictions equate the sensitivity of the island environments with Gateway's marshland areas so that only researchers and other permitted visitors can join the birds and seals: "All marshes and Hoffman and Swinburne Islands are closed to the public, except by special use permit. Marshes are fragile wetland habitats and the islands contain bird nesting sites. Remnant structures on the islands are in disrepair and present a safety hazard."[73] Built from dredged material, the islands were placed there, they have been trodden, they have collected additional waste, and they have hosted detritus of temporary programs and failed plans. Their original function of isolation placed their program of disease control away from land but in the middle of a watery threshold. Even if you were to arrive by ship, you might miss seeing them. Thanks in part to Moses, they do not have "welcome" written on them. Now they are off limits, isolated again, not by quarantine but by conservation.

Swinburne's dock is gone. Paddling from Staten Island's Midland Beach reveals a forest of pilings, barnacled perches for gulls and cormorants. They offer bearing for navigation, not structure. The granite boulders that surround the islands dwarf my kayak's precise but under-scaled approach. Planned by engineers, avoided by mariners, called "rip-rap" by both, this rocky fringe could have been placed by giants. Rip-rap is a reminder of the built character of these places. From the air, the material softens the hexagonal edges of the islands. But up close, from the kayak's low-slung perspective, this barrier to erosion protects from high seas and resists visitors as much as the administrative rules that shield its island habitats. On the passage from Swinburne to Hoffman, the long wave periods from container ships alternate with the shorter chop of weekend fishermen and tidal flux. Winds have also picked up. Even though this is the open water of the Lower Bay, it is as if the area is more closely bulkheaded, like Manhattan's rivers and harbors further north. Standing currents allow for welcome, though unsettling, pauses to hear the unseen cry of birds from Hoffman's jungled interior, and I wonder that this might also be the noise of quarantine.

NOTES

1. "Quarantine," *New York Times*, 10 July 1866, 4.

2. "The Great Mine Sprung: Flood Rock Shattered into Countless Fragments," *New York Times*, 11 October 1885, 2.

3. See Matthew Gandy, *Concrete and Clay: Reworking Nature in New York City* (Cambridge: MIT Press, 2003).

4. "The Great Mine Sprung: Flood Rock Shattered into Countless Fragments," *New York Times*, 11 October 1885, 2.

5. See Claude Rust, "Hellgate's Infamous Past: Part I. Fight against 'endless obstructions,'" *Military Engineer* (September-October 1971): 337–341, and "Hellgate's Infamous Past: Part II. General Newton and Conquest," *Military Engineer* (November-December 1971): 410–414.

6. "Twenty Feet from Death: The Barge Republic Stranded on Flood Rock," *New York Times*, 12 August 1880, 2.

7. "Hell Gate: The Nature of the Obstructions," *New York Times*, 19 August 1867, 2.

8. Some of the outcroppings were referred to as "nigger-head rocks," and others made references to American Indians. In such racist and derogatory references, some accounts seem to blame the rocks themselves for the fateful wrecks and accidents. See "Twenty Feet from Death: The Barge Republic Stranded on Flood Rock," *New York Times*, 12 August 1880, 2.

9. "By a Child's Gentle Touch," *New York Times*, 11 October 1885, 1. John Newton congratulated his daughter with the following: "There just look what you've done! You have probably created more commotion this morning than any other little girl in New York." Nine years earlier, 90,588 tons of broken rock were removed after the Hallet's Point blast, which occurred on September 24, 1876.

10. "By a Child's Gentle Touch," *New York Times*, 11 October 1885, 1.

11. William Hosea Ballou, "The Flood Rock Explosion," *The American Naturalist* 20 (February 1886): 137–140.

12. "The Ruins of Flood Rock," *New York Times* (October 12, 1885), 2.

13. "The Ruins of Flood Rock," 2. A reporter noted that the "whole run of the tides had been altered." Pilot Wicks of Steinway in Queens was quoted: "The explosion has changed the tides in a remarkable way." A whirlpool known as the "Turntable" that originally circulated southwest of Flood Rock was shifted two hundred feet where a "new turntable" formed near Astoria's shore. Most observers, fishermen, and pilots agreed that there was "everywhere evidence that the explosion had been a thorough one, and had wrought great changes."

14. Clifton Hood, *722 Miles: The Building of the Subways and How They Transformed New York* (Baltimore: Johns Hopkins University Press, 2004), 166-167, 295.

15. "Mayor Has Steinway Tunnel Work Stopped," *New York Times*, 20 September 1905, 16.

16. See "Draining the Belmont Tunnels in New York," *The Iron Trade Review* XLI (1907): 147. During this time period, the coffer-dam technique began with the construction of a sea wall around the preliminary piece of exposed land, which in the case of Belmont was less than three hundred square feet. Excavated material, some of it from the initial shaft, was then dumped between the outcropping and the sea wall to develop an island for continued excavation. See also Claude Rust, *Military Engineer* (1971).

17. "An Air Pipe Burst," *New York Times*, 16 January 1906, 2.

18. "Belmont Tunnel," *Railroad Gazette* XLIV (1908): 473.

19. Hood, 167.

20. Arthur B. Reeve, "The Romance of Tunnel Building," *World's Work* 13 (December 1906): 8343.

21. Hood, 166.

22. "Rush Belmont Tunnel Work," *New York Times*, 4 August 1907, 3.

23. "Walk Through the Belmont Tunnel," *New York Times*, 15 August 1907, 7.

24. "Belmont Tunnel Makes New Digging Record," *New York Times*, 15 November 1906, 18. In terms of transportation infrastructure, the Steinway Tunnels were anticipated as "an important link between the heart of this city and the vast network of trolley lines on Long Island," but the tunnels presaged the trolleys.

25. Sharon Seitz and Stuart Miller, *The Other Islands of New York City: A History and Guide* (Woodstock, VT: Countryman Press, 2011), 148, 195.

26. "Quarantine," *New York Times*, 10 July 1866, 4.

27. "Report of New York's Health Commission to the State Legislature in 1876," cited in The Citizens' Association, *A Typical American: Incidents in the Life of Dr. John Swinburne of Albany* (Albany, NY: The Citizens Office, 1888), 91.

28. The Citizens' Association provides one of the most detailed descriptions of the islands' original construction: "The two islands are of the same size and construction, with the exception of the buildings. The foundation [of Swinburne Island] is hexagon in form, two of the sides being two hundred and sixty feet in length, the other four one hundred and sixty-one feet. The exterior of the crib-work is thirty feet in width at the base, twenty feet at the top, and twenty feet in height, and is constructed of large timbers firmly fastened together and filled with small stones, the whole surrounded by a riprap of heavy stones . . . the area at the base of the riprap is over three acres; the extreme length at the top being five hundred and four feet, and two hundred and twenty-eight feet in width. To construct the island, nine thousand cubic feet of timber were required, and seventeen thousand cubic yards of stone in the riprap, five yards of stone in the crib, and fifty-six thousand four hundred yards of sand to fill the space enclosed by the crib-work." (82)

29. Frank Linstow White, "Barriers Against Invisible Foes," *Frank Leslie's Popular Monthly* 33 (June 1892): 662.

30. The Citizens' Association, 80–81.

31. In their account of New York City's islands, Sharon Seitz and Stuart Miller refer to these islands, particularly Swinburne, as "outcast islands" (83).

32. Seitz and Miller, 92.

33. *Staten Island Advance*, 17 July 1934, cited in Seitz and Miller, 79.

34. John C. Van Dyke, *The New New York* (New York: Macmillan Company, 1909), 392–393.

35. Patricia M. Salmon, *Realms of History: The Cemeteries of Staten Island* (Staten Island Museum and Staten Island Institute of Arts and Sciences, 2006), 367.

36. See the Report of the Committee of the New York Academy of Medicine, "The Present Condition, and Suggestions for the Improvement of the New York Quarantine Establishment," *The Sanitarian* 218 (January 1888). See also the Medico-Legal Society of New York published *The Sanitarian: A Monthly Magazine Devoted to the Preservation of Health, Mental and Physical Culture* XX (New York: A.N. Bell, 1888). The New York Health Commissioner and the Mayor asked the Academy of Medicine to make a report on quarantine facilities in New York in general and on Hoffman and Swinburne Islands in particular.

37. "Moses Assails Site for Summer Camp," *New York Times*, 2 June 1937, 25. See also "Unwanted $5,000,000 Island is White Elephant to Uncle Sam," *Staten Island Advance*, September 24, 1948, cited in Salmon, 368.

38. Moses includes this excerpt from the 16 November 1968 issue of *Newsday* in his book *Public Works: A Dangerous Trade* (New York: McGraw Hill, 1970), 111.

39. See Robert Caro, *The Power Broker* (New York: Vintage, 1975), 270. See also Elizabeth Barlow, "Robert Moses and Robert Caro Redux," *Sitelines: A Journal of Place* 3 (Fall 2007): 18.

40. Moses, 111.

41. Moses, 112.

42. Moses includes these quotations on page 100 of *Public Works.*

43. "City to Link 2 Isles for Park in Lower Bay," *New York Times*, 24 December 1956, 1.

44. "Mr. Baruch's Islands," *New York Times*, 25 December 1956, n.p.

45. Eric Sanderson, *Manhatta: A Natural History of New York* City (New York: Harry Abrams, 2009), 555.

46. Robert Moses in New York City Department of Parks report *The Future of Jamaica Bay* (1938).

47. Murray Schumach, "Park in Brooklyn Rising on Refuse," *New York Times*, 18 August 1957, 56.

48. Schumach, 56.

49. Schumach, 56.

50. A conference between the Harbor Line Board and the Riparian Commission brought capitalists and engineers together with commissioners to discuss how the existing harbor lines might be altered to allow for this linkage of island to mainland. The board determined that Edward Cragin, representing private business interests, would essentially be purchasing property from the Riparian Commission, the default owner of the submerged lands. Along with a Jersey shore landowner who called the project "outrageous and speculative," there were two main points of opposition: one based on the need for isolation and the other based on concerns about real estate speculation. The Chief Engineer for the United States Treasury Service Alfort Brooks Fry noted that the proposed dock should not touch Ellis Island because the "possibility of contagious disease demanded a channel of at least 500 feet." Opponents also noted that the constructed land would interfere with free access to property on the mainland, but the main opposition to the plan argued that granting the request "would establish a precedent for building islands which could be seized upon for speculative purposes." This latter position soberly forecast the rampant speculation and unfettered island-making that would soon occur further south over Biscayne Bay's submerged lands. See "Dock Island Project Meets Opposition," *New York Times*, 15 June 1900.

51. T. Kennard Thomson, "A Really Greater New York," *Popular Science Monthly* 88 (1916): 60–63.

52. Edison's proposal had circulated enough to garner response from popular journals of the time. The magazine *Indoors and Out* editorialized that "accretion is the evil of New York" and argued that only extending transportation infrastructure, rather than "destroying a natural and superb waterway," would remedy New York's current "physical difficulties" of transit and its "Nassau canyons." See *Indoors and Out* 2 (1906): 311.

53. In Thomson's plan, a central boulevard over railroad tracks extends the main axis of Broadway as the spine for a fanciful but anticipatory transportation scheme that alludes to Frederick Law Olmsted's Central Park—weaving diverse modes of transportation with separate elevated decks for cars, trains, and planes—and foreshadows the architect Le Corbusier's Ville Radieuse—with its elevated housing blocks.

54. Robert Moses, "New York Reclaims Its Waterfront," *New York Times*, 7 March 1948, 16. Moses also called New York City America's "peerless harbor city" and compared Manhattan to Venice.

55. Moses, *Public Works*, 27.

56. Wayne Phillips, "Park Plans for Two Gift Islands at Standstill after Three Years," *New York Times*, 10 September 1960, 23.

57. Charles Bennett, "Landfill Off S.I. Is Studied by City," *New York Times*, 7 July 1967, 1, 41.

58. *New York Times*, 8 February 1880.

59. *The Great Gatsby* (New York: Scribner, 1925), 23. Robert Caro cites the influence of this excerpt on Robert Moses' view of the Flushing Meadows and Riker's Island landscapes. In *Public Works*, Moses describes the "volcano" that was "over there on Riker's [sic]" and notes, in a lengthy but almost poetic footnote: "Just beyond Flushing Bay and blocking the entrance is Riker's island. For many years barges were floated to it with ashes and refuse which could not be dropped at sea in bad weather. Riker's Island in 1934 was a cloud of smoke by day and a pillar of fire by night. Fires broke out continuously all over the vast heap, and an inadequate water supply no sooner put out one than another began. The columns of odorous smoke drifted over to Flushing Meadow, mingling with the smoke and smell of the Corona dump and the miasmic sewage odors which at low tide arose from Flushing Bay. When the work of leveling the Corona mountain for the first World's Fair began, all dumping stopped at Riker's island. Fires were put out, and great mountains of refuse were leveled. A 70-acre tree nursery was installed by the Park Department." (541) Moses would later lament the loss of the trees, which died because of inadequate watering.

60. Moses quoted in Caro, 1084.

61. Thomas J. Deegan, Chairman, World's Fair Corporation, "A Statement by the Executive Committee of the New York World's Fair 1964–1965," *Flushing Meadows-Corona Park Pub-*

lic Ceremonies Commemorative Book, 3 June 1967. Moses includes the Executive Committee's statement in *Public Works*, 617–8.

62. Charles Bennett, "Nature-Lovers Lose Park Area To Landfill Forces in the Bronx," *New York Times*, 28 July 1967, 26.

63. As a coastal park, the legislation that enabled Gateway (Public Law 92-593, in 1972) includes Hoffman and Swinburne as well as "all submerged lands, islands, and waters within one-fourth of a mile of the mean low water line of any waterfront area" within the park's boundaries.

64. "U.S. Recreation Area Favored," *The Daily Register*, 12 December 1969, 2. Interior Secretary Hickel recalled his impressions from his May 1969 visit to the park areas.

65. In 1972, the Hoffman-Swinburne Island study reviewed the effects of two projects: an enclosure dike around the islands and a 3,900-foot dike at Norton Point. The latter proposal sought to protect Coney Island beaches from the Bay's polluted waters. The Hoffman and Swinburne portion of this project was absorbed by federal considerations in the Gateway National Park.

66. Douglas Robinson, "Gateway Still Just an Idea," *New York Times*, 14 April 1970, 49.

67. Kara Kirchhoff and Leah Stokes, Columbia University for Gateway National Recreation Area, *Gateway National Recreation Area: Long-term Resource Management Under a Changing Climate* (April 2009), 5.

68. National Park Service, *General Management Plan: Gateway National Recreation Area, New York / New Jersey* (U.S. Department of the Interior, August 1979), 97. The projected budget allotment for the "creation of new islands around Hoffman and Swinburne" was $37.5 million (National Park Service, 167).

69. National Park Service, 98.

70. National Park Service, 34.

71. John Leonard, "About New York," *New York Times*, 1 September 1976.

72. National Park Service, 96. Gateway's General Management Plan, originally targeted for 2009, is ongoing; and its release date was projected for Summer 2012.

73. National Park Service, Gateway National Recreation Area, "Superintendent's Compendium of Designations, Closures, Permit Requirements and Other Restrictions Imposed Under Discretionary Authority" (August 16, 2011): 9.

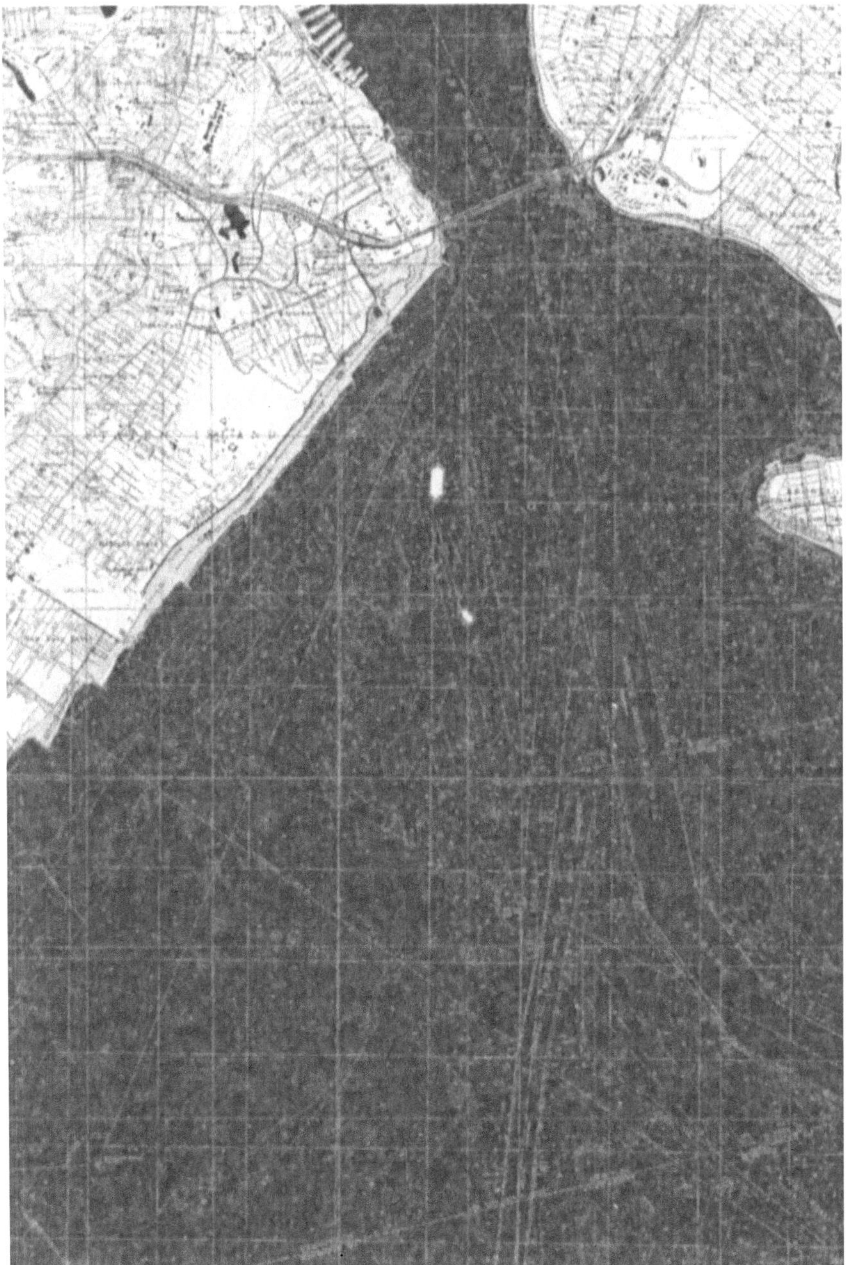

Rip-Rap

There is an iconic photograph of Robert Moses standing on a cantilevered beam over the East River. Arnold Newman composed the image in the summer of 1959 when Moses' tenure as Parks Commissioner was coming to its controversial end. Defiantly, Moses stands on the red beam, hands on hips but unnaturally high so that his elbows extend almost like a bird, maybe a sea gull. His left hand holds a roll of blueprints. This white sheaf of drawings, his white dress shirt, and black suit anchor the photograph's right side. To the far left is the United Nations complex. Its Secretariat building mirrors the palette of Moses' outfit and balances his figure. The rest of Manhattan's midtown skyline unfolds behind him, and the river's edge cuts a strong horizontal line. Above this line, skyscrapers press into a sky laced with dispersed jet-trails. Below this line, the waters of the East River roll and ebb. Low tide has exposed the northern edge of Blackwell's Reef, resting in the reflection of the Secretariat building. And the beam that supports Moses points south toward this reef and toward Belmont Island. The beam follows the middle line of the river, the line of islands that includes Man-o'-War, Belmont, Mill Rock, and Welfare. It is the southern tip of the latter island where Moses floats. In the decades to follow, this island would add ironies to the photograph's location. It would be renamed for Franklin Delano Roosevelt, who disliked Moses. In 1969, George Delacorte would donate a fountain to the city that would become a logistical problem for years to come. And here at the southern tip, the architect Louis Kahn, also Moses' nemesis, would design FDR's memorial Four Freedoms Park.

The staging of the photograph questions the stability of infrastructure and at the same time exaggerates its constructive potential. Newman painted the aluminum beam red and placed weights on its northern end to hold the cantilever in place. The connecting plate at the other end is cardboard with clay bolts. Making the plate allowed Newman to determine its precise form, and the arrow of the beam points to the rock outcropping, the natural infrastructure of the reef. If we imagine that the arrow points a bit further south, to Belmont Island, then this narrative of infrastructure is further complicated by the human-made island's presence. The "fake" beam becomes part of an extended infrastructure. Some of its underpinning is material that was cast aside long ago, like Belmont's spoils from rail construction. Some is residual reef structure, which has been left intact because of its minimal navigational risk. And some is perhaps related to current construction, so much excavated rock, dirt, and mud below the beam at Moses' feet.

Overall, this makes for a seemingly impossible scene. The connector plate cannot connect, its holes are filled, and its edge is awkwardly angled. Moses only stands thirty inches above the ground and a full twelve feet from the East River. Newman angled the shot to simulate Moses' floating above the water. In one sense, it is an allegory, one in which Moses, born on his beam vessel, floats on the East River's unstable waters to form his own island position, a new addition to Manhattan's archipelago. Moses agreed to the photo shoot for *Holiday* magazine, but he would later note: "I am still shaky as the result of perching on that steel beam" and "that photog [sic] made me stand on a steel beam cantilevered fourteen feet up over a pile of rocks. The horizontal beam shook and shivered. Hell of a way to take a picture."[1] Islands, like beams, are vulnerable, isolated, and sometimes humbling.

Newman's photograph is presumably about Robert Moses, master builder of New York. But it also captures the nature of infrastructure itself. Moses claimed he was not inventing nature, just moving it. The photograph is not really about Moses. It is more about what he is standing on, what Manhattan is built on, and how the city's watery context makes it a city of islands. In this photograph, Newman manages the elements of Moses' empire like so many raw materials. These materials cannot be exclusively rationalized from either technical or natural frameworks. Consequently, the photograph casts infrastructure as nature, and Moses might be perched above one of the beaches he moved. The photograph also shows that nature is infrastructure. The treed fringe of East River (FDR) Drive glows iridescent green in the late morning sun. Closer to us, Manhattan's rocky substrate is revealed, whether naturally formed or deposited in so many construction projects or waste disposals.

In Newman's original prints, another outcropping—more like a pile of rubble—can be seen below the beam. Moses hovers over this other component of infrastructure—rip-rap, unseen in the cropped photograph and often overlooked in everyday experience. At the edge of land and water, where so many of his plans tread, rip-rap is emblematic of the reciprocity of nature and infrastructure that spoil islands engender. Rip-rap is a cluster of rocks or rubble thrown together to create a boundary against erosion. Its first use as a technical engineering term can be traced to early American engineering handbooks, and it is believed to have been derived from an earlier nautical term that described rippling water over submerged outcroppings of land. Earlier uses of "riprap" denoted forces of impact, suggesting that the term might describe not only the image water rippling but also the sound of waves rapping on stony shoals and reefs. Rip-rap stabilizes a shore and allows artificial land to remain where it was not before. But the sound of "rip" and "rap" reminds us of water's power of erosion and submerged land's danger. Symphenomenal, it imitates the natural sounds of water. Like spoil islands, rip-rap's rocks are displaced natural materials. Origins for the name of this infrastructure are also to be found in nature. And the imitation works both ways. There is nature to be found *in* infrastructure.

NOTE

1. Robert Moses to Louis F. V. Mercler, October 16, 1959, and Robert Moses to C. C. Jones, December 5, 1959, in folder 8H, Box 46, Robert Moses Papers (RMP), Manuscripts and Archives Division, New York Public Library, Astor, Lenox, and Tilden Foundations, cited in Timothy Mennel, "A Fight to Forget: Urban Renewal, Robert Moses, Jane Jacobs, and the Stories of Our Cities," *Journal of Urban History* 37 (2011): 628.

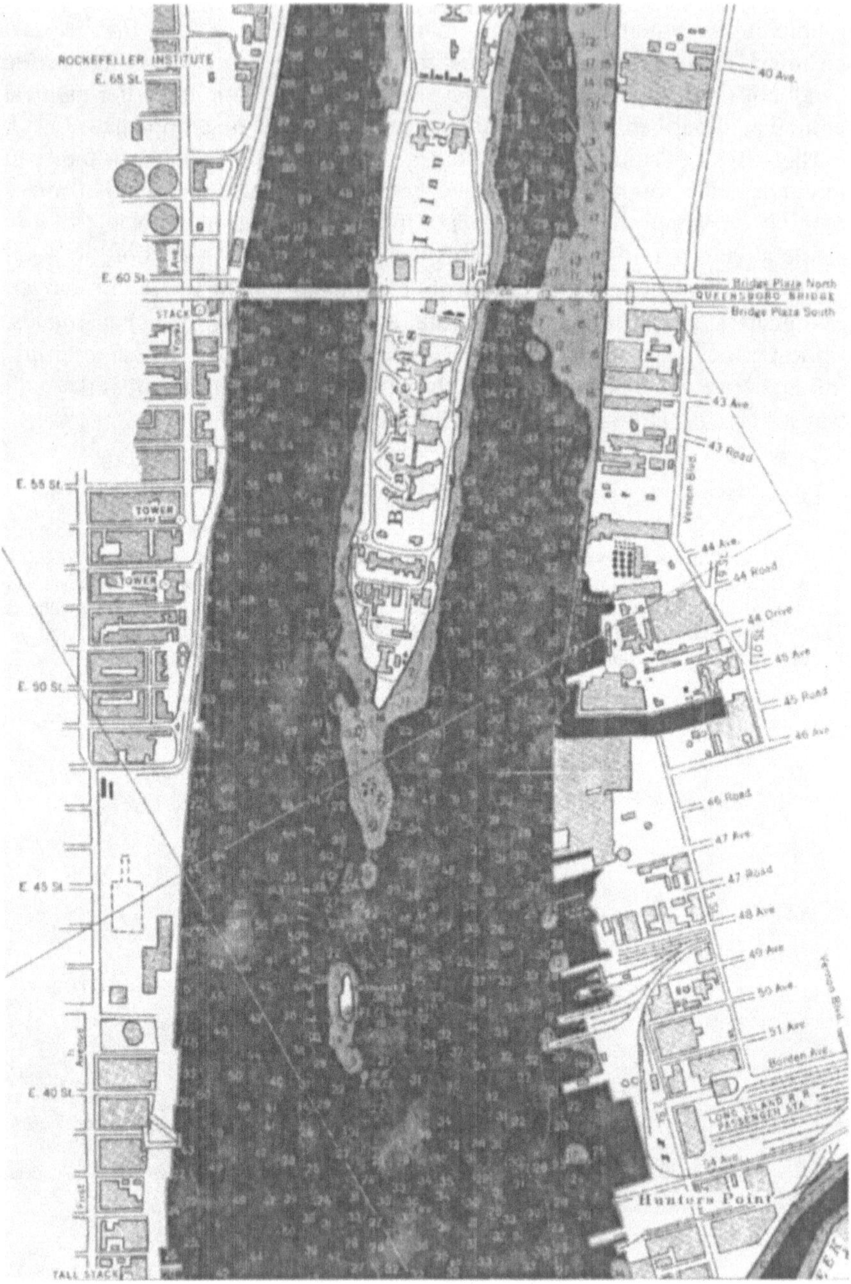
ROCKEFELLER INSTITUTE
E. 65 St.
40 Ave.
E. 60 St.
STACK
Bridge Plaza North
QUEENSBORO BRIDGE
Bridge Plaza South
43 Ave.
43 Road
E. 55 St.
TOWER
TOWER
44 Ave.
44 Road
44 Drive
45 Ave.
45 Road
E. 50 St.
46 Ave.
46 Road
47 Ave.
47 Road
E. 45 St.
48 Ave.
49 Ave.
50 Ave.
51 Ave.
Borden Ave.
E. 40 St.
LONG ISLAND R.R. PASSENGER STA.
Hunters Point
TALL STACK

Chapter Two

Spiritual and Infrastructural

U Thant's East River Island

VIEW OF AN ISLAND

Driving south, the view from FDR Drive is a film strip, edited by concrete and steel. Between columns and high guard rails, the island is there. Episodically pulsing like one of Muybridge's subjects, it appears to move, northward against the East River's tidal current. It bluntly cuts the water like one of the many barges that move up and down the strait. When these vessels pass, they dwarf Belmont Island in size and height. But the island stays put as the barges move on, and its flashing green Coast Guard light is a reliable beacon for river pilots. Set in motion by the FDR's structure that must also support dirt, sod, and the political weight of the United Nations directly above, the island is mobilized as another kind of infrastructure, one that vibrates with a spirit, difficult to discern so fleetingly but recognized by Secretary General U Thant from his vantage point in the UN. This attention recast Belmont as U Thant Island.

From the other side of the river, views of the island are also episodic and fragmented, but more static. Access works perpendicular to the shoreline rather than smoothly along it, and until recently it was difficult to see U Thant's island at all. Long Island City's industrial waterfront limited public views to glimpses over gates at the ends of east-west streets. In 1998, Gantry Plaza State Park began to open the waterfront, and tourists and New Yorkers now come to watch the boat traffic, gaze up and down the Manhattan skyline, or take a more introspective break in the gantries' generous shade. And now, they can't miss seeing U Thant Island, the anchored vessel that signals the nature of infrastructure.

Like Hoffman and Swinburne, Belmont is a human-made island largely off limits to humans. After its access shaft to the Steinway Tunnels was boarded up, there was very little reason to visit the tiny island. It was mostly ignored, except for thrill-seeking swimmers, ship pilots, and erstwhile thugs. On August 9, 1904, gang members left Lillie O'Neal on the Man-o'-War reef. After kidnapping the nineteen-year-old Irish woman, the "two thugs of the Mackerelville gang," rowed their victim to the island, where she was later rescued by police authorities.[1] The unlikely presence of this uninhabited island also made it an anonymous target, or stepping stone, for swimming contests. On July 28, 1932, James Bruno and Rocco Grasseo were rescued from Man-o'-War reef by a fireboat and the police launch *Gypsy* after they attempted a swim across the East River from 7th Street in Long Island City. The tide was coming in, and this part of the reef was six inches under water when they were finally picked up.[2] From the Manhattan side, George P. McGinnis drowned trying to reach Belmont Island. On October 10, 1949, after losing a game of checkers during their break as linotype operators at the *New York Daily News*, McGinnis challenged his co-worker Herman Plummer to the swimming competition.[3] More recently, on September 15, 2010, Belmont Island provided the backdrop for long-distance swimmer Marcos Diaz to complete his global swim in support of the United Nations' Millennium Development Goals. Diaz met UN Secretary General Ban Ki-Moon at Gantry State Park in Queens.

As an augmented reef, Belmont Island remained a hazardous, though more visible, obstacle to navigation. On the night of February 7, 1964, the Norwegian ship *Sigdal* ran aground on the island's submerged ledge.[4] The 33,310-ton, 669-foot tanker remained stranded on the island as rescuers attempted to dislodge the vessel and contain the spill. This plight recalled the island's other histories as an obstacle to navigation. On June 28, 1919, the wreck of the steamer *Calvin Austin* led to a court case that continued into the 1920s, and in July 1926, the reef damaged the *S-51* submarine as it was towed to Brooklyn's Navy Yard for salvaging.[5] Belmont was always meant to be seen—navigational beacons have generally assured that—but it gained a new visibility with the construction of the United Nations headquarters. Through its visual link with the 39-story Secretariat building, the island would become an object of diplomacy, protest, exhibition, conservation, and meditation. This increased exposure would also mean decreased—if not completely restricted—access to the general public. This link between building and island points toward a city's infrastructural nature. It also puts forward an infrastructure built on nature that is not displaced but placed at a distance. And it will then not be a surprise that this nature—represented by the island in the river—is in fact part of a human-made system.

Late in 1946, city officials, Robert Moses among them, negotiated the location of the United Nations in New York's Turtle Bay neighborhood along the East River. His plans for X-City having fallen through, developer William Zeckendorf sold the seventeen-acre property to John D. Rockefeller, who then donated it to the city. Wallace K. Harrison led a team of architects representing each of the founding member countries. As early as January 1947, this Board of Design Consultants was working from a design scheme with the Secretariat's tall narrow building in the center of the site located between Manhattan's 42nd and 48th Streets. Design discussions throughout the spring led the board to select Oscar Niemeyer's design known as Scheme 32, which placed the slab-like building close to its East River frontage and arranged the UN's other buildings so that they formed a central civic plaza that opened to First Avenue. Influenced by Le Corbusier's competing proposal in Scheme 23A, the board's final scheme placed the General Assembly building not on the periphery as Niemeyer had proposed but in a more central location, occupying what had been the open plaza. Finalized just as the site was being cleared in June 1947, this configuration pushed the public space to the edge of the East River.

A primary design concept in all of the schemes was the open space afforded by the East River frontage and the site's seventeen acres, an unusually large tract within Manhattan's density. The design board envisioned the UN Headquarters as the center of "an indefinite larger area" that could sustain a "certain distance from disturbing surroundings" and preserve the UN's "dignity and grandeur." The complex could then be the focus of the "whole world's common interests."[6] In this sense, the UN building was a global stage to be viewed symbolically. But the design schemes also addressed how the East River might be experienced from within the site. The north-south orientation of the Secretariat tower slab maximized views of the river. Proportions of the tower's height and footprint remained consistent throughout the design process, and the final dimensions of the Secretariat were 72 feet wide, 287 feet long, and 544 high. Expanded east and west facades meant increases in heat gain, but designers traded this climate control problem for the flood of light and the drama of views. Completed in Spring 1950, the Secretariat's Thermopane glass curtain wall offered panoramas of the surrounding boroughs as well as the river's tidal flow, its boat traffic, and its islands. One island in particular was most readily visible.

Four hundred yards to the east, on axis with 42nd Street, Belmont Island became a focal point for visitors, delegates, staff, and ambassadors. Oscar Niemeyer had also seen the island. It shows up in his sketch of Scheme 32 as a thin but prominent rocky mass, too big to be a boat and too small to be an extension of Roosevelt Island. In Niemeyer's sketch, the seventeen-acre site is itself an island, a rectilinear floating ground set within the East River's now graphically expanded waters where it is joined by Belmont's filial presence. Far more than Hugh Ferris' powerful renderings of the project, this sketch captures the idea that the UN joins New York City's archipelago, an island among united islands. In the UN's vision, this seventeen-acre island would also serve to unite the global archipelago of independent states—an idea that resonated with Buckminster Fuller's one-world-island, later modeled "geoscopically" on Belmont Island. Niemeyer's sketch also confirms the design board's idea that the East River was an integral part of the UN's landscape, even its site planning. And proposals, particularly those in the 1960s, considered the East River's opposite shore for expansion possibilities.[7] Whether formalized or not, the river space that extends eastward is a part of the UN's public zone in which Belmont Island remains as a focal point.

GEOSCOPIC ISLAND

The visibility of Belmont Island from the UN anchors the East River's event space. Its small size, the breadth of the East River, and the density of development along the river's embankments made the island otherwise difficult to see throughout the twentieth century, except by boat or with an oblique view from the river's other towers. In that sense, the UN is an auspicious witness to the island's cultural and natural histories, and the UN's Secretariat offers a privileged view of the island. Each year, as the UN prepares to open the General Assembly, adult sea gulls typically flock to the island. Ritual viewing of this event by UN staff members, diplomats, and ambassadors was already well-established a few years after the completion of the Secretariat building. At the start of the 1954 assembly, a news article noted that one of the "relaxations for delegates and others here is the placid contemplation of the birds on, above, and around the island."[8] Further emphasizing this connection between island and building, sea gulls have been known to fly from the island's shores to the top of the Secretariat to drop clams to the pavement below.

Protestors have taken advantage of the island's location in the shadow of the United Nations. On August 25, 1972, six members of the Greater New York Conference on Soviet Jewry occupied Belmont Island for two and a half hours. They renamed it "Soviet Jewry Freedom Island" and sought to "dramatize the plight of Soviet Jews," particularly decrying a practice that required Soviet Jews to pay for exit visas before going to Israel. They chose the island for its proximity to the UN and its symbolic value as a place of isolation.[9] Artists have also mined the island's presence. At dawn on the morning of the Republican National Convention in 2004, artist Duke Riley paddled a hand-made vessel from the ferry terminal at Long Island City in Queens to Belmont Island. When he arrived, Riley attached a twenty-one-foot long, glow-in-the-dark flag to the Coast Guard's tower. Riley's project sought to reclaim the island as public space. It also further established what he has called the United Islands of the East River, a hypothetical archipelago of resistance and free expression. Riley chose the site for its visibility to the UN, and he chose the timing for the heightened security during RNC's convention at Madison Square Garden. The East River around the UN is already a secure zone, and this context afforded Riley an opportunity to explore what he has called "individual behavior in conflict with institutional fear."[10]

Buckminster Fuller also understood the strong visual connection between the UN and Belmont Island. In 1956, he proposed a version of his Geoscope project, the Minni Earth, to be located on the island. Winslow Wedin's rendering of the project shows a globe floating above the stylized rock outcropping. Its axis tilted to match the earth's obliquity, the globe seems to have lodged on the top of the Secretariat building at a point where the curvature of its equator loops toward the thirty-eighth floor, which is, not surprisingly, where the Secretary General resides. In the drawing, we see Africa and most of Asia from a vantage point near the southern end of Roosevelt Island. The globe's orientation, or "posture" as Fuller called it, mimics that of the larger earth and places the United Nations, actually Belmont Island itself, at the precise zenith. The 200-foot diameter globe is tethered to the island and to lattice-frame towers so that its overall height of 400 feet would be on level with the UN building.[11] In a later lecture, Fuller described the globe's setting and the possibilities that the broader New York archipelago afforded:

> But realize, at back of the UN Building in New York in the East River is what used to be Blackwell's Island, and then it became Welfare Island, and then it had all kinds of insane asylums and city prisons on it, and gradually those were moved away. They built a new island of refuse up there, further towards the sound and all those things up there. That has now been renamed Roosevelt Island. At any rate, South of Blackwell's Island is Blackwell's ledge with bell buoys and so forth, a whole group of rocks in the water, and there right, actually out to the east of the United Nations building. And what I wanted to do was to build, then, a miniature earth, mounted from those rocks, having a mast, and mounted in cables it would weigh so very little that the cables would be really invisible and it would seem to be floating out there, look like a miniature earth that's come in close to our earth here, and it's, the United Nations building. . . . So, it would be a miniature earth really out confronting the representatives of the world.[12]

Belmont Island made Fuller's vision possible. In his project, he was expanding on an already augmented site, that "whole group of rocks" immediately to the east of the UN. Just as the buoys he mentions indicate ledges and reefs, his Minni Earth marked the island's location and from that modest position signaled the possibilities for global transformation. An oversized channel marker "confronting" the world.

At the time of his UN proposal, Fuller was developing a Geoscope prototype at the full two hundred foot diameter for the first time. He had previously worked with students at Cornell University in 1952 and would later work at Princeton in 1960. In all of these experiments, Fuller envisioned a tool for presenting data. And at the UN, he felt that the Minni Earth would be an informative, and persuasive, device in the decision-making process. In 1962, he presented the project to UN Secretary General U Thant, who apparently liked the idea and arranged a meeting with ambassadors at a New York hotel. Estimated to cost ten million dollars at the time, the UN project did not proceed, but its sophistication as a demographic and politically influential tool did. At Princeton, he had discovered that electric lights, rather than drawn lines, would allow more fluid mapping of data, and the Geoscope could now inventory "world resources wherever they occur."[13] Floating on air rather than water, this marker would address not only world representatives but the world itself.

The Geoscope also worked from inside out. Like a true planetarium, it could be occupied, and viewers could understand the globe's position as if they were in the center of the earth. Essentially, the viewer became the earth, and the interior of the scaled-down globe became the viewer's own sky. Fuller described the excitement of Cornell students who climbed a ladder and placed their chins on the tripod at the center of the sphere, where they could imagine themselves on any continent, with any star as its zenith. From inside, the sphere's interior surface became what Fuller called "your private sky."[14] Lodged within Fuller's "Idea of June 15, 1948," this concept foreshadows the Geoscope's capacity for individualized orientations and accurate visualizations of geography. Extending this idea to the UN site and to its island context, the Geoscope becomes a place of reckoning, not just in terms of location but for broader concepts of politics and society, even existence. Fuller compared the relationship between his Minni Earth and the planet to that of a lifeboat held by a davit crane on the *Queen Mary*. Big boat and little boat track with the same orientation, motion, and speed—not unlike the pairing of UN complex and its attendant island. Just as the lifeboat mimics the larger boat and saves those on board, understanding the smaller island might help steer the one-world-island. Landward versions of the ship's davits, the gantries on the river's east shore complete this image of the cybernetic ship of state, holding a smaller craft in reserve, to be accessed and moved by the kinetic industrial apparatus of the massive cranes that were previously used to lift barges between water and rail and now mark the location of their eponymous State Park.

Fuller's Geoscope carried a double meaning: from inside the sphere you could see the moon and stars, and from the outside you could see the globe as an object within a universal field. Decades earlier Fuller had discussed the concept of "point of view" that the Geoscope would later model. For him, point of view was not merely the angle of vision. It was more richly defined, closer to a story-telling model and the perspective of a narrative voice. Fuller's point of view concerned both introspection and external experience, a looking inward and outward. Fuller's hypothetical Geoscope is a globe in motion, a "great glass globe of the earth." Through its vitreous lens, it established relative position and mapped experience but at the same time provoked thought from "an abstract position" within the universe.[15] Overall, he sought a double movement in which perception could be phenomenally, and sensibly, defined and could also be "critical and intellectually independent."[16]

Islands in general and Belmont Island in particular have played similar roles in helping discover the human position in the universe—whether physically, spiritually, or philosophically. Another vision for this island, which would be called U Thant Island, also mixes the geoscopic with the anthropocentric, though cast within the framework of meditation and spirituality. Fuller's introspection is not unlike U Thant's meditation. Looking outward and looking inward. Islands are viewing stations as well as objects to be viewed, and looking at islands from a distance signals the opening up of space.[17] Although described in terms of real estate acquisition, this effect was described in a short news brief for the advertising business column in the *New York Times*. After being hemmed in on Lexington Avenue, advertising company president Alfred Seaman was "so joyous" over his new-found view after relocating to One Dag Hammarskjöld Plaza, where he looked down on the UN and the "little island in its shadow."[18]

ISLAND MEDITATION

While Buckminster Fuller was looking up at the United Nations with the Geoscope installation, UN Secretary General U Thant was looking down at the island from his office in the Secretariat building. In 1965, author June Bingham observed U Thant in his office: "Meanwhile, as the Sunshine Biscuit sign lights up across the East River and the beacons of the Long Island airports probe the darkening sky, the Secretary-General stands at his 38th-floor window. The view, he says, is conducive to meditation."[19] Neither Thant nor Bingham elaborate here, but the Burmese Secretary General was a practicing Buddhist, and we can assume that it was not Long Island City's Sunshine Biscuit sign that held his attention. It was the island, still visible at dusk, set within the East River's quiet evening repose, marked also by the red and green lights of passing boats, as the "winter sun has set . . . on its way to Burma."[20] During the month of April in 1970, U Thant invited Sri Chinmoy to be the UN's interfaith chaplain and to conduct twice-weekly meditation sessions for delegates, staff, and members of non-governmental organizations. Chinmoy, an Indian mystic headquartered in Queens, continued his work at the UN until his death in 2007—to some a controversial figure and to others a critical resource. The group named "The Peace Meditation at the UN" would meet throughout his tenure.

U Thant was continuing a program of spirituality that had begun with the United Nation's Meditation Room. In 1956, the same year that Fuller first proposed his own visualization environment with the Minni Earth, Dag Hammarskjöld directed the renovation of a small space that had been set aside for meditation in the General Assembly building. U Thant's predecessor as Secretary General believed that the UN should have a room dedicated to "silence in the outer sense and stillness in the inner sense" to complement the organization's mission of peace. When Hammarskjöld spoke about this room at its dedication ceremony, he could have also been speaking about the islands and insular spaces sought by Buddhist devotees:

> We had a very small space within which we sought to achieve a room of stillness. We must do everything possible in creating such a room to create an atmosphere where people could really withdraw into themselves and feel the void. In a sense, what I think we had at the back of our minds was something which is said, I believe, in one of Buddha's scripts, that the significance of the vessel is not the shell but the void.[21]

The room was meant as a place of retreat and isolation for silent meditation. In the renovation, Hammarskjöld exchanged chairs for benches, placed a six-and-a-half-ton rectangular block of iron ore in the center, and located the abstract mural *Infinity* painted by Bo Beskow on the narrowest wall of the trapezoidal room. Lighting from above illuminates the polished and "shimmering surface of solid rock." If the room is the silent vessel, the iron ore block serves as the meditative focal point, as the "weight and solidity of the everlasting," and symbolizes permanence in what Hammarskjöld called a world of "movement and change."[22]

The stone is like the island. From U Thant's vantage point atop the Secretariat, we can imagine the East River's surface unfolding in the view's middle ground. We might also imagine how its islands—Man-o'-War rock to the south, Roosevelt to the north, and certainly Belmont Island directly to the east—formed moments of stability and permanence within its choppy, racing waters. In such a view, the East River and its islets become a garden, not unlike Chinese rock gardens or Japanese *karesansui* gardens.[23] In their long traditions, these gardens initially included weathered stones or earthen mounds placed in pools of water and later constituted "dry landscapes" with rock outcroppings, suggestive of mountains or islands. Donald Keys, speechwriter for U Thant, believed that those who were meditating within Hammarskjöld's recently rededicated room used the "stone like a lens" to connect their thoughts with the global context.[24] For U Thant, the island was a similar lens, and its East River context was a garden. With this in mind, U Thant began a process of reshaping Belmont Island.

The placing of stones is an activity associated with landscapes of work as well as garden design. Displaced rocks, along with sediment, have long been a part of China's hydraulic landscape, where floods created *chou*, deposits of dry land, and where dredging for the Grand Canal displaced land for water on an immense scale. Displaced materials, particularly stones, came to connote a "geomantic magic" that could be talisman as well as sacred mountain.[25] Some stones continued to move as mountain islands that rested legendarily on the backs of swimming turtles in *horai* mythologies.[26] In these processes, fluid came first, and stones emerged from water and offered moments of stability. Stone-placing came to be associated with garden-making.[27]

Unmoving objects for garden design as well, individual stones are placed within landscapes of water, sand or gravel. In Japanese *karesansui*, patterns evoke waves, currents, and flows in a sea of raked gravel. These contours resonate with the remarkably textured water surface of the East River. Much like the dry materials of the garden landscape shaped by human agency, the water surface of the East River has been modified by human activity. The East River is almost completely contained by bulkheads. These human-made edges send waves back and forth within the river's field; and because the river is not really a river but a tidal strait, tides run up and down its length, adding to the intricate system of currents and the waves that mark them. Boat traffic adds another, more variable, set of wave patterns. New York's winds further texture the river's surface. U Thant Island interrupts these systems and in the process generates new patterns. This point of stability measures tidal fluctuations, offering a fixed point to visualize the racing tidal flow, which can be more than five knots with ebb currents, and to observe the tidal range, which is nearly five feet in the spring. With these currents, you might imagine the island to be a *horai* resting on the back of a turtle patiently negotiating the East River's rippling fluctuations.

As part of a scenic composition, rock-islands offer focal points and a steady structure within the overall field's emptiness and flow. Ryoan-ji, a well-known dry garden in Japan, exemplifies the elements of this composition and the connections to be made with the East River tableau. On the outskirts of Kyoto, the early sixteenth-century garden has been described as a "monument to stillness" and a "living landscape" that rises out of the abstracted water.[28] Though not the river's bulkheads or Manhattan's canyon-edge effect, the garden's surrounding walls closely meet the raked gravel's strong planar effect and provide a sense of enclosure. The garden's fifteen stones form an archipelago of five clusters that inspire variable meanings and associations. The field context greatly exceeds the size of the rocks, and viewers find the infinite in the infinitesimal, just as U Thant Island is almost lost in the middle of the half-mile-wide river. Its fifty-foot width makes it

New York's smallest island. Imagining either river or gravel as empty space also engages what is known as "spatial silence" in Japanese painting and garden design, where blank areas of the page, or ground, hold great significance in their contemplation. Garden stones and river islands also offer a sense of depth. Marking a location in the view, these objects help us to perceive relative distances and to scale relative sizes. In this sense and in its expansive context, the smallest island in New York's archipelago might give a sense of the nature of the universe.

Meanings of the rocks in these gardens vary widely. Reading them can be an open-ended process, but in other cases the rock-islands are specifically associated with paradise and immortality. In his early writing *Shiji*, Chinese historiographer Sima Qian described the three paradisiacal islands on China's east coast: Penglai, Yingzhou, and Fangzhang. This latter island translates to "ten square feet" and is suggestive of the limited space offered by islands of refuge and meditation.[29] Also known as the "Isles of the Blessed," these islands were simulated in an artificial lake built by Han Emperor Wu for his system of imperial parks. The isolated islands represent the "bold rock islands of immortals" found in the Eight Seas (*hakkai*) and thus tie into the Daoist idea of the island retreat as a remote place for rest and uninterrupted meditation. In the wider view, gardens themselves functioned similarly as places of retreat, isolation, reverie, and calm. These restful "hermitage" spaces might occur in the middle of urban environments and provide a "terrain of the imagination."[30]

Distance plays an important part in these gardens and their methods of meditation. Islands—whether imagined or real—can be remote places; and when they are objects of meditation, there must be distance between the contemplating subject and the island being contemplated. In this process, distances between subject and object within each meditation activity allow the solitary island's reshaping as an unfixed *topoi*, and it is no longer a commonplace part of the environment. Viewers of *karesansui* occupy elevated terraces that separate them from direct contact with the garden. The vantage point at Ryoan-ji is further limited within the raised veranda—not the four hundred foot elevation of U Thant in the Secretariat building but nonetheless a perspective that horizontally unfolds below the viewer. For other viewers of the U Thant Island, restricted public access highlights its isolated position in the river and requires that a safe, and meditative, distance is maintained. This restriction arises from its proximity to the United Nations and its unofficial designation as a bird nesting area. The UN meditation group's lease from the city has permitted some degree of island access, but otherwise the Safety and Security Zone around the UN limits activities, like the artist Duke Riley's attempt to establish an island state and restore public access.[31]

TINY ISLAND

The UN's interfaith chaplain Sri Chinmoy continued to work in this extended field of the East River's meditation garden with his formal adoption of Belmont Island. Soon after U Thant invited him to give meditations twice a week at the UN, Chinmoy proposed that the United Nations was a place without walls: "Please try to feel that the United Nations is not a mere building, but a place of worship, a place where all human beings can worship and pray." His meditation group modeled the "oneness-world-family" that he saw as the ideals of the UN, which Buckminster Fuller had also referenced with his Geoscope project.[32] Belmont Island became an official part of what Chinmoy called the "living shrine" on September 16, 1977, when the UN meditation group, led by Sri Chinmoy, dedicated the island to the memory of U Thant, who had died three years earlier. Thant's daughter Daw Aye Aye attended the ceremony and described the day as "rainy, damp and raw" but noted that "our spirit was joyous and fulfilled."[33] After New York Governor Hugh Carey commended its initiative, the group renamed the site "U Thant Island," planted trees and flowers, and inaugurated its official lease of the island from the state. The group periodically visited the island for ceremonies and maintenance, and on October 7, 1982, the meditation group dedicated the "oneness arch." A mock-up of the steel tubing lattice was assembled in Chinmoy's neighborhood of Jamaica, Queens, to test its structural performance before harbor patrol boats helped deliver the arch's components to the site. In addition to the thirty-foot-high arch, the memorial includes a buried vessel that is dedicated to U Thant and holds his statements about peace and spirituality, personal items, and the Secretary General's favorite tie clasp presented to the group by his daughter. The group also made official visits to the island for the U Thant Peace Award, which was offered by Chinmoy to WNBC TV reporter Russell Barber in 1986. Island visits became more limited in the 1990s as security around the United Nations was increased.

At its southern end, every four seconds, a U.S. Coast Guard navigational light blinks green from atop a skeletal tower. According to its USCG designation, lighted buoy 17 officially marks the reef rather than the island.[34] In 1999, the Meditation Group added to this informational content when it erected a sign on the island. Its lettering was difficult to read for unaided eyes on Manhattan or in Queens. Like the buoy, the signage was really directed to boaters who might approach the island. Within its six by twelve foot dimensions, the original 1999 sign reads: "U Thant Island: Compassion—Home. Dedicated to World Peace through Prayer and Meditation." The sign also includes the UN insignia, the attribution to "Sri Chinmoy: The Peace Meditation at the United Nations," and a quotation from U Thant—"Every man or woman should not only ask himself or herself what he or she is going to do in the world, but also ask 'Will there be a world in which I can live.'" Compared to urban and highway billboards, it is not large, but its relative size on the small island makes the sign a significant feature and recalls the billboards that Robert Moses feared for Hoffman and Swinburne. The sign appears to claim the island—a possibility reinforced by a photograph of Sri Chinmoy on the island, standing above the sign while leaning on a picket fence.

Another sign has been attached to the Coast Guard tower. Lacking the UN insignia, the sign faces west toward the Secretariat; it notes the dedication of the island and includes the poem Chinmoy wrote about U Thant and a brief biography of the Secretary General along with the following text: "Simplicity was U Thant's life. Sincerity was U Thant's mind. Purity was U Thant's heart. His was the approach of serene and illumined dignity." Both sign installations operate didactically—the first, to put forward U Thant's message, and the second, to put forward U Thant himself as an exemplary practitioner of peace. This signage underscores the island's deeper, though less legible, significance as an emblem of calmness for an audience that begins with U Thant, widens to the occupants of the Secretariat tower, and expands further to the possibility of a global collective.

Like other spoil islands, Belmont's scale and residual presence have allowed it to serve as a repository, a vessel that attracts meaning. In 1999, the *New York Times* hosted a competition for a time capsule that would be opened in the year 3000. When the *Times* asked competition participants and its readers a central question: "What would you take to a desert island circa 3000," it posited ideas of necessity and alluded to many of the specific features Belmont Island shares with desert islands. Deserted and desert-like, remote and arid, U Thant Island might resonate with many New Yorkers as Robinson Crusoe's Island of Despair.[35] An entry by the architecture firm Caples Jefferson made this connection with a proposed granite and metal obelisk for Belmont Island. The architects chose the site because "it was small enough so that it wouldn't be tempting . . . for somebody to build something in the course of the next thousand years, some practical real estate type of thing."[36]

The project reserves the island so that we might witness passing time. Wires extending into the river's brackish water hold in place the monument's nine slabs, which gradually fall away from their vertical position as the wires corrode and loosen. This molting process would be timed to reveal an inner glass core on the first day of the year 3000. The project brief cited the "beauty of decay" as a driving concept, and architects Sarah Caples and Everardo Jefferson asked the question: "How do you feel time?" For them, the changes of the monument would allow viewers to visualize the passage of time, as the corroded metal plates expand and press out on the granite slabs and the wires eventually release their hold. The project's entropy calls to mind Robert Smithson's interest in landscapes of decay, and the use of the island as a data visualization tool parallels the informational sphere of Fuller's Minni Earth project. This combination of monument and island becomes a reserve of data that engages inevitable temporal shifts. But the island itself already does that. It is in effect already the time capsule, the "thousand year visible monument."[37] Too small for development, too solid—and now too meaningful—for removal. Its granite gneiss already shifts and slowly erodes to measure time.

Belmont Island was never meant to be occupied. It originated as a platform for the mine's power equipment and the temporary access point for the middle shaft and its ventilation. Now, it stands as a reserve and consequently holds unlimited potential for meaning, imagination, and meditation. This potential ranges from the functional to the ephemeral. For the former, city officials proposed the relocation of the Delacorte fountain from the southern end of Roosevelt Island to Belmont's rocky ledge. Its six hundred foot spray caused problems on the larger island; and in 1974, and again in 1987, some argued that the fountain donated in 1969 by George T. Delacorte "really has no business being on an island that people use" and "it wouldn't bother anybody" on the smaller island. David Ozerkis, chief engineer of the Roosevelt Island project, proposed the fountain's installation on Belmont Island to mitigate the constant problems of its spray. Delacorte responded that Belmont Island was "'rather impractical' because a power line would have to be run to it and there was no way to get there, and the foundation must be serviced daily."[38]

For Sri Chinmoy and his meditation group, the island also held meaning as a potential "heaven on earth." In one of his many statements that mix didactic content with poetic form, Chinmoy wrote:

> God alone knows
> When the kingdom of Heaven
> Will finally descend on earth.
> In the meantime,
> God is begging His aspiring children
> To try to create
> At least a tiny island
> Which He can call Heaven on earth.[39]

Written after the dedication of Belmont Island to U Thant, Chinmoy's poem proposes a "tiny island" as the site of peace and harmony espoused by Buddhist principles. Chinmoy reserved Belmont as this place. The idea of island has long served metaphorically in UN discourse, and it has been applied specifically to U Thant. New York Mayor Abraham Beame noted: "In stormy times, he was an island of calm in a sea of controversy. His meditative ways helped him to maintain the neutrality so necessary to sustaining the confidence of differing nations."[40]

In December 1968, U Thant reiterated the organization's mission that the global community could not function with the isolation of individual nation-states: "Gone are the days when each nation was an island unto itself. Today, questions of human rights are a matter of international concern."[41] Thant implies a singular island community that is, in his mind, hopefully united in peace. This idea can also be found symbolically rendered in the UN's insignia, visually uniting the continents with its polar vantage point. This symbolic map recalls Buckminster Fuller's 1943 Dymaxion project of the Air Ocean World that unfolds the globe in a series of polyhedral components, which reduce distortion of the land masses and can be viewed from any position, removing the bias of orientation. Fuller generated the Minni Earth, planned for Belmont Island, from this map's polyhedral construction and its 2,160 triangular components. Both projects illustrate his "one island" concept. For him, the "one-world-island-in-one-world-ocean" format demonstrated the earth's geomorphic links as well as the migration patterns that unite the global community.[42]

Such an island proposes a new model of the world, a state that achieves balance and equilibrium. Buckminster Fuller suggests a similar method when he writes: "You never change things by fighting the existing reality. To change something, build a new model that makes the existing model obsolete."[43] The tiny East River island inspired such transformation. For Thant, it marked fallibility but also stability. For Chinmoy, it could be "heaven on earth." And for Fuller, it anchored his message to the UN and the world. It offered a place for "awareness" that required a necessary distance from its viewers but still provided linkages by modeling ideas that might change the world.[44]

ISLAND INSIGHT

In the second chapter of his memoirs titled *View from the UN*, U Thant emphasizes the role that meditation played in his daily life and in his role at the UN. In one passage, he describes the four features of Buddhist meditation. It is the fourth feature *upekka*—referring to equanimity, equilibrium, and detachment—that relates most directly to what his literal "view from the UN" offered:

> *Upekka* connotes the acquisition of a balance of mind, whether in triumph or tragedy. This balance is achieved only as a result of deep insight into the nature of things, and primarily by contemplation and meditation. If one understands how unstable and impermanent all worldly things and conditions are, one learns to bear lightly even the greatest misfortune that befalls one or the greatest reward that is bestowed on one. This lofty quality of even-mindedness or emotional equilibrium is the most difficult of all ethical virtues to practice and apply in our hectic world. To contemplate life, but not to be enmeshed in it, is the law of the Buddha.[45]

U Thant's island was a means for participating in the world but doing so at a distance by contemplating its presence. The island allowed for the contemplation of nature and the nature of things without becoming "enmeshed" in its worldly conditions. Even if U Thant did not know that the island originated as the residue of human activity, he knew of its fluctuations. And it was this impermanence that made the island a suitable, if not ideal, object of contemplation and meditation. The island made *upekka* possible.

The small island in the middle of the East River is a piece of infrastructure that is inherently unstable. It represents the precariousness as well as the triumphs of human intervention. Later versions of Chinese rock gardens provided similar landscapes, which came to be appreciated for their human-made—rather than nature-made—qualities. Such garden-making shifted design from adaptations made within natural landscapes to largely artificial constructs that held within them the ideas and interpretations of naturalness. The human agency and scenic composition of stone-placing became key elements in garden-making. Along similar lines, meditating on the artificial island, U Thant could contemplate human vulnerability and agency through a landscape element that was not only a tiny outcropping but also the waste product of human activity. This residual nature made it a particularly compelling tool for understanding impermanency and instability. From U Thant's perspective, the East River garden was the "mock landscape" that allowed consideration of not just human interpretations of nature but also human construction itself—systems of "worldly things" and the waste they generate. The island was about the nature of infrastructure.

The island's limitations are also its potencies for contemplation. In Daoism, which U Thant espoused, the engagement of environments such as islands allows for *dao*, the "way of nature," to be revealed. Dry landscapes inspire meditative immersion in the garden as a way to "experience the world in its fullness."[46] U Thant Island resists habitation, and its lack of development makes for indeterminate scale. Consequently, its emptiness can be a refuge for the mind, and its exceedingly small size can take on macrocosmic significance—a condition that has been described as "thousand miles in an inch."[47] U Thant Island also resists definable shape. The East River's tides, light reflected off of the water and nearby buildings, even accumulations of garbage and detritus, constantly change its demeanor. Without shape and with its form also difficult to ascertain, the island inspires the construction of a mental image apart from physical constraint. Such representations of nature—whether above Ryoan-ji or above the East River's textured waterscape—offer mediums for insight into the nature of things and the nature of being.

U Thant Island rests at the intersection of two axes. Along a north-south axis, it bisects the river's navigational infrastructure. Its position also reveals the hidden east-west axis of rail transportation that links the boroughs. Within these two systems, the island is lodged between differing views of nature—as impediment and expedient. North-south, it is hazardous land in water; and east-west, it was a means to an end, the by-product of infrastructural construction. Both axes share a fundamental view that nature serves human endeavor—that "human skills" will triumph over "nature's resistance." Nature is also a means to an end—whether cast within the process of island destruction or island formation. But there is a third axis.

U Thant's view of the island overlays a meditative sight line in which artificiality and nature collude to form an infrastructure that does not simply facilitate worldly pursuits but assists contemplation. U Thant makes the case for this sensibility when he proposes the ethical virtues of even-mindedness. And that is one of the lessons of New York's artificial islands. Recognizing that human-made landscapes are mock landscapes at the unstable nexus of human and nature—rather than one or the other—provides insights into the nature of things. Such insights facilitate further connections between what is made, where we make it, how we preserve it, and how we continue to contemplate it. Conservation and meditation. Islands are reminders that nature is in us, just like artificial islands are reminders that we make nature. Contemplating such places means that restricted space meets meditation, infrastructure engages wildness, and naturalness comes to the city. These ideas were latently expressed in a planning commissioner's statement as New York City considered the Gateway project's future: "We are challenging ourselves to reinvent the city and the harbor, and these islands show there is room for serious environmental features, that nature doesn't have to be elsewhere."[48] With such islands for infrastructure, neither nature nor spirituality has to be elsewhere.

NOTES

1. See "Thugs Marooned Girl on 'Man-o'-War' Ledges: Lillie O'Neal, Waitress, Cruelly Beaten and Left to Die," *New York Times*, 9 August 1904, 12.
2. "2 Boys Rescued from Reef Engulfed by Tides," *New York Times*, 29 July 1932, 17.
3. "Printer Drowns in River," *New York Times*, 11 October 1949, 33.
4. George Horne, "Tanker Teeters on Rock near U.N," *New York Times*, 9 February 1964, 90.
5. "*Calvin Austin* wrecks on Man-o'-War rocks," *Federal Reporter*, 15 July 1921, 3; and "Salvaged submarine S-51 hit the Reef while being towed to Brooklyn's Navy Yard" *New York Times*, 8 July 1926, 1.
6. UN Document, A/311, pages 79–80, cited in Norman J. Padelford, "United Nations Headquarters: Alternatives for Expansion," *International Organization* 21 (Autumn 1967): 778–779.
7. See Padelford, 779. As it was noted at the time and as a 2010 proposal demonstrated, the connection would require an expensive bridge, and it would be a "hazardous walk in inclement weather."
8. "Return of Gulls Awaited at U.N," *New York Times* (August 22, 1954), 7.
9. "Isle at U.N. 'Occupied' in a Jewish Protest," *New York Times* (August 26, 1972), 4. Eight years earlier, on December 11, 1964, a 3.5-inch bazooka fired a shell from Long Island

City lot at 12:10 p.m. during Ernesto "Che" Guevara's speech at the UN General Assembly. Police searched for clues on the island, which was in the line of fire. See Homer Bigart, "Bazooka Fired at UN as Cuban Speaks," *New York Times*, 12 December 1964, 1.

10. See Duke Riley's website: "Belmont Island Situation, Mission, Execution, Administration, Command (SMEAC)," accessed March 1, 2012, www.dukeriley.info. See also the documentation of "United Islands of the East River," Duke Riley's show at Sarah Lawrence College in Winter 2003.

11. Buckminster Fuller's Geodesic Dome for the Montreal Expo '67 was a similar size—250 feet in diameter and 200 feet high. It was also placed on an island, Ile de St. Helene, which was augmented with fill and was located next to Isle de Notre Dame, which was created entirely from fill.

12. Buckminster Fuller, Session 9, Part 7, "Everything I Know" lectures (January 1975), accessed January 17, 2012, http://www.bfi.org/about-bucky/resources/everything-i-know/eik-session-9/eik-session-9-part-7.

13. Buckminster Fuller, Session 10, Part 1, "Everything I Know" lectures (January 1975), accessed January 17, 2012, http://www.bfi.org/about-bucky/resources/everything-i-know/eik-session-10/eik-session-10-part-1.

14. Buckminster Fuller, *Your Private Sky*, eds. Joachim Krausse and Claude Lichtenstein (Baden: Lars Müller Publishers, 1999), 344–352. Inside the sphere, sections of the world map form the private sky-dome and accurately provide orientation to cardinal directions and to the vertical calibration of zenith so that the inhabitant can "see his geography correctly."

15. Fuller, *4D Timelock* (1928), 31, cited in *Your Private Sky*, 346. Buckminster Fuller wrote: "The point of view, through introspection, unlimited to the segmental area of our temporal eyes, is our abstract position in the center of the universe, looking or building from the inside out, as from the center of a great glass globe of the earth. Through this globe may be viewed the progression of relative positions to the starry universe, looking along the time lines in all directions. The separate paragraph thoughts are only connected by their common truths, which are the material crystalline spheres of sensible and reasonable fact, through which the radial time lines of individualism must inevitably pass in their outward progression towards the temporal infinity." (346) In a formulation reminiscent of Lucretius' skin-like "films" that are released from objects, Fuller extends the model to include a series of "material crystalline spheres of sensible and reasonable fact."

16. Fuller, 344.

17. There are many examples of these island experiences. Exiled to the island of Patmos, St. John visualized content for his writings from an insular position. For a discussion of this island context, see my article "Florida Porch Reverie," *The Florida Historical Quarterly* 90 (Winter 2012): 324–330. Herman Melville masterfully describes the effect of islands viewed from a distance. At the start of "Norfolk Isle and the Chola Widow," his eighth short-story sketch from *The Encantadas*, Melville captures the isle's mystical isolation: "however insignificant to most voyagers, to me, through sympathy, that lone island has become a spot made sacred by the strangest trials of humanity." See "The Encantadas, or Enchanted Isles," accessed July 12, 2012, http://www.melville.org/encant.htm.

18. Philip H. Dougherty, "Seaman's Island," *New York Times*, 8 January 1974, 51. The article notes that Seaman, president of SSC&B, "claimed" the island from his forty-second-floor office and called it "Seaman's Island."

19. June Bingham, *U Thant: The Search for Peace* (New York: Alfred A. Knopf, 1966), 281.

20. Bingham, 281.

21. These excerpts come from the brief text Hammarskjöld wrote as an introduction for visitors to the room. The UN's website includes this text on a page titled "A Room of Quiet," accessed December 18, 2012, http://www.un.org/Depts/dhl/dag/meditationroom.htm. The Meditation Room originally opened on October 14, 1952.

22. Hammarskjöld began the document with the rationale for the room: "We all have within us a stillness surrounded by silence. . . . It has been the aim to create in this small room a place where the doors may be open to the infinite lands of thought and prayer." He concluded with the following: "There is an ancient saying that the sense of a vessel is not in its shell but in the

void. So it is with this room. It is for those who come here to fill the void with what they find in their center of stillness."

23. This latter term sometimes appears as *Hiraniwa karesansui*, which is translated as "planar dry landscape." With Japanese as well as Chinese influences, the rock garden can be traced back to sixth- and seventh-century garden configurations.

24. Donald Keys, "Spirituality at the United Nations," accessed November 9, 2011, http://www.aquaac.org/un/sprtatun.html.

25. Wybe Kuitert, *Themes in the History of Japanese Garden Art* (Honolulu: University of Hawai'i Press, 2002), 148.

26. Kuitert, 127, 196. *Horai* are goddesses of the seasons.

27. Kuitert, 51–52, 148. Underscoring the connection between island and garden, the word *shima* translates to "island" but was also an early synonym for garden.

28. Meher McArthur, *Reading Buddhist Art* (London: Thames and Hudson, 2002), 203.

29. Helmut Brinker, Hiroshi Kanazawa, and Andreas Leisinger, *Zen Masters of Meditation in Images and Writings* (Honolulu: University of Hawai'i Press, 1996), 68. *Penglai* translates to "Dishevelled Goosefoot," and *Yingzhou* to "Ocean Continent." This latter space finds the world in the island as with Buckminster Fuller's Geoscope. Wybe Kuitert also provides examples of island typologies, drawn from his translation of *Sakuteiki* ("Notes on Garden Construction"), the eleventh-century Chinese manual on garden construction. He includes The Field Island, The Rocky Island, Hill Island, Forest Island, and the Ebb-tide Style as idealized types "that serve as a mental image to inspire the senses." (33)

30. Edwin T. Morris, *The Gardens of China: History, Art, and Meanings* (New York: Charles Scribner's Sons, 1983), 68. Such spaces are not unlike the United Nations' Meditation Room.

31. The "Safety and Security Zone: New York Marine Inspection Zone and Captain of the Port Zone" is listed under Title 33 Code of Federal Regulations (CFR) 165.169, established in East River. Note A on the U.S. Coast Guard's navigation chart 12335 designates the "Specific Regulated Navigation Areas and Limited Access Area." The island remains as a public space only to be seen. But it is in fact hard to see. From Manhattan, East River tours have noted that the north end of Tudor City offers a glimpse of the island; and from the Queens side, the pier at Gantry Plaza State Park now provides a good vantage point. Otherwise, it is easily missed.

32. Sri Chinmoy, "A Living Shrine," March 1, 1974, Sri Chinmoy's address to the Meditation Group after the March 1st meeting, accessed November 3, 2011, http://www.srichinmoylibrary.com/books/0295/1/4. See also Sri Chinmoy, *My Meditation-Service at the United Nations for 25 Years* (Jamaica, NY: Agni Press, 1995), 38. Many of Chinmoy's books, some published at his own Agni Press, look reflectively on his work at the United Nations. These include the following: *My Meditation-Service at the United Nations* (1995), *The Garland of Nation-Souls: Complete Talks at the United Nations* (1995), and *The United Nations: The World's Oneness-Home* (2005).

33. "Daw Aye Aye Thant," Sri Chinmoy Centre News, accessed July 3, 2012, http://www.srichinmoybio.co.uk/news/daw-aye-aye-thant.

34. See U.S. Coast Guard, *Light List: Volume 1: Atlantic Coast: St. Croix River, Maine to Shrewsbury River, New Jersey* (Washington, D.C.: Government Printing Office, 2012), 244. The light's number is 27315, its location is 40-44-47.450N by 073-57-51.980W, and atop the steel skeleton tower is a square green dayboard with a reflective border. The light has a "higher intensity beam up and down stream."

35. Robinson Crusoe laments his landing on the "dismal unfortunate island," which he names the "Island of Despair." See the castaway's journal entry dated September 30, 1659, in *Robinson Crusoe* (New York: Oxford University Press, 2009), 60.

36. Colby Devitt, interview with Sarah Caples, *New York Times* on the web, September 29, 1999.

37. Colby Devitt, interview with Sarah Caples.

38. Dembart Lee, "Follow-up on the News: Fountain," *New York Times*, 19 May 1974, 35.

39. Sri Chinmoy, *Ten Thousand Flower-Flames* (Jamaica, NY: Agni Press, 1983), 93.

40. Sri Chinmoy, *U Thant: Divinity's Smile, Humanity's Cry* (Jamaica, NY: Agni Press, 1977), also excerpted in "U Thant," accessed July 2, 2012, http://www.srichinmoylibrary.com/books/0303/1/45/index.html.

41. U Thant, General Assembly statement commemorating the anniversary of the Universal Declaration of Human Rights, 9 December 1968 (S-0885-0003-20 in UN archives).

42. Buckminster Fuller and Anwar Dil, *Humans in Universe* (New York: Mouton, 1983), 101. See also the idea of "one island world" in Fuller's *Critical Path* (New York: St. Martin's, 1982).

43. Buckminster Fuller, Session 2, Part 8, "Everything I Know" lectures (January 1975), accessed January 17, 2012, http://www.bfi.org/about-bucky/resources/everything-i-know/eik-session-2/eik-session-2-part-8.

44. Islands worked that way for Fuller. In the second session of his "Everything I Know" lectures, he noted: "I want to think about a rather simple model here of, I'm going to have an entity, and I'm going to make it a spherical entity. It's an island entity, and there are no aberrations or forces operating on it, so it tends to be spherical. And there is otherness, but it, for the moment, is not aware of it."

45. U Thant, *View from the UN* (New York: Doubleday, 1978), 22.

46. Brinker et al, 80.

47. Kuitert, 84. He quotes fourteenth-century poet Sesson Yubai: "A little group of fist-sized stones creates the effect of a thousand miles." In this section, he also notes that the Chinese term *kasenzui*, or "mock landscape," entails the "formation of the scenic garden" (84–85).

48. Omie Medford, quoted in Seitz and Miller, 4. Medford was a project manager in the "Waterfront and Open Space" division of New York City's Department of City Planning. See also *The New Waterfront Revitalization Program* (September 2002).

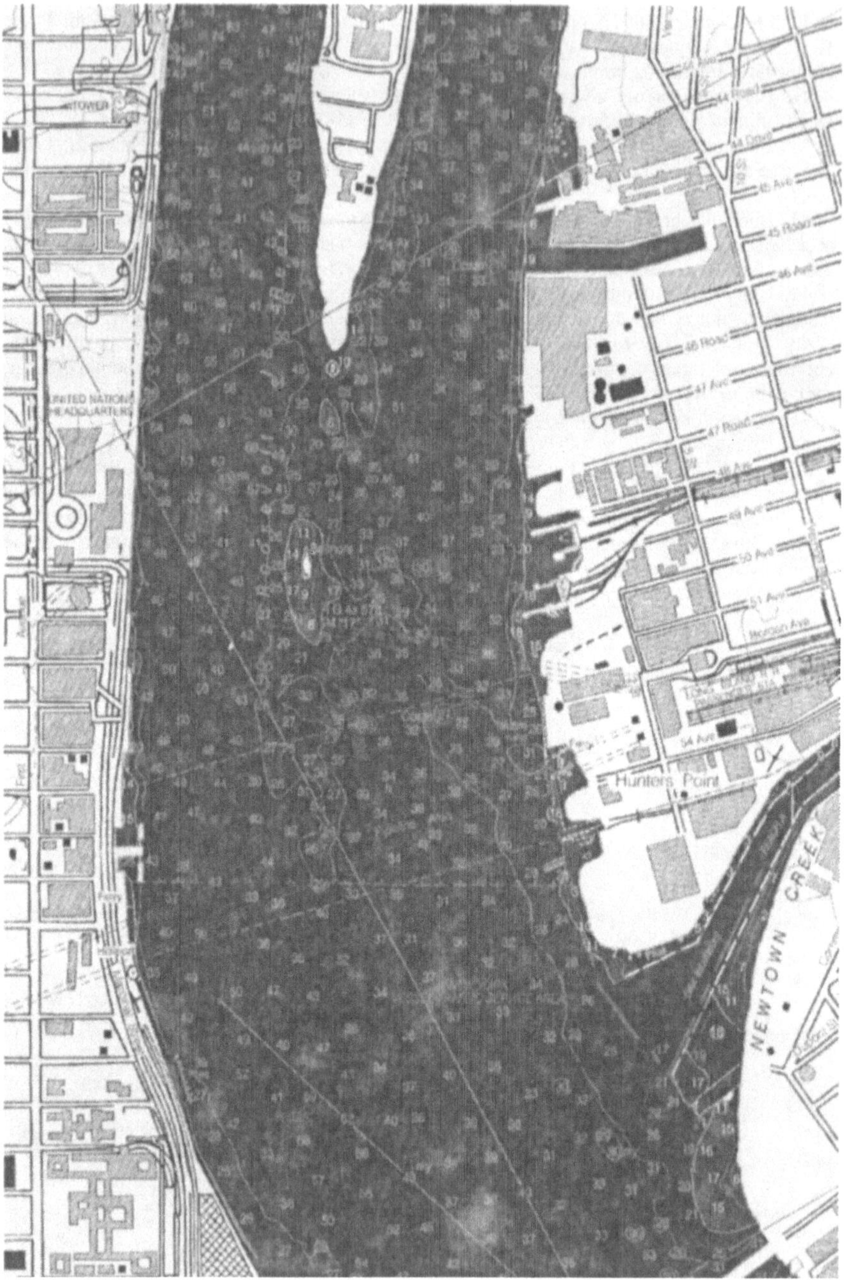
UNITED NATIONS HEADQUARTERS
44 Road
44 Drive
45 Ave
45 Road
46 Ave
46 Road
47 Ave
47 Road
48 Ave
49 Ave
50 Ave
51 Ave
Borden Ave
54 Ave
Hunters Point
NEWTOWN CREEK

Mound

At 10:28am on September 11, 2009, Jacksonville firefighter Skip Smith raised an American flag above a mounded island along the Intracoastal Waterway. One minute later, he lowered it to half-staff to commemorate the time when the second tower of the World Trade Center fell. He had seen the mound of earth on his daily commute across Atlantic Boulevard's bridge: "It looked like an unclaimed island just calling for a flag." Smith claimed the hill, erected a flagpole, and named the spoil island Hill 343 in honor of the 343 New York firefighters who died during 9-11.

The hill was a makeshift memorial built on what was destined to be a temporary mound. In the 1960s and 1970s, dredgers created the seven and three-quarter acre island with spoil from the Intracoastal Waterway. Its official name—FIND Island—is an acronym for the waterway's administrator—Florida Inland Navigation District. Over time, grasses colonized wetter areas of the island, and its uplands hosted other vegetation, including exotics like tallow trees and natives like red cedar, wax myrtle, black cherry, and yaupon holly. Four decades later, a mitigation project sought to redistribute the long-forgotten spoil material. In exchange for the environmental impacts from its construction of Wonderwood Expressway, the Jacksonville Transportation Authority (JTA) was required to restore a wetlands area. JTA chose FIND for the project and began clearing the island in early spring 2007. Citing habitat destruction and high costs, opposition halted the transformation of the island and the construction of an earthen access causeway on July 4th. Manager Wendy Morrow summarized JTA's position: "We're sensitive to the fact that there would be a change in the type of wildlife there. But still, the salt water marsh will have a habitat of its own. It was a spoil island. It was created. It wasn't something that was there before. Originally, it was a saltwater marsh area. So, we're actually restoring it to what it originally was."[1]

At the center of the debate was an island, but the sides were really arguing over spoil. Residents of Neptune Beach had grown accustomed to their view of the island's upland marsh habitat. To them, it was a resource, and to JTA it was an opportunity—and then a contractual obligation—to correct an error. But the historic dredge material would soon be re-formed, and the upland's slight elevation would become one of the highest points for miles. The original plan to remove the spoil material from the area was determined impractical. As public opposition grew, costs from the delay mounted. Permits to barge the material over the marshland were denied, and JTA decided to consolidate seven acres of material into a mound that covered the remaining three-quarters of an acre. This transformation from spoil island to salt marsh yielded a thirty-foot high mountain of material, its angle of repose matching the slope of a car's windshield and its base approximating a narrow football field with rounded ends.

In the first week of September 2009, Smith and other members of Jacksonville Fire and Rescue Local 122 paddled shovels, sand, and 750 pounds of concrete to the mounded island. They also ferried solar panels and equipment for two solar-powered spotlights to illuminate the flag. On September 8th, the firemen poured the concrete and installed the twenty-five foot flagpole and four-foot by six-foot flag. In the weeks following the commemorative event on the 11th, Smith paddled out to the island each morning and night to raise and then lower the flag. The lights had proven inoperable, and federal code states that the United States flag should not fly unless it is illuminated. With donated funds, the firemen installed a new lighting system and a larger flag, which would later be sent to Ground Zero for a temporary exhibition. The flag on the hill, illuminated nightly, became a focal point of emotions as well as politics. Neptune Beach Mayor Harriet Pruette thanked the firefighters, the Jacksonville Beaches Exchange Club gave an award, and JTA praised the effort, even though they had not been aware of the project's planning.

In spite of the memorial's popularity, opposition to the mound remained. Planted spartina grass failed, and the hill was brown. One consultant to JTA's engineering report called the mound's memorial an "attractive nuisance."[2] This oxymoron captures the aporia of the mound. In the public domain as well as the public eye, the spoil mound was caught between public use as the firefighters' memorial and public purpose with the general benefits of an environmentally compatible landscape. It was impossible to ignore the memorial's evocative power, and it was equally difficult not to admire the local firefighters' vision and their dedicated effort to build and maintain the project. While the memorial elevated the memory of a significant event, the mound that supported it obstructed residents' views. Although at first the memorial stabilized public opinion about the mound, the memorial's presence undermined JTA's conservation work, even though those efforts targeted an artificial landmass. Conservation easements required by JTA's permit from the water management district disallowed access to the island, and foot traffic continued to prevent vegetation growth. Regrading started in February 2011, and by April the larger mound was redistributed into smaller tumuli, which have remained as tributes to the 9-11 memorial.

Mounds by definition are human-made. They are heaps of earth that have been artificially created.[3] When we call naturally-occurring elevated areas "mounds," we are comparing them to human-made piles, banks, fortifications, boundaries, or obstacles. The comparison does not usually work the other way, and the spoil mound looked out of place in the intracoastal wetlands on Jacksonville's eastern edge. Before JTA's project, many considered the island to be a natural part of the salt marsh. The project interrupted that vision and made visible the site's latent artificiality. Making the mound gave form to the spoil's presence and its history. The loamy sand recalled—and quantified—the Corps of Engineers' channel-deepening, but it was no longer the formless by-product of that project. In a short period of time, the 23,000 cubic yards of spoil gained additional meaning. To some, it was a displaced natural resource. To others, it was the substrate for a solemn memorial. In logistical terms, it was previously excavated material that could again be redistributed. Turnbull Engineering of Yorktown, Virginia planned seven smaller mounds, which are clearly visible in aerial views. These nearly perfect seventy-foot diameter circles are sprinkled across what was previously an island and clustered around what was previously a larger mound. To mitigate environmental damage elsewhere, an artificial island was spoiled and then remade. Now, its progeny of smaller islands provide perfect geometries to measure the next forty years of natural adoption.

NOTES

1. Drew Dixon, "FIND Island removal is put on hold," *Florida Times-Union*, 4 July 2007, L-1.
2. Engineering Design Management, Memorandum on "FIND Mitigation Site," Robert Franques to John T. Davis, 31 March 2010, 6.
3. Etymologically, the origins of "mound" are ambiguous. In one respect, they emerge from "mundus" and connote a model of the earth. In another respect, as "mund," they are works of protection that engage the "manus" of hand.

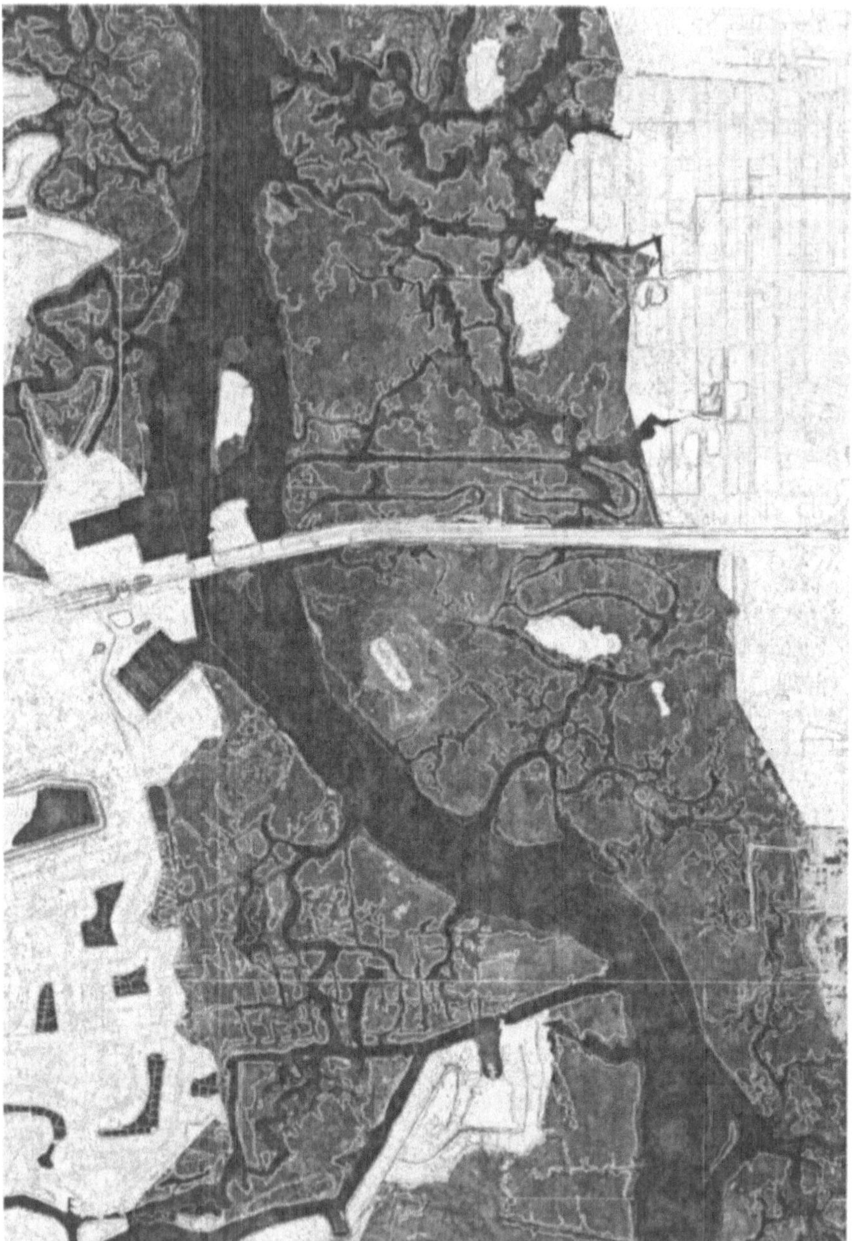

Chapter Three

Public and Private

The Common Wildness of Indian River's Linear Archipelago

He was revolted by the thought of known places and dreamed strange migrations.[1]

COMMUNAL WASTE

Collections of waste indicate shared undertakings. This residue can also constitute a communal space. When waste remains near the origins of its production, the landscape becomes a legible canvas of resource exchange. Gather and consume, give and take, cut and fill. Derived from a common resource, such waste material continues as not just a record but a source of communal life. In some cases, it is repurposed for more narrowly defined, even individualized, uses. Either way, it remains as a site with transformative potential. It might be claimed, reused, or conserved. Sometimes it is ignored, but when the waste is a mound or anomalous island, it remains within public perception. Rising above tidal flatlands, the middens and the dredged islands of Florida's east coast both offer landscapes that narrate waste as a shared resource.

One landscape extends thousands of years into the past, and the other is more recent, occurring in the last century. Both demonstrate how the residual spaces of waste record shifting values of public and private use. In their perhaps unexpectedly parallel histories, these coastal waste spaces move from communal and public projects to sites of more individualized experience and then largely return to a status as common places. Both shell midden and spoil island began as public—or at least communal—infrastructure before becoming idiosyncratic sites of experience and emergent places of habitation. Development and increased use have resulted in a third phase that manages and sometimes codifies these sites as commonly held spaces. Such residual waste sites become cultural and natural reserves, even monuments, within the public sphere.

Their communal waste also suggests a common wildness. Common not because it is always shared but because it is open and available. And wild not because it is distant or untamed but because it offers room for play, rest, privacy, loneliness, even madness, like that of Sweeney's topographic movement along "strange migrations." A less scientific version of the naturalist William Bartram and a more mystically inclined observer than John Muir, Sweeney immerses himself in the marginal landscapes of Ireland and western Scotland. In one of his "strange migrations," he swims the wildland's rivers from pool to pool. One might also traverse Florida's Indian River, not from pool to pool, but from spoil island to spoil island.

No longer remote, these islands are still largely unknown. Without the long natural and cultural histories of the British Isles' coastline, these islands maintain a wildness born out of waste and obscurity. Extensive in number and depositional rhythm, Indian River's islands constitute an unprecedented archipelago that translates waste to wildness and residual space to public commons. To move island to island is to navigate histories of human intervention and, like Sweeney, to discover timeless experiences of water, rock, and sand. In the former, our own activities of dredging locate us. In the latter, the resulting islands displace us. These islands raise us to an unusual height above the waterscape, and they maroon us on a small piece of ground made vulnerable by currents and winds. They also remind us of other human activities. Spoil islands and shell middens are both instances of waste as public space.

MIDDEN

More than two thousand years ago, Indians settled on the Florida peninsula where they built an extensive system of earthworks. Mounds served as the basis for ritual and burial practices. As their coastal habitation became more sedentary, the Indians also left behind middens formed from shells discarded after extracting their contents. These middens, often confused with burial mounds, were refuse heaps whose contents depended on the harvest—clams in the fresher water, oysters in higher salinity. A function of the bountiful shellfish in its estuaries coastal tidelands, and wetlands, Florida's east coast shell middens are some of the largest refuse mounds in the United States. Turtle Mound rises fifty feet above sea level, the Castle Windy midden is seventeen feet high and three hundred feet long, and Seminole Rest—though depleted by mining operations—is twenty feet high.[2]

The middens of the Indian River Lagoon are the residue of communal resource extraction. They measure a compliant landscape's bounty and potential. The network of middens, most of them now destroyed, also illustrates the limits of these resources. When a particular area's yield of shellfish was exhausted, shelling activities were shifted to a new location. The resulting matrix of refuse heaps contradicts assumptions that the native population lived on sustainable practices.[3] In this archaic context, limited human populations could overharvest a locale without overtaxing the region as a whole. But the midden network also represents a collective focus on seasonal harvest and cooperative living. Longer-term occupation occurred in base camps around "kitchen" middens, and other middens were used as transitory sites for shelling particular harvests. Larger permutations of these heaps are thought to be more centralized processing stations.[4]

Centuries after the zenith of these bountiful extractions but only decades after the final expulsion of Indians from the Atlantic coast, the refuse earthworks of spoil islands will join these abandoned middens in a landscape altered more and more by human intervention. The artificiality of each will be sometimes contested and sometimes forgotten. The magnitude of the mounds might seem to exceed human capacity. In 1908, Florida Fish Commissioner J. Y. Detwiler recounted in an article for the state's historical society that an "intelligent person" asked him what "upheaval of nature's forces must have occurred to produce this vast pile of shells."[5] In the following decades such large-scale displacements of resources and unearthed materials will become common, so that the natural might be more broadly identified—and confused—with the human-made, and the blurring of what we have made and what has been made around us becomes more ubiquitous.

TURTLE MOUND: GREEN DESERT AND DEAD GARDEN

In the spring of 1921, John K. Small plied the waters of Florida's East Coast Canal to reach Turtle Mound. Likening himself to a pioneer who goes south rather than west, the horticulturist and botanist was travelling through Florida on a four-week trip to report on the state's largely overlooked "floristics" for the New York Botanical Garden. Small had arrived in Florida at the heights of natural drought and human-made drainage. In the title of his report, Small captures the paradoxes of Florida's landscape, both natural and human-made. For him, "green deserts" are the region's sandy deposits—its dunes, hammocks, and prairies—that support an extraordinary range of flora. As one example, he recounts how the Everglades emerged from the sea as a sandy desert.[6] In contrast, "dead gardens" are areas lost to human development.

Small observes with deep concern the effects that the state's network of canals and drainage ditches have had on its wetlands, particularly the Everglades and the lands around Lake Okeechobee. From the account of his travels, it is clear that Small feared human development as much as he worried about the destruction of fire. He attributes these fires less to the natural drought and more to the protracted artificial drought caused by the drainage projects. Natural drought prevented the "green desert" from becoming the "painted desert," with its array of flowers that he expected. But the artificial drought might transform Florida's "green desert" to "dead garden." At the conclusion of his article, he notes that these "fires, if frequently repeated, will transform the Florida peninsula into a desolate waste and not as we found it in the spring of 1921, a Green Desert."[7]

Small's destination on that particular spring day rose fifty-five feet above the barrier dunes that separate the Atlantic Ocean and the brackish waters of Mosquito Lagoon. He noted that the view from the top of the mound was unique along the entire coast for its "extent and variety." For Small, Turtle Mound is caught between green desert and dead garden. The midden is a desert because it augments the coastal dunes' sandy deposits with shell. It is also a garden because migrating birds have dropped—and, according to Small's hypothesis, Indians may have planted—the seeds that yielded the mound's diverse flora. Much of the midden's greenness is foreign to the surrounding context, and its foundation in shell makes it vulnerable, not to natural forces such as hurricanes or fires, but to human intervention. At the time of his visit, Small learned that the county had recently purchased the mound of shell for road repair projects.

ISLAND ESTATES

Small's 1921 passage from Jacksonville to Turtle Mound took him along the Florida East Coast Canal. This navigation project formed a part of the broader land engineering projects that Small decried in his essays about Florida development. Dredging for this waterway had been completed nine years earlier, but the private entities responsible for its navigability had not maintained the canal, which in many places was beginning to silt up. While Small inventoried botanical wonders, engineers planned improvements to the waterway. On his southward trip, Small might have passed the surveyors charged by Congress to study the canal's current widths and controlling depths and to consider the possibilities for federal purchase of its right of way. Following a previous survey that found no commercial justification to purchase the canal, this renewed activity initiated by the July 1920 Rivers and Harbors Act eventually found that benefit exceeded expense. According to the Secretary of War and the U.S. Army's Chief Engineer in 1926, an inland waterway dredged eight feet deep and seventy-five feet wide along Florida's east coast would lower transportation costs.

The Florida East Coast Canal was one of many waterway projects that eventually became incorporated as the Atlantic Intracoastal Waterway (ICW). In this context, the Chief Engineer's report and its incorporation into the River and Harbor Act of 1927 initiated the modern history of spoil islands along the coast of Florida. The Act stipulated that a local agency would acquire the canal's right of ways, which would then be transferred to the federal government. It also made this agency—established as Florida Inland Navigation District (FIND)—responsible for the placement of spoil: "local interests shall furnish suitable areas for deposit of dredged material."[8] FIND remains today as the state sponsor for the inland waterway. As the "special taxing district for the management and maintenance" of the ICW in Florida, FIND's land-use easements cover most of the waterway's spoil islands.[9]

Nine years after Small's journey, Herbert Hoover steered the presidential yacht *Sequoia* along the circuitous waterway. Near the headwaters of the Matanzas River, where the natural channel becomes an artificial cut, the waterway originally took an almost ninety-degree turn to the east in what was known as Fox Cut. According to local stories, as the *Sequoia* negotiated this turn's coquina shores, President Hoover yelled "somebody straighten this damn thing out" as he was thrown from his captain's chair. By 1933, the Corps of Engineers created a second canal that provided a smoother turn and a wider channel. This alternative route at Fox Cut created an island that was built up with spoil from the dredging operations. The old and new waterways parallel each other for two and a half miles and make legible the evolution of this waterway from a modest, but quickly antiquated, channel five feet deep and 50 feet wide to a full-fledged waterway eight feet deep and 100 feet wide. Today, the older eastern canal remains less than this originally planned width, while the larger western canal offers breadths that range from 200 to 300 feet. This latter waterway contains the ICW's controlling depths of twelve feet and the required minimum channel width of 125 feet—dimensions that were established in January 1944 after favorable reports from state investigations and the Corps of Engineers' federally funded studies. Dredgers completed Flagler County's section of the waterway in 1951, and overall work between Jacksonville and Ft. Pierce was completed in 1965.

Severed from its mainland by the two canals, a residual island, which would become known as Island Estates, first hosted the town of Ocean City before many of its residents moved east and south across the waterway to Flagler Beach. The island was notable for its dangerous hunting after this uprooting of its settlers and before Rayonier and Levitt Corporations—the latter known for Levittown's development—purchased the island's 242 acres in 1967. Topographic maps from this time show a lumpy terrain of rolling tumuli with alternating crests of 13, 16, 19, and 22 feet. Its name naturalizing the artificial deposits of this topography, Hammock Dunes laid out preliminary development plans in 1973, but Island Estates' housing was not built until the 1980s.

Waterborne transportation imperatives had created the territory for Island Estates, but lack of automobile access hindered its growth. On paper, this parcel of land was readymade for development but its insularity proved too restrictive, even for Levitt's development machine. For almost two decades, Island Estates languished as one of Florida's many "unfinished cities," poetically illustrated by E. B. White in the 1970s: "Although I am no archeologist, I love Florida . . . for the remains of her unfinished cities. . . . I love to prowl the dead sidewalks that run off into the live jungle, under the broiling sun of noon, where the cabbage palms throw their spiny shade across the stillborn streets and the creepers bind old curbstones in a fierce sensual embrace."[10] But development finally caught up with what engineers had wrought, and the island's real estate potential was realized as a housing development. Its insularity now provided security, and the artificially created island became a community gated by its geographical conditions.

Like Island Estates, the spoil islands of Florida's east coast waterways are the residue of the technical rationalization that produced the transportation network with its channels, canals, and cuts. Throughout the twentieth century, the ICW was widened, straightened, and deepened to accommodate larger vessels and to keep pace with commercial growth. Spoil islands have not attracted the same development interest as sites like Island Estates. In addition to their intransigent public status, they pose a host of other problems for developers: most are too small, their terrain is variable, and their inextricable connection to the waterway's channel complicates bridgework and makes their relative proximity to the mainland appear irreconcilable. Held in reserve, the spoil islands remain as a common space and are left to measure other conditions of the rationalized landscape.

INDIAN RIVER LAGOON

> The view from the top of the mound is unique along the Florida coast, both in extent and variety, for it is the highest isolated elevation, and commands a view of the ocean towards the east, the coastal dunes stretching north and south, and the lagoon and the mainland to the west fading away into the distant pinelands.[11]

When John Small scaled Turtle Mound, he was surveying a diverse and far-reaching landscape. The Indian River Lagoon system extends for 155 miles. It begins at Ponce de Leon Inlet near New Smyrna Beach—about ten miles north of Turtle Mound—and extends southward to Jupiter Inlet. Its biological and topographic diversity reflects the difficulties of its classification as both a lagoon, opening to the sea, and a bar-built estuary, mixing fresh and salt water. Small's view from Turtle Mound also included the earlier histories of the ICW. Before commercial interests called for the waterway's expansion, the Second Seminole War prompted the U.S. Army to carve a canal through the central area of the Indian River Lagoon. Shallow waters hindered the transport of supplies, even by flat boats, to the front lines of battle with the Seminole Indians. Many years before, the Indians made a similar cut, and the Army was retracing this waterway infrastructure when they dug a canal eight feet wide and two feet deep across the isthmus that separated Mosquito Lagoon and the historic headwaters of Indian River. These transformations evidence the logic of a modern transport network. In some sections of its route, Florida's ICW exploits the natural courses of rivers, creeks, and lagoons that host the north-south passage from Jacksonville to Miami; but interventions like the Haulover Cut do not follow a natural system and instead impose a new system to link previously disconnected water bodies. Throughout the twentieth century, the Corps incrementally increased its controlling depths and widths. Projects, once considered "improvements" that expanded already-present attributes, instead became effectively independent of what the natural context offered. And Indian River's dredged waterways exemplify the illogic of dredging extremely shallow bodies of water.

From the midden, Small witnessed a wholly distinctive view. On the midden, he also discovered tropical vegetation not typically found in the area, even within the Indian River's diverse environment. The elevated mound of shells supported papaya, wild plumbago, torchwood, snowberry, wild coffee, mentzelia (known as "poor man's patches"), and wild orange at all stages of growth. On top, nettles, commonly found in waste areas, grew to unprecedented heights of five feet as if to call further attention to the mound's ruderal underpinnings. Many of these plants grow there today and attest to the human-made earthwork's continued support for natural but also exotic vegetation. Later studies identified Turtle Mound as the northernmost extent of eight species of subtropical plants, confirming the anomalous status of this landscape.[12]

Going southward, Small travelled from midden to midden, where he found more instances of atypical vegetation. He described these human-made hammocks as "new stations," and he marveled at their "floristics . . . of a type wholly foreign to those of the surrounding country."[13] Spoil islands also support anomalous vegetation. Shell middens are conducive to wild pepper, and spoil islands sustain Brazil pepper. Just as Small found some botanical links between dunes and shell middens, the sandy material dredged from the waterway's bed approximates dunes that have moved inland to a more protected context. The result is an upland environment in an unlikely place, supporting improbable—and many times exotic—growth. At their edges and their sufficiently inundated and partially submerged interiors, spoil islands also provide readymade footholds for mangrove colonies to continue their production of new land. These human-made mangrove habitats have combined with a warming climate to extend further an ongoing northern progression of the mangrove.

The Indian River Lagoon system is exceedingly shallow. The average depth of this system is between three and four feet, with broad areas less than three feet and only a few areas naturally reaching depths of six feet. Cast within the Indian River's sensitive estuary environment, this relative shallowness exaggerates shifts in elevation. The upland conditions generated by the spoil islands—ranging from one foot to almost twenty feet above high tide—demonstrate the effect of increased elevations on a particular location's microclimate, and unnatural heights support non-native vegetation. Shallowness yields not just higher islands but also more islands, and the Indian River's archipelago permits a visualization of human intervention. The channels are hidden below the surface, but most of the spoils are not. They reveal the give and take between human engineering and natural conditions, between logic and its application, between cut and fill. From above, you can read the enlargement of cuts and inlets through the mass of land that has been displaced and deposited to make channels navigable and harbors and ports accessible.

In the overall pattern of islands, you can also read differences in the natural depths of rivers, creeks, and lagoons where the waterway's route has been placed. Typical spacing along this section of the east coast is a half mile between dredge deposits. In Mosquito Lagoon, with its lower average depths, spoil island deposits are closer together and usually occur less than one thousand feet apart. And in some cases, an almost contiguous line of spoil approximates a newly formed barrier island or spit along the eastern side of the ICW. Near Sebastian, in the lower section of the Indian River's waterway, nine spoil islands—six of them in rapid succession—form a geometrically inspired archipelago. Imagine six equilateral triangles, with each side about four hundred feet long and with each apex pointing toward the mainland slightly south of due west. Spits of sand—each also about four hundred feet long—extend this geometric gesture to within nine hundred feet of the mainland.

Spoils allow the reading of currents, and these Indian River islands are the windsocks of the water. There would be three more triangular islands in this series, but these deposits have had the misfortune of being located at the waterway's confluence with the Sebastian River's outflow, and remaining material lies submerged and amorphous, a spectral residue of the dredging process. Elsewhere, other patterns emerge. Further south a series of five diminutive islets, officially known as IR (for Indian River) 15 through IR 19, disclose a slightly deeper context for dredging. And their near circular form—each about 150 feet in diameter—draws a calmer hydrographic chart.

Most of them more than four decades old, these Indian River islands can be read as experiments in displaced material. Like landform buoys that designate channels, test air and water, and move with changing tides and currents, these marks in the landscape remind us of our own susceptibility to these natural forces. Spoil islands are so many Franklin kites flown unwittingly in waters that are themselves only now becoming fully understood. They demonstrate, quite dramatically, the fluid relationship between our interventions and the natural systems, which they interrupt. They also inspire poetic readings. Their geomorphologies recall tadpoles, fish, snakes, as well as familiar geometries—circles, ellipses, triangles, and crescents.

In the Indian River region, proximity of shell midden and spoil island provides a rare opportunity to investigate how waste materials gauge a landscape's submission and resistance to human intervention. The ICW works as a continuous line that cuts through a highly diverse, mostly invisible, submerged landscape. The volume and frequency of spoil islands measure the landscape's resistance to technical rationalization. If spoil islands evidence a place's opposition to the necessities of an imposed navigational system, then shell middens record a particular place's resource potential, its compliance to nomadic systems of subsistence. These two extractive processes—dredging and harvesting—differ in purpose, but they both yield waste products that create distinct places, similarly formed as earthworks unique to their context.

A shallow waterscape's resistance to channeling generates islands, and a bountiful lagoon's shellfish population gives rise to middens. Each waste product becomes a new resource. Midden and island create spaces for habitat and habitation. Plants, birds, and people come to live there. Homesteaders on middens and squatters on spoil islands follow ruderal vegetation—both exotic and native—in the colonization of this new landscape. Small's enthusiasm for these landscapes—evident in his breathless, if slightly hyperbolic, accounts—suggests an Edenic environment, a heaven on earth not only for botanists but also for homesteaders. Middens rise above mosquitoes and swamp miasma, and spoils offer an alternative island paradise. Waste generates aberrant but habitable spaces, and many of its denizens will seek to preserve these anomalous sites.

HOMESTEADING

Florida's mounds and middens provided auspicious homesites. If not the same paradise recounted by Small, the area's numerous earthworks were useful to a burgeoning population of homesteaders. Mounds rose high above the miasma of adjacent swamps, and middens resisted the tidal fluctuations.

The high ground—hard to come by in much of the peninsula—also served as a defensible position for mid- and later-nineteenth-century homesteads. After the turn of the century, the economic value of the earthworks' material supplanted their topographic significance. Increased populations required more roads of better quality, and midden shells became a valuable resource for road-building in a landscape of sand and muck that provided very little hardscape. Many owners whose property included shell middens sold the mining rights to construction companies building and improving Florida's road systems.

As early as 1905, New Smyrna Beach promoted its network of shell roads as the largest in the state and the "pride of its populace."[14] Thousands of cubic yards were redistributed throughout the area. Two thousand rail cars and four months of excavation flattened one of the many middens along Mosquito Lagoon's western shore. This repurposing of Indian earthworks was not new. Estimated to be seventy-five feet high in the eighteenth century, Turtle Mound served as a navigational marker for centuries of Atlantic sailors, and New Smyrna Beach's mounds provided raised foundations for some of its first commercial structures, while many others bolstered civic pride with road material.

The revaluation of one culture's refuse as another's material resource—from landfill to road-fill—placed private landowners in an economically strategic position that outreached what the mounds might have offered to earlier pioneering homesteaders who faced Indian attacks. Midden owners found that their topographic anomalies—not yet widely acknowledged for their historical significance—were reserves for public infrastructure, and the contents of the mounds were more valuable than their overall form. Archeologists, botanists like John K. Small, and others certainly recognized the earthworks' historical importance, but development continued. Visitors were often confused whether the mounds were artificial or natural, and accounts of mound homesteads sometimes extolled nature's work in the remarkable artificial tumuli without recognizing their provenance. In 1891, a writer for *The Topical Sun*, Lake Worth's first newspaper, described Oak Lawn Hotel's setting as if it were a typical, even though well-appointed, part of the Florida landscape:

> The house is situated upon a picturesquely wooded mound—of the kind very common in Florida. This is peculiarly divided into two portions by a wide, deep ditch intersecting it. Since Judge Heyser settled the place eight years ago, many old Spanish relics and curious objects of interest have been unearthed by the residents. It is indeed a most interesting and delightful spot. A well-appointed establishment, a fine garden, a profusion of dairy products, fine fruit trees, well kept premises, clever genial people make it just what it is—a pleasant home-place for rest and ease. It is nearer the Inlet than any other hotel on the lake and is also on the county road leading to Jupiter and northward.[15]

But property owners who may have inadvertently settled on the mounds would quickly learn the economic value of their shells. Even Paul Kroegel, avid protector of Pelican Island's bird population, sold material from his midden homestead. The first homesteader in the Sebastian area of Indian River's middle section, Gottlib Kroegel, Paul's father, had moved his family from Germany and settled on the mound known as Barker's Bluff in 1881. It is said that the elder Kroegel favored the contours of the mound because it reminded him of the hills of his homeland Chemnitz in Germany. Six years after being named warden of the Pelican Island refuge, Paul Kroegel sold his mound's shells for the construction of a paved road between Micco and Stuart in 1908.[16]

Not all landowners capitalized on their earthworks' material reserves. In 1911, Wesley Snyder purchased what became known as Seminole Rest. Previously owned by Lord Christopher Hatton Turner, the property included two dwellings—a caretaker's house and a main house built before 1888. While mounds and middens up and down Florida's Atlantic coast were being destroyed, Snyder's Mound, as it was also called, would remain intact. Snyder appreciated the aesthetic value of his homesite more than its economic worth: "When they were building the roads or railroads, they wanted to buy the shell. Granddaddy said he wouldn't sell. He said it would spoil the beauty of the property; that's why we have the elevation here."[17] While Kroegel saved the pelicans but lost the midden, Snyder preserved this land for its environmental attributes.

Its raised position captured breezes and protected the houses from the lagoon's tides and storms. Quite simply, and perhaps ironically, the refuse heap provided aesthetically pleasing terrain and beautiful prospects with its artificial elevation. And the homesteader became caretaker of this beauty, at risk of being spoiled. At once fortuitous and complex, this collision of archeological site and home-site also created a dilemma for preservationists after the 1988 acquisition of the property by the state of Florida. How to maintain the property's houses while not harming its equally significant mounded substrate became even more complicated with requirements of public access and education. The changing valuations of this land—from waste, to private homestead, then to cultural resource protected for public purpose—frame the significance of private tactics for settling and, in many cases, taking care of Indian River's spoil islands, which are even more numerous than its middens.

SQUATTING

> Often considered to be "no man's land," the islets are, in truth, the property of the state, although some have proved to be more than temporary campgrounds. . . .[18]

From no man's land to everyman's paradise, the spoil islands of Florida's east coast have provided sites of makeshift settlement. Neither truly camping nor officially homesteading, those who settled the islands practiced versions of domestic life and tactics of resistance, all within the ICW's very public zone. The islands and those who used them would occupy this contested space for more than a half century. Nowhere was this transition from the waste of no man's land to the perception of the islands as paradise and natural resource more protracted and more diverse than in the coastal areas of Brevard, Indian River, St. Lucie, and Martin counties. Along this ninety miles of coastline, 137 spoil islands form an unprecedented archipelago of public land, some of which was privately claimed and then later publicly organized.

On maps, this island chain resembles a necklace laid carefully into the jewel box of the east coast's protected waters. They make up yet another version of the geomorphic patterning along the Indian River's island chains. From far above, their form appears as consistent as a string of pearls, but on closer inspection, they have been molded and transformed by local currents and disturbed and at times submerged by weather events. In segments with consistent tidal conditions, there is a pattern to their island form, with easterly spits of sand in the lower Mosquito Lagoon, amorphous deposits near Titusville, westerly spits near Sebastian Inlet, and substantial uplands, typically six hundred feet to a side, north of Ft. Pierce. The largest island is seventy-five acres, but most are closer to an acre. Some have disappeared, but most have persisted as foundations for mangroves, upland vegetation, and further deposition since the 1920s. Many of the islands served again as places of deposit during dredging operations between 1953 and 1961, but have been left alone since then because of environmental regulations on the deposit of spoil.

Indian River's spoil islands became the setting for an exceptional range of experiences and uses. The dual character of these sites—as waste and island—created a unique interaction of public and private. As waste, the islands were for a long time disregarded by the agencies that formed them, and these by-products of public infrastructure were claimed for private use. As islands, they afforded seclusion and solitude, and their insularity meant privacy. These spoil islands consequently draw from the full meaning of private space—*within* as well as *outside* the public sphere. The islands are contracted private areas within public space, but they also remain very close to mainland life. In some cases, only six hundred feet separate island from developed—often urbanized—shore. Not always readily accessible but chronically visible, this chain of islands remains part of the image of the linear city that has grown up along Florida's east coast. Their sheer number offers multiplicity. Not one island one person, but diverse locations for human activity and multiple points of dispersed responsibility for the agencies that oversee them.

Though isolated and multivalent, St. Lucie County's spoil islands formed an ad hoc community. Reports on the "spoil island squatters" suggest a shared existence, a kind of aqueous neighborhood with an ICW address. One family lived for more than twenty-five years on a spoil island that would later be designated SL16 (SL for St. Lucie County). Prompted by residents' complaints to the Internal Improvement Fund (IIF), a 1963 survey found that, starting in 1938, Luther Bass raised his children just above North Bridge on the spoil island. Another spoil island resident taught high school in Ft. Pierce. Jean Park, a teacher at West Wood High School, lived on the spoil island for five years until her cabin burned in late 1985. Her father had built the cabin in the early 1950s. There was a sense of permanence to this neighborhood, and the assistant executive director of FIND was struck by the sophistication of many island dwellings: "It's amazing the structures they've built. Some are pretty elaborate—just short of two- and three-bedroom homes."[19]

The study found more than a dozen residents—some without mainland addresses—living on islands in the Ft. Pierce section of the Indian River. From their initial creation in the 1920s to drafting of spoil island policies in the 1970s up to evictions in the 1980s, these spoil islands existed in a regulatory gap that allowed for an idiosyncratic negotiation of what "private" means when it exists in opposition to "public" and when it grows from within public space. These long-lived environments tell us how the private—in terms of isolation and non-public—might precede and even inform public management and how individual practices frame conservation practices. A combination of public strategies of conserving and private tactics of "making do" define the makeshift ecology of the Indian River's spoil islands.

For some, the spoil islands offered places of escape. A diverse set of occupants found private spaces of seclusion that were clearly defined by each island's physical boundaries but were still held, even if ambiguously, within the public life of the nearby town. This combination allowed movement *from* the city and *to* paradise. In Ft. Pierce, the islands became a part of weekly rituals that moved between nearby town and readily accessible recreational waters. Residents visited these "weekend paradises" for cookouts, picnics, fish fries, and family reunions.[20] A 1972 article noted that "hippies" had also found the islands to be a "world away" and that the islands had become a "handy place to shun establishment culture without being too far away from its conveniences."[21]

Further up the coast, a spoil island became the site to escape the law and to resist authority. In court documents, John Wesley Davis listed his home address as "Marker 146, Indian River." At the time, Davis was facing conviction for his involvement in the 1970s railroad strikes during which federal authorities accused him of trying to dynamite a bridge trestle. A folk hero known as the Barefoot Picket, Davis took his pet pig Ollie to court with him before his conviction was overturned on appeal. In the meantime, his residency on the island—officially designated IR34—placed him 165 yards northwest of his postal reference at navigational marker 146A and only six hundred feet east of Vero Beach. Other spoil islands have similarly been associated with legal exception. Legends tell of Al Capone hiding out on one. Capone's island experiences also include lawful, if part-time, residence on the artificial but purpose-built Palm Island in Miami's Biscayne Bay and his attempt to purchase what would become Deerfield Island in the early 1930s. In the 1970s, Pinellas County police arrested a man who listed one of the county's spoil islands as his home address, which was confirmed when the manager of Caladesi Island State Park retrieved his belongings from the island camp.[22] Back in the Indian River in 2005, John Devore claimed ownership of a series of St. Lucie and Martin County spoil islands, which formed an "extraterritorial constitutional democracy" that he named the Republic of Anodyne.[23]

Spoil islands are also for hermits. At the age of seventy-two, Charlie Brick moved to Seven Pine Island with his dog Duke in 1962. On this spoil island four miles north of Ft. Pierce's town center, Brick went without clothes and grew a long white beard. Brick typically went to town twice a month—"two times too many"—and he relied on thirty araucana chickens for food and company. He would say they were the only chickens with barnacles on their legs.[24] Brick was born in Calcutta, Ohio; and before retiring to his island hermitage, he lost his first business in the 1929 crash and then worked as dishwasher, merchant seaman, marine policeman in Great Lakes, handyman, and yacht mechanic in Ft. Lauderdale. Interviewed in 1972 on his Ft. Pierce spoil island, he was living in a small wood-framed building with tar and paper sheathing. Brick kept a .22 caliber rifle to shoot the chickens, and a prized possession was a Christmas card from President Nixon with, as he noted with pride from his remote island home, the return address of the White House. Even though he moved into a high-rise apartment for senior citizens—"I got too sick for roughing it"—after eleven years, the island was home; and he planned a concrete block shrine with his ashes in a coffee cup to be returned to the island after he died.

For others, the islands represented spaces of entitlement. With island deposits as leftovers from costly dredging exercises, some local picnickers, campers, and settlers saw the spoils as charitable upshots and happy accidents to be enjoyed, while their adjacent waterways played host to a broader vacationing public who came from afar—the northern or midwestern United States—to take advantage of the area's natural resources. Even in the early stages of Florida's intracoastal waterway, it was clear that a deeper channel would benefit recreational yachts as much as, if not more than, commercial interests. After FIND was formed and the Rivers and Harbors Act approved deeper and wider channels, the *New York Times* reported news of the waterway's 1930s construction progress in its sports section. An avid readership of yachtsmen, sports fishermen, and even hunters watched for updates as they planned vacation trips southward. One report from October 22, 1933 extolled the channeling efforts, noting that nine dredges were at work on the Florida segment of the intracoastal waterway.[25]

Along the coastline's shallows, described later by President Johnson with an almost religious respect, the Corps of Engineers based dredging operations on commercial use, but it was soon apparent that the ICW was more valuable for recreation. In 1935, the completed waterway was promoted for use by "watercraft of all kinds," but the new channel depth assured access for deep-draft pleasure boats and yachts that became common during the region's burgeoning pre-war development. Before these dredging projects, even vacationing presidents were not immune to the problem of shallow waters. On March 12, 1923, President Warren Harding's houseboat *Pioneer* ran across and Indian River sand bar. Commodore A. H. Brook had to rescue the President when his boat was again grounded the next day, twelve years before the dredgers arrived. Such events became less frequent with the deeper waterway. Channels joined railroads and other infrastructure projects that had been approved by the IIF following the state's encouragement and its offers of vast areas of land in exchange for private investments toward the transportation networks. Railroads brought vacationers from the north just as they also afforded increased trade for Floridians. Amidst public scrutiny, it seemed that the dredged waterway only brought yachts, and the spoil islands that remained were the "poor man's paradise."[26] They were considered spoils of a culture war.

These feelings that the waterway was the province of the wealthy persisted and remained evident in the 1980s when the state of Florida redoubled its efforts to evict those who had settled on the spoil islands. This juxtaposition of external prosperity and local everyday life fed a narrative that opposed island freedom to mainland rules: "People on the mainland can do no wrong, but a small group living on an island, everything they do is wrong." With renewed evictions, many of the spoil islanders assumed the islands would become luxury developments. Ft. Pierce resident Mike Evans narrated a populist version of the waterway's foundational history: "These islands were created because rich people wanted to get their yachts up and down the Intracoastal and it was dredged out. These islands were just useless little clumps of sand until a few people started getting a little enjoyment out of them and then somebody suddenly saw some value in them."[27]

The state's evaluation of the islands had started many years before—with bird sanctuary designations and parkland proposals in the 1960s—but Evans' comments frame the islands in terms of public use as opposed to public purpose. In this sense, Ft. Pierce residents discovered ways of using waste material, "little clumps of sand" that no one wanted. Here, claiming land takes the form of finding use where it is not expected. Evans and others filled a gap of ownership and adopted the spoil islands, noting that the "state never claimed responsibility for them." Not all the islands were adopted and maintained, and the trash and overuse that state officials feared was a problem at some sites. But the connection that the island settlers made with these residual spaces suggests a more nuanced relation to mainland living than the simple resistance to authority. This opportunism translated to the sustained—and for the most part careful—cultivation of land that had previously been ignored.

Many settlers became caretakers of the islands. Self-appointed to the role of maintaining these environments, they claimed the spoil islands for their own use but maintained them as public space. They argued that their localized tactics of conservation were "doing more to take care of the islands" than the state's broader strategies were able to do. On their Ft. Pierce island, Evans and his family built two cabins with screened porches, set up picnic tables, and plotted a garden. T. J. Evans took on the role of custodian for some of the other islands' part-time residents. His description of this work inscribes ordinary activities of domestic landscaping within the island environment. The islands extend the quotidian, making these practices out of the ordinary because of their marginal location and the broader purposefulness they hold. For Evans, eviction meant loss of home, loss of resource, and a ceding of paradise: "It's like taking your home away and putting you in jail—somewhere you don't want to be. I'm over there practically every day, raking the yard and cleaning up the pine straw. I don't know what I'll do to get away from all this racket and hassle now." The Evans family also hosted visitors to the island. In the face of eviction, they argued that boaters could use the dock they had built, families could picnic in the areas they had cleared, and overnighters could sleep in the spare bed. Another resident Philip Smith, who bought a cabin on a spoil island locally called "Ambergis Island," summarized this indefinite ownership in practical terms of cleaning versus claiming: "Nobody's tried to claim the land—we've just cleaned it up."[28]

Officials with FIND earlier echoed this line of thought that the islanders' tactics had largely improved an otherwise wasted landscape. When it became evident in the early 1970s that most of the Florida waterway's islands would no longer serve as spoil sites because of environmental restrictions, Astor Summerlin, the chairman of FIND's state board during this time, proposed that his agency might negotiate leases with the island settlers. He noted that patrolling the islands was an "impossible burden" and eviction would only make room for another wave of squatters. By leasing the island, "public use could be insured and maintenance of the islands' appearance could be enforced without evicting cooperative people."[29] This model would set up private contracts for public interests. After the 1960s survey, the Chief Deputy Sheriff reported that there had been no further orders for eviction from the IIF and that "as long as the questioned shelters are open and free to the public, and the islands themselves fully available for general use, there would be no need for further action on his part." At the same time, another resident Herman Summerlin had already formalized this policy with a sign placed at the entrance to his cabin built on a Ft. Pierce spoil island: "This is for the use of the public—please use and leave it clean."[30] Such nuanced claims of ownership contrast with more absolute declarations of private property. Just as middens attracted homesteading claims like Wesley Snyder's Seminole Rest and Paul Kroegel's lagoon dwelling, newly formed islands also elicited opportunism. It was not uncommon for dredgers to claim newly formed spoil islands as their own. In the 1920s, J. F. Lofton dredged the city of Ft. Myers' turning basin and then claimed the resulting island as his property. J. L. Hunt later bought the island, and by mid-century he had constructed a house on the island in its rapidly urbanizing context.

In these discussions, the complex relation of waste and island came into contact with negotiations of public and private. On these islands, it was argued that not just human endeavor in general but particularly private activities outside of government sanction might improve the environment. In 1972, Astor Summerlin acknowledged that some squatters had abused the sites, but "for the most part the people have provided sanitation facilities and made the islands nicer than they were in their natural state." Summerlin's use of "natural state" elides the islands' human-made origins. But having originated as waste products, outside of conventional zones of conservation, the spoil islands constituted a "natural state" deemed imperfect. By this logic, with their natural attributes already disturbed, the spoil islands could be improved by human agency. On these spoil islands, caretakers were both gardeners and guardians. A few miles south of Charlie Brick and closer to Ft. Pierce, two residents transformed their spoil island into a garden. In 1955, Otto and Molly Walthers built a concrete block house with an indoor toilet, water storage tanks, and screened windows on their island. For seventeen years, they cultivated a garden of blooming flowers, cacti, and "carefully tended paths."[31]

The Indian River context also holds traditions of protection that have had national significance. Not a spoil island, but similar in its geomorphic instability, complexity of management, and national attention disproportionate to its size, Pelican Island is located in the Indian River north of Vero Beach near Sebastian Inlet. Like Ft. Pierce's spoil islands, this island was first cared for by a self-appointed guardian. Beginning in 1881, Paul Kroegel watched over the pelicans and other wildlife that frequented the sand spit less than four acres in area. When President Theodore Roosevelt officially declared the waters around the island as the nation's first Wildlife Refuge, Kroegel's role was formalized as the preserve's first game warden. From his vantage point atop Barker's Bluff, Kroegel challenged boaters and hunters who nearly decimated the declining population of brown pelicans.

Kroegel recognized the natural resource but not the cultural resource. The warden guarded the pelicans but did not hesitate selling the shell material that supported his homestead. Wesley Snyder's homesteading recognized the site's aesthetic and climatic value but at the expense of what was an archeologically sensitive site. Preserving the island is also changing it. Both midden and spoil island hold complex mixtures of natural, cultural, technical, and aesthetic attributes. In Kroegel's case, the midden was there for the same reasons the pelicans were there—the rich harvest of sea life. The spoil islands were artificially introduced but quickly became upland habitats—footholds for vegetation and ready-made nesting sites for birds and, not surprisingly, for people. The waste of middens and spoil islands are uniquely multivalent spaces, knots of value, that require multiple frameworks for understanding their intersections of human agency, habitat, and habitation. The development of spoil island policy was the first attempt to loosen this Gordian knot without severing its bonds.

GREEN RELIEF

For state officials, spoil islands represented both problem and prospect. These officials soon recognized that spoil islands formed a distinctive system of lands within the state and in the 1960s and 1970s began to develop a policy that might regulate their "public purpose." Fashioning this policy evolved gradually and its formulation was reactive and speculative, rather than proactive. IIF's Board members responded to reports of squatting and addressed the Corps of Engineers' diminishing use of the islands on a case by case basis. As dredging guidelines became more restrictive, the board began to reconsider the islands as natural rather than human-made resources. In a June 1962 meeting of the IIF Board, Florida Governor Cecil Farris Bryant proposed the use of new spoil areas for public recreation. Bryant was concerned about the effect of spoil easements on the state's title to the land after the USACE District Engineer for Jacksonville applied for an additional spoil area on submerged land in the Gulf of Mexico near the mouth of the Suwannee River in Levy County. After the governor directed the board to "explore the possibility of using such areas for parks and recreation," board members expressed agreement that spoil islands might have "great value and be an important potential resource for public purposes."[32] The attorney general then proposed a study to find islands that were no longer "required for continual deposit of spoil." These discussions marked a transition from spoil islands as places of deposit—a functional network of waste areas—to a valued resource with recreational potential.

The IIF held the official role of defining the emergent landscape of spoils. Created in 1854 as an agency of the Florida government, IIF managed public lands granted to the State by Congressional acts.[33] The agency's Board of Trustees oversaw the sale and development of these lands, and projects included transportation networks, canalization, drainage, and reclamation. Their decisions effected expansive territories and natural systems. The IIF Board's deliberations to organize these spaces and systems often went beyond technical rationalization to include the social and cultural contexts, as new infrastructure came to rest on layers of previous occupation and resources. The board also contended with indefinite geographic boundaries within a landscape that was often at the shifting intersection of water and land. The public nature of land, particularly in Florida, has always been a contentious subject, particularly when many of the development projects reviewed by the IIF were privately funded, mostly by northern industrialists. Redoubling this complexity, questions arose about the waste areas and new public lands that were by-products of the projects it considered. The Board's minutes reveal a growing recognition of these areas, and nowhere are their contradictions more evident than in the deliberations about spoil islands.

References to spoil areas during the IIF meetings demonstrate an initial ambivalence followed by an increased consideration of the islands as distinct places. These areas were variably called spoil banks, deposits, and muck. Insular deposits were not called "spoil islands" until 1946.[34] In these cases, the board's concerns lay more with satisfying their directive to provide a continuous public domain for the waterway. The spoil areas became a contentious part of this mandate for "perpetual" easements because they were subject to tidal changes, natural disturbances like hurricanes, and public interference. To clarify their official inclusion within the waterway's easement, "perpetual" spoil areas became known as "spoil islands." By mid-century there was a growing awareness of their value as actual places.

Even if state sovereignty of submerged lands applied, these sites emerged as newly visible focal points of community, identity, and aesthetics backgrounded by the state's burgeoning development. Valuation of spoil islands ran the gamut of public perception. They could be "useful," "undesirable," and "unauthorized."[35] As early as 1942, the IIF Board addressed the question of squatters who were "taking possession of the spoil areas thrown up by dredging the Intracoastal Waterway" in Dade County's Biscayne Bay.[36] Offers from individuals to buy the islands were denied, while offers from companies or organizations that might be in the "public interest" were approved. But defining what constituted "public interest" became increasingly difficult as approvals for other purposes, such as an ammunition dump in 1943, gave way to questions about alternative uses, such as for-profit concessions on Miami's picnic islands.[37]

Following Governor Bryant's and the attorney general's interest in spoil islands, board discussions in the 1960s reflected a growing awareness of spoil islands, not only among board members but also within the general population. At the IIF's August 4, 1964 meeting, a series of Indian River islands were proposed as public recreation areas to be managed by individual counties. Representing Martin County, District Engineer Frederick L. Bell requested the dedication of spoil areas M-1-A, M-1, M-2, and M-3 for "public recreation and preservation of wildlife." The board reserved the dedication of the submerged lands around the spoil areas but approved exclusive public use rights for the spoil islands to county management. Though he did not cite specific projects, IIF's Director and Chief Engineer William Kidd noted that similar spoil areas were being used effectively for recreational purposes, "some left in the natural state and others with facilities for the public."[38] In one respect, the board divested itself of the vexations of at least four spoil islands in Martin County and would be relieved of responsibilities for their management. This provision is clear in the approved motion, but the broader implications are for an emergent system of public spaces. Also at this August meeting, the Florida Audubon Society requested the designation of a spoil island in the Gulf of Mexico, just south of the St. Joe Paper Company plant, as a bird sanctuary.[39] The Society asked for and received the same designation for spoil areas in Pinellas County. The board's approvals were still subject to the reservations granted to the United States and "to the right of the public to use the island for recreational purposes," but this action set a precedent and two more spoil islands were added to the Pinellas County bird sanctuary in 1967.[40]

Not until 1970 did the IIF Board outline its first official policy for spoil islands. On August 11, the board adopted the following two-part guidelines:

1. Spoil islands should be left in their natural state unless a greater public purpose would be served by development. If non-state development is to take place, it should be undertaken under lease by the Trustees, rather than sale and consistent with guidelines set forth in Section 253.111 Florida Statutes.
2. Proposals for public development of any spoil island should be authorized only after consultation with appropriate state agencies to see if any better public purpose might be served.[41]

The policy set a "natural state" as the default condition of the islands. Future deliberations would further define what this state constitutes, but this first presentation of the policy had its origins in the Division of Recreation and Parks' recommendations for future use of the spoil islands. Requested by the Secretary of State and authored by Director Ney C. Landrum, this report "recognized the aesthetic and recreational value of spoil islands as they are, with no development."[42] In April of that year, Secretary of State Thomas Adams recommended that the Division be notified of the board's recent deliberations about spoil islands, and he suggested that the agencies should work to make the islands "more useful to the public as marine parks."[43] According to the state's findings, not only were the islands useful for recreation and leisure activities, but they were also beautiful. Spoil had not been deposited on most of the islands for more than a decade, and vegetation—native and exotic—had rapidly taken over the islands' largely upland habitats. Along the central and southern coastlines, mangroves had also begun to colonize the spoils. This policy set in motion the paradoxes of trying to maintain the spoil islands in an "unspoiled" state, and sanctioned development could occur only through lease agreements—not sale—and only if a "greater" or "better" public purpose would result.

Leading up to this policy, spoil islands brought into focus for the board the vexing and often conflicting issues of private use and public purpose. Squatters were continuing to occupy the spoil areas. At the August 11th meeting, Adams interjected during the board's approval of the spoil island policy that IIF Trustees should proceed immediately to inventory all of the spoil islands so that the state could take swift action to evict the islands' squatters. At the same time, private groups were continuing to develop public lands. Adams also noted in June that he had recently become aware of Florida developers who were platting "sovereignty bottoms"—submerged land and spoil areas—that they did not own. Some developers were also distributing brochures promoting the sale of these lands even before they applied for permits. Florida had a long history of these activities throughout the century; but in this case, Adams linked the problem directly to the issue of spoil islands. At the August meeting, he reminded the IIF Board of his request for a study of spoil islands. This time including the Natural Resources Department, he called the archipelago a "system" of marine parks that would encompass the east and west coasts of the state.[44] These reports provided the foundation of the spoil island policy, but it was a more localized issue that had sparked Adams' interest in the spoil islands.

Further south along the ICW in Biscayne Bay, the spoil islands of rapidly growing Miami and North Miami had begun to feel the pressures of private enterprise. At an April 1970 meeting, the city of North Miami requested the approval of a license to sell concessions on Spoil Island No. 1 in Biscayne Bay. This spoil deposit was one of nine islands that the IIF Board dedicated to North Miami in 1965 and 1966 for "public park recreational use." These islands constituted nearly one hundred acres, and they were officially referred to as "permanent maintenance spoil areas" but came to be popularly known as Biscayne Bay's "picnic islands." As landward urban growth focused on housing and commercial uses, North Miami looked waterward to these islands for much-needed public space.

Private entities also targeted the islands, particularly after the islands' informal success for recreational use had been officially recognized. Miami's sailing club wanted to open a concession stand on Spoil Island No. 1. Selling soft drinks and hot dogs to famished boaters would be a relatively small-scale and innocuous operation, but the approval of a for-profit activity would also set a precedent for the rest of the spoil island archipelago. IIF's Executive Director James Apthorp framed the issue as a policy question of whether the board wanted to lease spoil islands. Along with his staff, he recommended approval of North Miami's request, noting his belief that North Miami would keep "tight control" on the concession's operator. Representatives of the Sailing Club, who would receive the license, emphasized the concession's accessibility to the general public. But there was opposition at the meeting. Mayor Sheppard Broad of the Town of Bay Harbor Islands argued that this model allowed a private party to profit from land that had been dedicated as a public space for parks and recreation. He summarized his disagreement by noting that the spoil islands should not be "despoiled for private profit."[45] In IIF meeting minutes, this astute play on words directly preceded—and its sagacity might have inspired—Secretary of State Adams' call for the marine park system of spoil islands.

The IIF Board deferred ruling on the case of Spoil Island No.1 at its May 1970 meeting and again two months later, while it waited for the policy report on using spoil islands. After approving the spoil island policy in August, the board reviewed North Miami's application at the November meeting, where at first it recommended denying the lease agreement for concessions and rescinding the dedication to the City of North Miami. The mayor of Bay Harbor Islands had again driven up to Tallahassee for the meeting, and his comments again framed succinctly the paradox of the spoil islands. He supported the board's recommendation to deny the licensing request but showed support for the city to retain the dedication of the islands as long as they remained in their "unspoiled state." This caveat provided one interpretation of the new policy's "natural state" and suggested that "natural" denotes a condition apart from development, specifically development that involves private profit.

Citing documents provided by the Division of Recreation, the mayor also pointed out that the spoil islands would provide much-needed "green relief" to the bay that had become confined by concrete embankments and seawalls.[46] This discussion posits the islands as a visual element of public perception, with the bay as a painterly canvas. The spoil islands are a reserve for visual—if not directly physical—release. The mayor's statements acknowledged the islands as a significant part of the urban image, and consequently the urban identity, of greater Miami. The comments also foreshadowed Christo's *Surrounded Islands* project, which highlighted those elements of "green relief" with pink fabric.[47] The November 11th board meeting closed with approval of the motion to amend the 1965 dedication, which had "contemplated recreational development, boating, beach and other facilities or construction," with a more restrictive policy that left the islands in their natural state unless a "higher public purpose" were to be established by the state. The parks division had said it was unfeasible to link spoil islands to state park development, and there was no further discussion of what that greater purpose might be.

In spite of this official deliberation, informal use of the islands persisted. Squatters, particularly those in the Indian River, continued to live on their spoil islands. In 1973, the IIF Board reaffirmed its 1970 policy and added the following provision:

> 3. Dwelling houses and structures not owned by the State of Florida or any of its agencies that have been constructed on the spoil islands should be removed, either by the individuals claiming a possessory interest in the structures, within a reasonable period of time, or by appropriate state agencies with assistance from local governmental officials. Continuous habitation of any spoil islands should be discontinued.[48]

While the city of North Miami's proposal for concession stands tested the bay's waters for private commerce, Indian River's spoil islanders continued with their own experiments for living. As their dwellings became less isolated, their settlement of the island became less reconcilable with public interests and newly-formed policies. They had constructed an informal network of cabins, picnic tables, and docks that serviced the residents of Indian River. The specificity of this 1973 amendment contrasted with the breadth of the spoil island policy's first provision. The earlier provision was more universal in its ruling, and the later stipulation targeted a very specific condition of occupation. This policy amendment clarified any doubt that, in the state's management framework, individual occupation of the islands contradicted the "natural state" outlined in the policy's first provision.

SPOIL CABINS

As preliminary studies were being made of Florida's east coast spoil islands, another set of spoil islands—nearly two thousand miles along the waterway—had come up in discussions at a public hearing in Corpus Christi, Texas. In March 1972, the building superintendent for Nueces County was arguing for the closure of a Park Service road that, as it turns out, came close to his spoil island cabin. The superintendent's cabin was one of an estimated 115 "fishing shacks" within the boundaries of the Padre Island National Seashore, which had been laid out in 1966 with its western edge at the Intracoastal Waterway. More than one hundred other such structures could be found on spoil islands outside the Seashore. Environmentalists argued that the Seashore's human-made islands were state land under the additional jurisdiction of the Seashore's master plan and—lacking sanitary facilities—were potential health hazards. Fishing shacks had been a part of the landscape since the early 1900s, even before spoil islands; and cabin dwellers countered with their case of ownership that preceded the formation of the park. Dredging operations from 1939 into the 1940s had deposited many of the Laguna Madre's spoil islands, which were sandwiched between two vast land territories: King Ranch to the west and Padre Island to the east. The former was one of the world's largest private ranches, and latter is the world's longest barrier island. Padre's northern section, which includes the National Seashore, is also considered one of the longest stretches of undeveloped barrier island environments.

Between 1972 and 1975, the National Park Service cleared the Seashore's spoil island cabins. During that time, in 1973, the Texas legislature approved the Coastal Public Lands Management Act (CPLMA), which authorized the General Land Office to issue recreation permits for the cabins that remained. This Act officially codified the previously unauthorized shacks as state-owned structures that could be leased for five-year periods. The 1950 construction of JFK Causeway had formalized transportation connections between Padre Island and the mainland, allowing many more residents easier access to the coastal waters and to the nearshore spoil islands. In the early 1970s, USACE testified that the shacks were navigational hazards, and Parks and Wildlife believed occupying the spoil islands threatened water quality; but the Land Office did not have the funding to remove the illegal structures and county agencies looked on evicting spoil island residents with trepidation. As a result, what some have called "The Free Republic of Spoil Islands" was unofficially sanctioned, and the status of its cabin dwellings was officially formalized with the Cabin Program.

The Cabin Program now administers 287 permit holders, significantly less than the original 612 permits first issued in August 1973. Many of the cabins have been built on pilings, rising above the islands' mix of sand, mud, and shell. Strict regulations and attrition by decay and storms are slowly returning the spoil islands to uninhabited conditions. If the structure is within one thousand feet of privately owned littoral residential property, the owner of that property must agree to the leasing of the cabin.[49] If they are damaged by hurricanes, the cabins can only be rebuilt with special permits and under strict guidelines. In a 1973 report, the Texas Land Commissioner noted that the program provided recreational rather than essential services and only benefitted a small proportion of the state's population—twenty-one thousand Texas citizens for every recreational cabin.[50] To avoid drawing from the state's limited general revenue, financing of the program is based on a user fee approach—sixty cents per square foot of the structure's area, regardless whether it is roofed or unroofed. [51] In its description of the program, the General Land Office makes very clear that the cabins are state-owned structures on coastal public land.[52] They are not private residences, but permits can be renewed every five years and the names of cabin dwellers are often presented alongside their cabin permit or "PC" numbers, which stand in for the lease-holder's address and sometimes attest to membership in the "republic." The program has its own newsletter, the "Cabin Connection." Introduced with a message from the Texas Land Commissioner, issues discuss cabin histories, provide updates on hurricanes and cabin permit bid offerings, and present recipes for "Seafood Salad" and "Curry Shrimp."

In the tradition of self-build programs, the Texas spoil island archipelago was inscribed with the self-driven activities of individuals. Like the Florida east coast's spoil islands, it became a site for independent practices within public space. Some came temporarily to escape mainland rules, others to find a permanent home, and still others to colonize what they felt was rightfully theirs. In the Florida context, some spoil island colonists believed their islands were waste products from a waterway serving industrialists and yachtsmen. Along the Texas coast, the spoil islands supported continuing traditions of fishing, hunting, and trapping. All of these activities skirted the margins of land use. In the context of Colin Ward's makeshift landscape, these spoil islands are geographic anomalies, within the interstices or on the margins of broader programs. The Texas and Florida cases offer divergent programmatic goals. On the Texas coast, the cabins are the resource, providing a fishing community's focus. On Florida's east coast, the spoil islands themselves become the resource, providing "green relief" in the midst of urban growth. The former privileges recreational use, and the latter emphasizes natural conditions. Neither cedes an overarching public mandate.

SPOIL TO NATURE

"Help protect your islands." Brochures for spoil island use in Brevard, Indian River, Martin, and St. Lucie counties remind readers of their role as owners and stewards of this extensive archipelago. With increased awareness of the islands, management shifted from a loosely defined array of private caretakers who operated with little state oversight to environmental agencies who sought "optimal use and control" of the islands. Today, the islands remain as public infrastructure that, like roads, can be maintained by volunteerism. You can adopt a spoil island just as you can adopt a road.[53] Published in 2007 and 2009, the pamphlets came out of the Spoil Island Project, an effort by FIND to manage 128 of its 137 spoil islands in a four-county region of Florida's Atlantic coast. There are forty-nine spoil islands—more than two per mile of shoreline—in Indian River County, forty-one in Brevard, and thirty-eight in St. Lucie and Martin.[54] The brochures list four general rules: no littering of any kind, short term camping only, no building of any structures, and no cutting any living trees or other live plants. Noting that these policies are consistent with regulations for state parks, a 1990 management plan, from which the Spoil Island Project originated, specifies that camping should not exceed a period of fourteen days.[55] The brochures contain maps that show each island's location and differentiate spoil islands from "natural" and "upland" islands. The maps also show the location of surfaced boat ramps and ICW navigational markers. An index further locates the islands in terms of distance from channel markers, includes each island's designated use, and shows the locations and depths of approaches by boat.

A 1980 report, sponsored by the Indian River County Board of Commissioners, made the first steps toward the use designations and management policies for later plans. *The Spoil Island Study: Indian River County* noted that "water was of the upmost importance" and made the case that "traditional land use planning has only secondary influence" in the management of coastal zones.[56] The biological sensitivity of these predominantly aqueous areas was rapidly coming into greater contact with extreme development pressures. The study hinted at a degree of immunity for spoil islands from these pressures, noting that spoil islands present "significant development limitations" although people "have been known to inhabit them, even to the extent of constructing permanent residences."

The study also framed spoil island management as a specific form of coastal management and noted that the value of these islands includes education, research, recreation, nature, continued deposition of dredge spoil, and beauty. This latter valuation was placed in the context of the public domain: "the natural beauty of their physical existence serves to provide us with aesthetic value which increases the quality of life in the community."[57] The authors of this study hoped that the pilot project—named the Spoil Island Management (SIM) plan—might facilitate Indian River County's "efforts to secure control over the spoil islands through lease negotiations with the state." Finally, the study recommended classification groupings to ease the complexity of managing more than one hundred islands in the public sphere. These three categories—conservation, passive recreation, and limited development recreation—provided the basis for subsequent policy groupings.[58]

The Spoil Island Project came out of this report and also drew heavily from the 1990 *Spoil Island Management Plan* for Indian River Lagoon. This plan coincided with the clean-up program initiated by FIND soon after the last wave of evictions in the mid and late 1980s. Biological surveys of the islands were undertaken from October 6, 1987 through January 31, 1989. The management plan developed three island designations "based on physical structure, ecological importance and historical use."[59] These categories—conservation, education, and recreation—script public use. On conservation islands, the plan discourages human use that might disturb nesting birds, threaten vulnerable species, or damage shallow-water ecosystems. Deemed sensitive and fragile, these habitats are not open to the public and require special permission for access. Education islands, which typically exhibit diverse and mostly native habitats, are recommended as environmental classrooms and veritable teaching laboratories for school children and other groups such as scout troops and continuing education classes. Recreation islands can be readily accessed, show clear evidence of human use, and do not provide habitats for breeding bird populations or animal species that are endangered or threatened. Islands for recreation are further divided into two sub-categories: passive and active. Spoil islands for passive recreation are "small or densely vegetated islands that cannot support development of structures" and are typically used for daytime recreation. The original spoil management plan recommends passive use of the islands for "picnics, sunbathing, and primitive overnight camping."[60] In contrast, active recreation islands are typically larger and able to support permanently built structures—docks, picnic shelters, brick fireplaces, grills, and even sanitary facilities. Most of the project's officially sanctioned campsites are on these active recreation islands.

Specific management plans for each county work from this management template. In the cases of Indian River County and Brevard County, each county's islands are grouped into two distinct management areas. For Brevard County, geographic separation as well as the islands' topographic characteristics determine the groupings. To the north, a series of lower and smaller islands occur in open water and as a result have suffered from erosion. To the south, islands located in a narrower section of the lagoon have been protected from winds and strong tidal action so that their areas are larger and their elevations higher. As a result, these islands support more diverse vegetation and native trees. The specific management plan for Indian River County divides the islands into two groups defined by population centers—Sebastian to the north, and Vero Beach to the south. The management plans have accounted for these geographic distributions and included the full range of use designations so that each distinct area has opportunities for conservation, education, and recreation.[61]

The Indian River Lagoon Aquatic Preserve Management Plan set up the conservation framework for these spoil island policies. Adopted in January 1986, as many of the lagoon's last squatters were being evicted and as the clean-up programs were nearing completion, this plan provided four policy directives that influenced subsequent spoil island management. First, "essentially natural conditions" are to be maintained to ensure the healthy propagation of fish and wildlife and to support opportunities for public recreation. Two other directives focus on the development of resource inventories with maps and studies so that specific habitats, particularly those of threatened or endangered species, can be identified and protected. The fourth and most specific directive is to manage state-owned spoil islands within the aquatic preserve as "bird rookeries and wildlife habitat areas." And the fifth seeks to encourage the preserve's public use "consistent with the continued maintenance of its natural value and functions."[62]

An earlier report from USACE's Dredged Material Research Program noted dire consequences for coastal bird populations if spoil islands were not maintained as nesting areas.[63] Often cited in later spoil island management plans, this report worked from the assumption that "human intrusion" and development would soon overtake nesting sites on natural islands. Spoil islands would become the final reserves for coastal bird communities. By this logic, the 1978 report argued that spoil islands, referenced almost euphemistically in the document as "dredged material islands," were taking the place of disturbed natural island habitats. Researchers advised restricting access to the spoil islands.[64] With these recommendations, waste from a public infrastructure project gained meaning as a natural preserve for avian communities. Natural islands—many of them not owned by the state—faced pressures of development not felt by the archipelago of spoil islands. And it seems birds prefer spoil islands, even if protected natural islands are available as alternatives. The spoil island, known as Hobart Island, recently surpassed Pelican Island, the historic centerpiece of the nation's first Wildlife Refuge, as a nesting site. In 2007, studies found 760 bird nests on the spoil island.

MIDDEN TO MONUMENT

> The fate of this natural garden—with a rich flora almost wholly different from that of surrounding territory for many miles—seemed so surely sealed that we put it in our list of dead gardens. However a year later we learned that some public-spirited citizens had succeeded in rescuing the mound and intended to preserve it as one of Florida's monuments.[65]

Three decades before the buffer around Cape Canaveral's NASA facilities and four decades before the establishment of Canaveral National Seashore would officially protect Turtle Mound, local citizens saved the site from road projects of the early 1920s. For John K. Small, the human-made mound not only supported an extraordinary "natural" garden but also remained as a monument to the region's past and future. In his terms, it was one of Florida's greenest deserts. Its granulated tumulus encapsulated the peninsula's origins in the uplifted seafloor, so much sand dried by the sun but supportive of "rich flora." It was also a hammock that gathered the region's paradoxes. For Small, it was the quintessential garden. A human-made substrate for exotic plantings—some delivered as if by chance from birds' over-flight, the outsized refuse heap remained a monument to industry, cultivation, and coastal habitat.

The translation of midden to monument demonstrates how the waste site becomes a part of public agency. In one respect, Turtle Mound became a monument because it was different. It was rescued by "public-spirited citizens" because it was "wholly different" as an unlikely garden hammock and because it collected the memory of the other, largely erased, mound landscapes. By definition, some monuments memorialize singularity of event, person, or place. Those seeking to preserve Turtle Mound might have recognized its commemorative value as one of the last remaining substantial shell middens of the region. They also might have found natural value in its diversity of flora and vegetation—Small's reading of Turtle Mound as an exotic garden. Put in the terms used by Alois Riegl in his essay "The Modern Cult of Monuments," Turtle Mound is a historical monument commemorating "precedents" of human activity and their "irreplaceable" events.[66]

But the midden originated as waste. It is a remnant, not unlike what Riegl described as a "torn off slip of paper." Such residue also entails value. Middens—and by extension spoil islands—are public earthworks that have been created by humans and then turned over to nature. Riegl's characterization of a monument's age value follows a similarly dynamic process. Monuments are reminders of singularity and difference, but they are also indicators of change over time. The romance of the ruin measures time in the face of nature's transformative process. In his commentary on age value, Riegl notes that "from nature working over time . . . we expect the dissolution of completeness as a symbol of an equally necessary and lawful decay."[67] Over time, the midden has eroded and adapted to natural change, so that its age value is also a "decay value."

Turtle Mound is a counter-monument. Its "decay value" comes from its physical impermanence and from the contingency of its meanings.[68] Empty shells are traces of a vanished culture, the eroded landform recalls natural attrition and borrowed roadwork material, and exotic plants allude to remote places. Middens redirect the construction of memory far afield—across the lagoon and its depleted shellfish resources, along mainland roadway networks built with crushed shells, and down avian flyways that transport exotic seeds to the anomalous midden uplands that support them. Such attributes run counter to forms of conservation that seek stability of both form and meaning. Middens, like spoil islands, gain meaning from change. After they have been waste sites, they become places of defense, protest, escape, and foreignness. Their unexpected presence and idiosyncratic use elicit a reflective process. Though mound and island offer some degree of isolation, this process is rarely private and internal. It is expansive and public, exposing nature and culture just as it has been uprooted spatially and temporally. Middens and spoil islands are monuments to entropy.

In 1966, the same year that Miami's spoil islands were recognized as picnic islands, the artist Robert Smithson outlined a "new kind of monumentality."[69] In the essay, Smithson presents the work of sculptors, Robert Morris, Dan Flavin, Donald Judd, and others, as explorations of entropy. He argues that the artists have created monuments to the present, momentarily lodged between an "inactive history" and a forgetting of the future. To illustrate this idea, he quotes Vladimir Nabokov: "The future is but the obsolete in reverse."[70] He opens the possibility that something obsolete or ruined can also be returned to use or built up as it decays or passes into disuse. Entropy, then, is not merely the movement from order to disorder but the negotiation of creation and destruction. In other projects, Smithson observed what he called "ruins in reverse" where buildings occupied a dynamic present and decay offered opportunity for rethinking past and future.

For Smithson, entropy occurs in this transformation between states, in a landscape that is temporally as well as spatially charged. Midden and spoil island operate in transitions from water to land, from tidal moment to tidal moment, and from disuse to purpose. In this framework, waste is not measured by obsolescence but by purposefulness. Midden and spoil island recall histories of work, index geologies, and map longer-term environmental change, but they also exist in their present condition. We find them, and we read them on their own terms. They come to us as ready-mades, appearing initially as ordinary islands in places that are sometimes out of the ordinary—unexceptional sites made remarkable by topography and placement. Smithson finds entropy here as well, in the gap between the inexact and the measured.[71]

Though quantifiable as displaced material, middens and spoil islands do not recall the precision of the activity that formed them. They recall the activity itself—harvesting and eating or dredging and depositing. Cast off as waste from these activities, they then become tools of measurement for tides, climate change, even public sentiment. Smithson recognizes that entropy occurs within specific, even if extensive, contexts; and he casts his entropic monuments into this broad landscape. His essay's epigraph describes one such landscape: "On rising to my feet, and peering across the green glow of the Desert, I perceived that the monument against which I had slept was but one of thousands. Before me stretched long parallel avenues, clear to the far horizon of similar broad, low pillars."[72] Resonating with Small's "green deserts," this setting is not unlike the Indian River's network of spoil and midden that places entropy in the public view and makes it a common condition.

COMMON WILDNESS

In the Indian River's Mosquito Lagoon, the ICW passes between midden and spoil island. To the west, the remnants of shell middens push out into the waterway, some as slight topographic bumps interrupting the shoreline and a few as significant bulbs of land reshaping the water's edge. To the east, often less than one thousand feet away, spoil islands mirror their landward counterparts. This alleyway of midden and island runs through Canaveral National Seashore, where island campsite Number 8 places you directly across the ICW from Seminole Rest, where more than a century ago Wesley Snyder set up his homestead on the midden that extends into the lagoon. From across the water, the homestead stands apart. Its wood-frame dwellings contrast with the travel trailers on another protrusion to the north, its oaks shade the now subtle rise in terrain, and Spanish moss cloaks the site in a gray fleece that helps recall the ancient site's full height.

When you camp on the spoil islands of Canaveral National Seashore, you receive the Backcountry Use Permit. According to National Park Service Protocol, Form 10-404 must be fastened to your pack, saddle, boat, or tent. The assumption is that you are roughing it. Pack in and pack out. This is your tag and you are in transit, because NPS backcountry is the primitive and undeveloped area of the park. In its expanded classification, backcountry is a "general condition of land" that has the potential to be managed as wilderness. And spoil islands are wilderness understudies.

To reach this backcountry one blustery Sunday afternoon, we parked our car in Riverbreeze Park, unloaded our boats, and paddled two thousand feet to campsite 8 on the spoil island known as Bissette Bay. If the weather had been better, we might have paddled farther to reach one of the Seashore's other spoil islands along the ICW. Two miles would have taken us to campsite 9 on Middle Dredge or to campsite 10 on South Middle Dredge. But in each case, only the width of the waterway—about the length of two football fields—would separate our primitive camp from the Florida mainland and Route 1's line of development.

On first glance, these spoil islands easily fail the Park Service's test of wilderness. The first sentence of the Wilderness Act's definition is interrupted with the clause "in contrast with those areas where man and his works dominate the landscape." With spoils, humans and their works are the landscape. These areas also fail the test of size. There are no spoil islands on the east coast that approach the Act's recommended minimum of five thousand acres, and IRL's total spoil island acreage is a fraction of that. The core definition, which follows the interjected clause, is a challenge even for the more remote islands: "an area where earth and its community of life are untrammeled by man, where man himself is a visitor who does not remain."[73] With spoil islands, man's lingering presence is more indirect. The dredge material remains, where most visitors would not.

There are other characteristics of this technically defined wilderness that also provoke ambiguity in the classification of spoil islands. The Act calls for wilderness in undeveloped land that retains its "primeval character." The

Indian River Lagoon's spoil islands are not old, but they resonate with the morphology of the area's Indian middens, and their geological disposition is hauntingly ancient. Coral fragments and chunks of coquina form the islands' visible substrate. This accident of archeology is not the primeval landscape the Act's authors had in mind, but it does hold latent values of wilderness—what the Act alludes to when it says that wilderness "may also contain ecological, geological, or other features of scientific, educational, scenic or historical value." What is underfoot reveals normally hidden stories of geology and human history. What can be seen from the island offers a scenic view and discloses the context of island formation. Spoil islands are unique pedagogical vehicles for understanding these frameworks. Wilderness defined by the Act must also have "outstanding opportunities for solitude or a primitive and unconfined type of recreation." In spite of their proximity to the mainland, these spoil areas offer the isolation of islands and the recreational opportunities of the lagoon's waterscape. Spoil islands are typically placed in open, often expansive, locations where neither islands nor humans would normally occur.

The sense of isolation on these islands is short-lived. In most areas, the ICW is a ribbon of public infrastructure, a highway of boats in which the islands are its occasional rest areas. The sheer number of islands in Indian River's archipelago offers a multiplicity and a variety of experiences unique among the east coast's human-made islands. Overall, the Indian River Lagoon system of spoil islands is a shared backcountry that falls short of what the NPS defines as wilderness. As waste, the islands have roughness and a dredged-up primitiveness, but this potential wilderness is ultimately too close. They are not wilderness, but they have wildness. Spoil islands occupy a middle zone, a cultivated—we might even say invented—wild space, not unlike the common lands that Gary Snyder located between deep wilderness and the "private places of the farmstead."[74]

The public nature of spoil islands means that they exist as a shared resource. Each island—and each midden—contains historical evidence of communal waste, but their public status means that they have also continued as the "common pool of resources" described in Snyder's analysis. For birds, they offer refuge and nesting grounds. For plants, they offer unique microclimates. By extension, for John K. Small, the middens were exceptional monuments of horticulture. For others, the islands are "green relief." Their common resources have also afforded ritualized movements between city and country. Further afield, the resource commons extends across the lagoon's and the river's waterway to include the ancient middens, with their histories of collecting shellfish and homesteading the land.

Spoil islands offer a common wildness. Unearthing spoil inserts wildness within the confines of the cultivated. Traditionally, the commons were areas not suitable for crops and thus held in reserve for—or simply given over to—

the community. But Snyder takes issue with the simple definition of the commons as "the undivided land belonging to the members of a local community as a whole." Commons is not just land; it is also a communal institution that "determines the carrying capacity for its various subunits and defines the rights and obligations of those who use it."[75] The commons takes its meaning from practice, not from definitions that tend to oppose concepts of natural and human-made. Recognition of their hybridity is more important than delineation of their differences. Part of the islands' wildness derives from the fact that they are spoil. When Bay Harbor's mayor described their value as "green relief," spoil was a function of their natural appeal. They were human-made waste, already spoiled, and yet they represented a natural state. Their "unspoiled" state was also not to be "despoiled" for private profit.

This common wildness works between "habitat" and "inhabit." The process of inhabiting by its very nature entails uncertainty: each inhabitant will have his or her own interpretation of what it means to dwell on the island, as in the case of the spoil islands' squatters and caretakers. In their sheer breadth, when management plans take on an "abstract and functional character," habitat falls back to its "purest form."[76] There is the burden of constraint rather than the openness of possibility. In some respects, the spoil island management plan installs "habitats" at the expense of habitation. To impose deliberateness on these unintentional monuments also runs counter to their value as a "common pool of resources" and to the spirit of public agency. Spoil islands' lineage from waste to natural and cultural resource demonstrates the potential for truly ecological monuments. The Spoil Island Project captured one part of this formulation when it referred to the goal of balancing "overall environmental quality and aesthetics."[77] But both beauty and environment are multivalent, and spoil islands are part of a much more complex public discourse. To understand the islands is to take into account natural systems and contingencies of meaning, to study their industrial and biological histories, to balance habitat and habitation, and to reflect on their popularity as well as their entropic tendency. These ecological monuments must retain their common wildness.

NOTES

1. Seamus Heaney, *Sweeney Astray* (London: Faber and Faber, 2001), 8. Sweeney's journey takes him through paradisiacal and hellish topographies, vegetated with thick briars as much as wild strawberries. In and out of madness, cast adrift by war, Sweeney finds this landscape alternately harsh and homely, much like the tidal margins he recounts here: "A year until last night / I have lived among dark trees, / between the flood and ebb-tide . . ." (12).

2. Susan Parker's *Canaveral National Seashore Historic Resource Study* (National Parks Service, 2008), edited by Robert Blythe, provides a good summary of these middens and the documents related to their more detailed study. The distinction between mound and midden

also helps distinguish between purpose-built earthworks and those that have emerged as waste, by-products of shellfish harvests. It is interesting that Hill 343, as a mound created with dredged waste, constitutes a third but related type (see the previous transitional piece "Mound").

3. Parker, 9.

4. Elizabeth A. Horvath, ed., *Final Report of the Archeological Investigations at the Seminole Rest Site* (Tallahassee: National Park Service, 1995), 140. See also Parker's *Canaveral National Seashore Historic Resource Study* (2008).

5. J. Y. Detwiler, "Origin of the Shell Mounds," *Publications of the Florida Historical Society* 1, no. 1 (1908): 14–17.

6. John K. Small, "The Everglades," *The Scientific Monthly* 28, no. 1 (1929): 87.

7. John K. Small, "Green Deserts and Dead Gardens: A Record of Exploration in Florida in the Spring of 1921," *Journal of the New York Botanical Garden* 24, no.286 (1923): 247.

8. River and Harbor Act of 1927, U.S. Congress, Act of January 1, 1927, 44 Stat. 1040, chapter 47. See also "History of FIND," Florida Inland Navigation District, accessed April 5, 2012, http://www.aicw.org/history.jsp.

9. For an exhaustive account of the canal and its related projects in Florida, see William G. Crawford, *Florida's Big Dig: The Atlantic Intracoastal Waterway* (Cocoa, FL: Florida Historical Society Press, 2007).

10. E. B. White, "On a Florida Key," in *Essays of E. B. White* (New York: Harper and Row, 1977), 140. For further discussion of the "unfinished city" in the context of semi-permanent camps, see my discussion in *Campsite: Architectures of Duration and Place* (Baton Rouge: Louisiana State University Press, 2008).

11. Small, 204.

12. Eliane Norman, "An Analysis of the Vegetation at Turtle Mound," *Florida Scientist* 39, no. 1 (1976): 23.

13. Small, 200, 205.

14. Robert H. Weeks, *New Smyrna, Volusia County "The Land of Flowers"* (History Facsimile Series East Florida, 1905), 1, cited in Parker, 56.

15. Cited in Lynn Brink, ed., *A History of Riviera Beach, Florida* (City of Riviera Beach: Bicentennial Commission of Riviera Beach, Florida, 1976), 7.

16. See Mary Beth Herzog, "Sebastian Dates back to the 1840s," *Press Journal*, 22 June 1975, 1.

17. Interview with Jacqueline Snyder Stevens and Marian Porta, August 27, 1992, by National Park Service personnel, typescript transcription at main office of Canaveral National Seashore, cited in Parker, 78.

18. Jud Deakins, "Spoil Island Squatters," *Palm Beach Post*, 6 September 1963, 1.

19. Teresa Lane, "Spoil Island Squatters To Be Ousted," *Palm Beach Post*, 16 June 1987, 1.

20. Lane, 1.

21. Eric Sharp, "Island Squatters Lack Nothing but a Deed," *Sarasota Herald Tribune*, 5 November 1972, 1F.

22. Alicia Caldwell, "Poor Man's Paradise is Close to Home," *St. Petersburg Times*, 30 July 1989, 1.

23. Suzanne Wentley, "Fire on Island Reveals Hazardous Materials," *Stuart News*, 16 December 2005, 2. Devore wanted to use one Martin County island, locally known as Commodore Island and Crab Island, for a small camping business near the St. Lucie Inlet. He argued this function's compatibility with the passive recreational use designated for the five-acre island. Devore filed a quit-claim deed for the island, which he named Sea Horse Island. Later in 2005, the State of Florida filed a lawsuit against Devore after a fire broke out on Thanksgiving from hazardous waste left on the island.

24. Sharp, 1F.

25. *New York Times*, 20 October 1933, 9:1. See also Nelson Manfred Blake, *Land Into Water—Water into Land* (Gainesville, FL: University Press of Florida, 1980), 149.

26. Caldwell, 1.

27. Lane, 1.

28. Lane, 1.

29. Sharp, 1F.

30. Deakins, 1. In 1988, Summerlin and his construction company Seven-Seas would hold the contract for cleanup of the spoil islands. He removed four barge loads of materials from the islands over a period of four weeks. See Teresa Lane, "Cleanup of spoil islands nears end," *Palm Beach Post*, 15 March 1988.

31. Sharp, 1F.

32. June 16, 1962, *Minutes of the Trustees of the Internal Improvement Fund* 33 (Tallahassee, FL: 1962): 553.

33. The Florida Legislature created the Internal Improvement Fund in 1854 to "facilitate the drainage and conversion of those lands the federal government conveyed to Florida." The language of the legislation used "swamp and overflowed lands" to "justify the disposal and conversion of those lands." The full designation of the board as the "internal improvement trust fund" helped legislators make the argument that "public interest" was served by the group, which effectively assisted railroad ventures led by David Yulee and his associates in the acquisition of land from the state. The board was originally called the Trustees of the Internal Improvement Fund and later renamed the Board of Trustees of the Internal Improvement Trust Fund. See Sidney F. Ansbacher and Joe Knetsch, "Negotiating the Maze: Tracing Historical Title Claims in Spanish Land Grants & Swamp and Overflowed Lands Act," *Journal of Land Use and Environmental Law*, 17 (Spring 2002): 355–356. I have used the acronym "IIF" throughout this text.

34. December 8, 1945, *Minutes of the Trustees of the Internal Improvement Fund* 25 (Tallahassee, FL: 1946): 489.

35. October 4, 1960, *Minutes of the Trustees of the Internal Improvement Fund* 33 (Tallahassee, FL: 1960): 73; and August 14, 1962, *Minutes of the Trustees of the Internal Improvement Fund* 34 (Tallahassee, FL: 1964): 31.

36. June 25, 1941, *Minutes of the Trustees of the Internal Improvement Fund* 23 (Tallahassee, FL: 1942): 187.

37. March 11, 1942, *Minutes* 23, 449–450.

38. August 4, 1964, *Minutes of the Trustees of the Internal Improvement Fund* 35 (Tallahassee, FL: 1964): 27.

39. August 4, 1964, *Minutes*, 27.

40. February 21, 1967, *Minutes of the Trustees of the Internal Improvement Fund* 27 (Tallahassee, FL: 1967): 158.

41. August 4, 1964, *Minutes of the Trustees of the Internal Improvement Fund* 35 (Tallahassee, FL: 1964): 42–43.

42. August 11, 1970, *Minutes of the Trustees of the Internal Improvement Fund* 37 (Tallahassee, FL: 1970): 30. The Division of Recreation and Parks' report is dated June 19, 1970, and additional correspondence included a July 23, 1970 memorandum from James W. Apthorp, Executive Director of IIF.

43. April 7, 1970, *Minutes of the Trustees of the Internal Improvement Fund* 37 (Tallahassee, FL: 1970): 639.

44. June 2, 1970, *Minutes* 37, 704–705.

45. April 7, 1970, *Minutes* 37, 639.

46. November 17, 1970, *Minutes of the Trustees of the Internal Improvement Trust Fund* 38 (Tallahassee, FL, 1973): 124–5.

47. For further discussion of picnic islands, "green relief," and *Surrounded Islands*, see chapters 5 and 6.

48. March 20, 1973, *Minutes of the Trustees of the Internal Improvement Fund* 39 (Tallahassee, FL: 1973): 334.

49. For details on the permits, see Texas Administrative Code, title 31, part 4, chap. 155, subchapter A, rule §155.4, available at http://www.glo.texas.gov/what-we-do/state-lands/_documents/professional-services/Rule_Cabin_Permit_Tex_Admin_Code.pdf

50. Gary Mitchell and Duane Windsor, *Regulatory Effectiveness Study for the Christmas Bay Coastal Preserve*, Publication GBNEP-14 (Galveston Bay National Estuary Program, 1991), 106. In Florida, with a 2011 population nearly 20 million, the ratio of residents to Indian

River spoil islands approached 100,000 to 1. In 1970, with a population of 6.8 million, the ratio was closer to 30,000 to 1—on par with the Cabin Program's ratio.

51. Mitchell and Windsor, 110.

52. See "Cabin Program," Texas General Land Office, accessed January 19, 2012, http://www.glo.texas.gov/what-we-do/state-lands/leasing-and-easments/coastal/cabin_program.html.

53. See "Adopt-a-Spoil-Island," Spoil Island Project, Indian River Lagoon, accessed February 4, 2012, http://www.spoilislandproject.org/adopt.htm.

54. The State of Florida owns 124 of the islands, and four of the remaining islands have been dedicated to Martin County. The other 13 islands are owned by private interests, the United States Government, or FIND.

55. Nancy Brown-Peterson and Ross W. Eames, *Indian River Lagoon Spoil Island Management Plan* (Bureau of Aquatic Preserves, Division of State Lands, Florida Department of Natural Resources, 1990), 5-5. This plan notes the importance of this policy to help enforcement officials distinguish between "campers and trespassers."

56. Indian River County Board of County Commissioners, *Spoil Island Study: Indian River County* (Vero Beach, FL: Indian River County Board of County Commissioners, 1980), i.

57. Indian River County Board of County Commissioners, ii.

58. Indian River County Board of County Commissioners, 6.

59. Brown-Peterson and Eames, 5-1. The cover illustration drawn by Charissa Baker shows stippled lumps that recede dramatically toward the horizon and starkly render the depositional qualities of the islands. To the left of this procession of amorphous deposits, a sailboat leaves a small wake as it traverses an unmarked field.

60. Brown-Peterson and Eames, 2–10.

61. In a few cases, designations overlap and reflect what the Management Plan calls the "unique multi-use classification of these islands." Three islands received dual classification as education and conservation, and three others were classified as both education and recreation.

62. See chapter 3 of the Bureau of Historic and Environmental Land Management's *Indian River Lagoon Aquatic Preserve Management Plan* (January 1986), 23–27. A revised management plan was issued in January 2004. This document expands the preserve's objectives to neighboring buffer areas, acquired through Florida Forever and other sources. The overall goal is to extend what has been called the "blueways" of the Atlantic coast.

63. Ralph W. Schreiber and Elizabeth A. Schreiber, *Colonial Bird Use and Plant Succession on Dredged Material Islands in Florida. Vol. 1: Sea and Wading Bird Colonies*, Technical Report D-78-14 of the Dredged Material Research Program (Washington, D.C.: Office of Chief of Engineers, 1978).

64. Schreiber and Schreiber, 2. This report also noted that "the relationships between 'natural' and 'dredged' habitats are poorly understood, but the former are rapidly being destroyed and the latter will thus become increasingly important." (6)

65. Small, 204.

66. Alois Riegl, *Der moderne Denkmalkultus, sein Wesen, seine Entstehung* (Vienna, 1903), trans. K. W. Forster and D. Ghirardo, "The modern cult of monuments: its character and origin," *Oppositions* 25 (1982): 20–21.

67. Riegl, 24.

68. For more discussion of "counter-monument," see James E. Young, "Memory/Monument," *Critical Terms for Art History*, eds. Robert S. Nelson and Richard Shiff (Chicago: University of Chicago Press, 1996), 234–247.

69. Robert Smithson, "Entropy and the New Monuments," *Artforum* (June 1966), included in *Robert Smithson: Collected Writings*, ed. Jack Flam (Berkeley: University of California Press, 1996), 10–23.

70. Smithson, 12.

71. Smithson excerpts Henri Bergson to make this connection: "'The history of philosophy is there, however, and shows us the eternal conflict of systems, the impossibility of satisfactorily getting the real into the ready-made garments of our ready-made concepts, the necessity of making to measure.'" (19)

72. Smithson draws this excerpt from John Taine's novel *The Time Stream* (10).

73. The Wilderness Act, September 3, 1964, Public Law 88-577 (16 U.S. C. 1131–1136), 88th Congress, Second Session.

74. Gary Snyder, *The Practice of the Wild* (San Francisco: North Point Press, 1990), 30. These places also parallel what Henri Lefebvre called the "far order" and the "near order." See his discussion in *Writings on Cities*, trans. and eds. Eleonore Kofman and Elizabeth Lebas (Oxford: Blackwell Publishers, 1996), 109, 113–114.

75. Snyder, 38.

76. Lefebvre, 79.

77. "About the Spoil Island Project," Spoil Island Project, Indian River Lagoon, accessed December 11, 2011, http://www.spoilislandproject.org/about.htm.

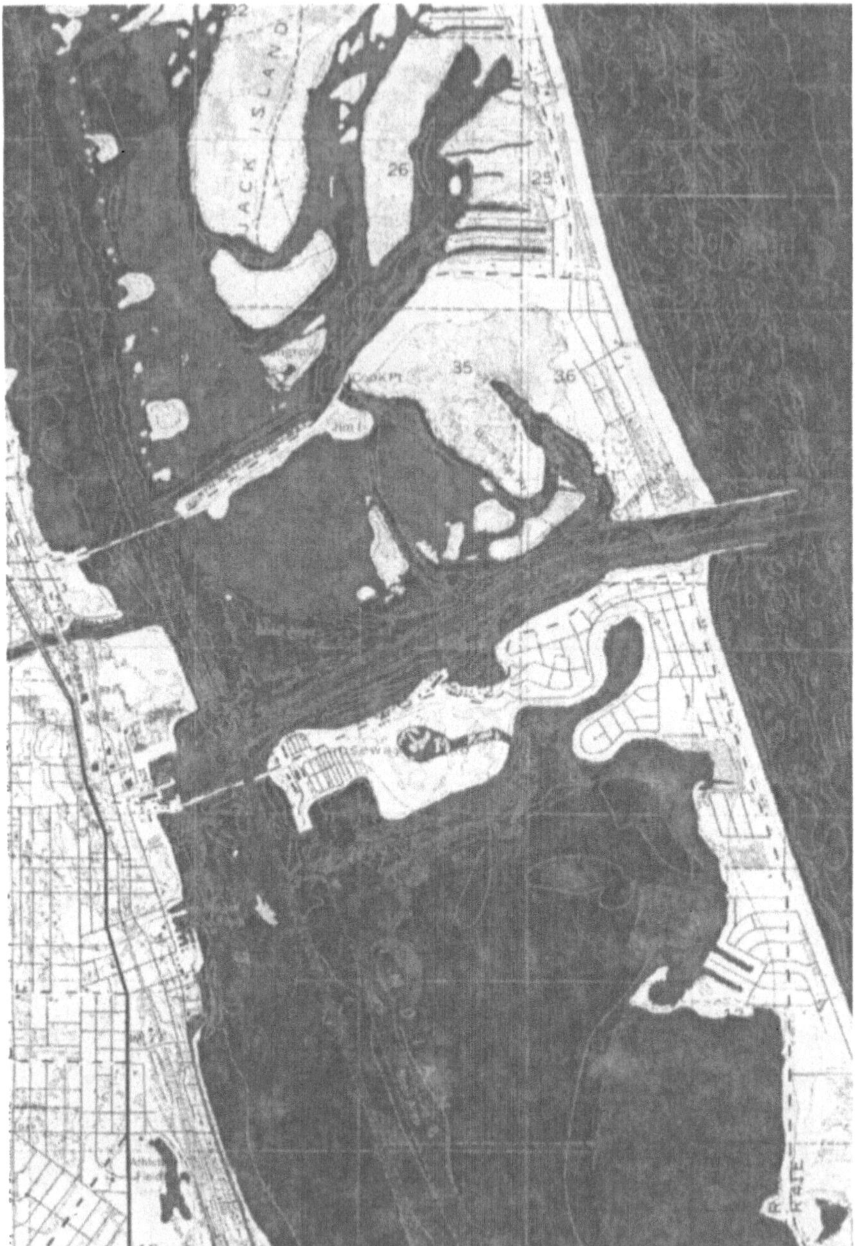
JACK ISLAND
26
25
35
36
15
18

Nettle

East of unlighted day beacon #215, you will catch sight of a huge trailer park on Nettles Island.[1]

Gonzalo: Had I plantation of this isle, my lord—
Antonio: He'd sow't with nettle seed.[2]

On Nettles Island, site-built amenities have supplanted trailer-park densities, but from the ICW's northern approach, the navigator's mistake can be forgiven. Along with a few obstinate tag-along trailers, more than fourteen hundred houses cling to the sand and sod ground of Nettles Island, which opened as a resort for recreational vehicle (RV) enthusiasts in the early 1970s. Such parks have not always been viewed affectionately, and to some the trailers and RVs may seem to have spread like the weeds that flourish on the marginal sites of spoil material. In the middle of the Indian River, the one hundred acre island avoided mainland attention and barrier-island costs. Across from the mainland town of Eden, it promised Florida's idyllic climate and lush vegetation. But like its namesake and like the spoil island that hosts it, the resort's paradise is imperfect. Nettles sting as they also cure. They embody the flawed promise of *The Tempest*'s island—Shakespeare sets Gonzalo's dreams of utopia against threats of his adversary's weed dystopia. Nettles Island represents burden and opportunism. Formation of the island overstepped the bulkhead lines that define shore construction, and colonization of the island exceeded density limitations. Temporary settlement became permanent, and conventions of gated community normalized the island's ruderal nature, but its contradictory underpinnings remain.

Although residents have noted the weed's presence, their island was named for the man, not the plant. In October 1889, John M. Nettles cleared a few acres of the forty-acre sand bank, planted five hundred banana trees, and built a small house. In April 1890, Nettles applied to have his new island estate surveyed, but he met with unexpected results. The island was considered public land, and ordering surveys prompted advertisements for public sale. After their inspection, government surveyors proclaimed the island's acreage "fit for agricultural purposes," even though its highest area was only two feet above high tides.[3] Five years later, Nettles tried to gain formal ownership of the island under homestead laws, but his claim was again rejected.[4] Future actions are not well documented historically, but the name has remained. In subsequent years, Caribbean rum-runners reportedly buried their treasures on the island, and other stories narrate its ownership by various Lord Mayors of London.

Un-surveyed boundaries and dredging accidents haunted the island's development. In 1956, its owner Gene T. Dyer purchased fifty acres of submerged lands to the west of the island. Six years later, when he applied to the IIF Board for a permit to extend his island westward, IIF staff members noted discrepancies with the bulkhead lines that define how far dry land can be extended into water beyond the shoreline. Nettles Island lay in an indeterminate zone where some bulkhead lines had not been established and others overlapped. To retract nearby Hutchinson Island's western bulkhead line by 550 feet, Dyer agreed to exchange the westernmost half of his plot for equivalent landward acreage on the east side of Nettles.[5] In 1969, the island's new owner Outdoor Resorts of America began dredging and filling to expand its territory and raise its elevation. By Fall of that year, spoil material reached heights of twenty feet, but as the project neared completion, its dredger sank. Replacement equipment was delayed, and the company's permit expired. While the project was underway, section A-10 of Indian River's aquatic preserve had been established, and dredging permits required environmental reports that the project failed. But the IIF Board was sympathetic and considered the dilemma a "matter of equity." With a tinge of parenthetical irony, the Board noted that ". . . had the work been completed on time (had the dredge not sunk), it would have been prior to the creation of the Aquatic Preserve."[6]

Despite news of the delays, Nettles' new residents began to arrive. A makeshift settlement hosted fifteen RV families on the barrier island while power was connected and slabs were poured. There was no grass yet, and sand blew across the island as tenants positioned trailers and contractors distributed the promised concrete tables and benches. Officially opened on December 19, 1971, the community was Outdoor Resorts of America's first experiment in a series of "luxury RV resorts" that continues in other locations today. Nashville country music personality Bobby Lord, first president of ORA's board of directors, performed with friends from the Grand Ole Opry in the island's newly completed clubhouse. Along with Lord's salesmanship and with gimmicks like two English double-decker buses—one with the erstwhile destination "Prison" still placarded in its window—serving as shuttles to the oceanside beaches, its population steadily increased and soon filled the resort's 1,585 lots. The island gained more autonomy three years later when its residents incorporated as an association.[7]

With its narrow plots and relatively wide roads, Nettles' plan uncannily resembles a hedge maze. Roads accommodated the girth of RVs, and many of the plots were pull-through for the easy exit of large rigs. The island's overall plan approximates a square, one-third of a mile to a side, with the eastern edge angling out to the southeast so that it parallels the natural shoreline, which can be reached across a causeway. Like an outsized parking lot, property lines angle at thirty degrees from streets so that the standard lot is a parallelogram, 34.64 feet on the short side and 65 feet on the long. To maximize site-built construction, the houses that replaced the RVs follow this geometry, with neighboring roofs coming within three or four feet of each other. To get a sense of this peculiar but strangely congruent layout, imagine a series of skewed, crowded houses that have been extruded from four angled parking spaces. What is not roof or asphalt is manicured grass. The communal area at the island's center includes horseshoes, mini golf, a swimming pool, playground, tennis and bocce courts, shuffleboard, sand for volleyball, a picnic area facing the Grand Canal, and, of course, a parking lot, which is evocatively metonymic with the larger housing lot.

The island is packed with program, and the population density easily exceeds its nine-unit per acre zoning designation that allows for typical two-bedroom and two-bath dwellings.[8] In spite of its seven-foot elevation held in place by sea walls, the island is vulnerable. For some it might be Shakespeare's weed-sown island. For others it could be a nationalistic emblem of ingenuity, encapsulating geography in streets named for far-flung streets where many of its residents—Florida non-natives—originated. But to view its position next to the ICW is to rethink how we live near the coast. The island is unabashedly artificial, it is efficient, and its density is in marked contrast with the undeveloped green spaces across from it on Hutchinson Island. Nearby, "Holiday Out" development carves old land and wasted space on the barrier island. If nettles are plants "whose virtues have not yet been discovered," then perhaps spoil is a material whose value lies in experimentation, and Nettles is a test case whose trailer park houses float on a sea of possibility.

NOTES

1. Claiborne S. Young, *Cruising Guide to Eastern Florida* (Gretna, LA: Pelican Publishing Company, 1996), 302.
2. William Shakespeare, *The Tempest* (New York: Simon and Schuster, 2004), 61.
3. S. V. Proudfit, ed., *Decisions of the Interior and General Land Office in Cases Relating to the Public Lands* XIV (Washington, D.C.: Government Printing Office, 1892), 458.

4. S. V. Proudfit, ed., *Decisions of the Interior and General Land Office in Cases Relating to the Public Lands* XX (Washington, D.C.: Government Printing Office, 1895), 407–408.

5. May 28, 1963, *Minutes of the Trustees of the Internal Improvement Fund* 33 (Tallahassee, FL: 1963): 218.

6. *IIF Minutes*, April 7, 1970, page 647. The IIF Board's director disagreed, in principle, with the motion to permit resuming the process of dredging: ". . . there are equities, but the staff could not recommend granting a dredge permit to get fill material within an Aquatic Preserve, but only to dredge navigation channels." The staff felt that it could not "violate the philosophy and policy adopted in connection with establishment of the aquatic preserve," which had been officially designated as A-10.

7. Mary H. Tifft, "The Story of Nettles Island," accessed November 17, 2012, http://www.nettlesislandcondo.com.

8. Along with areas of Hutchinson Island, Nettles' zoning designation is "Hutchinson Island Residential District" (HIRD).

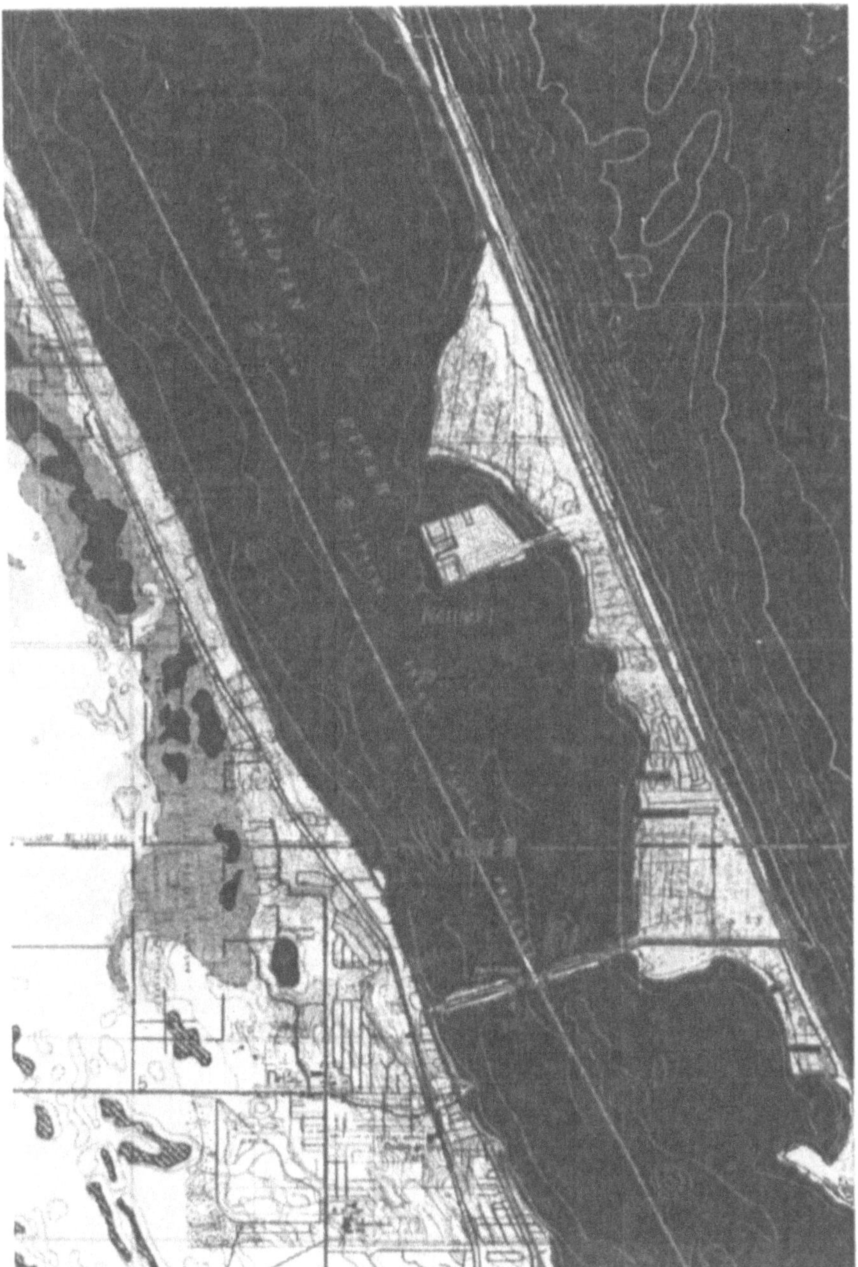

Chapter Four

Useful and Cultural

Peanut Island's Mutinous Landscapes

"I don't want Peanut Island." (John D. MacArthur in 1958)[1]

MUTINY

You might suppose it to be a developer's dream. Eighty-six acres of island, eight hundred feet from Palm Beach, circumnavigated daily by cruise ships and yachts, rising above Lake Worth's high tides more than twenty feet—as high as forty-seven feet in the past, augmented by eight decades of layered dredge material from one of Florida's most reliable inlets, opening to the Atlantic, to the Gulf Stream's closest approach, and to the Bahamas only fifty-seven miles away. But even now, well into the twenty-first century, most of Peanut Island is empty. A thin ring of green space glazes the island's edge, intermittently sprinkled with distinct interventions: a historic Coast Guard station, John F. Kennedy's bunker safe haven, an artificial lagoon, campsites, a fishing dock, and a narrative nature trail—all of these remain where numerous other projects have failed. They cluster around the central void that is the island's enduring site for the deposit of dredge spoil. Environmentally tenuous but logistically persistent, this place is more like a modern landfill than a dredged island. It has adapted where other spoil islands failed. With its perimeter event space for informal and formalized activities, Peanut Island is also a park. It complicates the lineage of other waste places like Staten Island's Fresh Kills that have been carefully translated from landfill to parkland. Peanut Island has always hosted dredgers and picnickers, the latter dodging the former's spoil buckets soon after the island first appeared above Lake Worth's azure-brown waters. This tightly-wound complexity in its cramped position between barrier shores and mainland has attracted the public while it has repulsed private investment.

Peanut Island is a mutinous space. Captain Cook's HM Bark *Endeavour* came to port in March 1998, and it is fitting that a replica of the HMS *Bounty* docked here for a few months in 2010. Just as the original *Bounty* could not sustain a common objective—to collect breadfruit saplings for the West Indies—amidst the multifarious concerns of its crew, Peanut Island has resisted master plans. Like the ship, the island has remained closed in on itself, apart from totalizing schemes, but open to experience. Each proposal, including an oceanographic institute in the 1960s and a full-blown high-rise development in the 1970s, was spurned by financiers or rejected by public outcry in spite of the island's auspicious location, ample size, and impressive rise in assessed real estate value. And yet, like the ships that pass it and like both versions of the *Bounty*, the island has been an "instrument of economic development," even if by default.[2] Up until recent work to restore the island, this economic development has occurred irregularly, as a waste site and an open-ended recreational area. On paper, the restoration project, initiated with studies in 2000, appears to be the master plan the island has resisted for so long. But Peanut's heterogeneous, layered histories—not to mention the byzantine complexities of its management overlays, jurisdictions, and easements, which are themselves cultural traces—inhibit conventional top-down plans. Additionally, the site's continued use for spoil deposition has ensured that this landscape of work, also a historic component of the place, is not frozen.

In these respects, Peanut Island is a heterotopia. Tracking along the type of "other space" described by Michel Foucault, the island is a zone of illusion that "exposes very real space," while it is also a site of compensation that reconstitutes another real space. For the latter, Peanut Island now organizes the waste it once accumulated haphazardly. In the center of this loosely formed spoil island, berms, liners, and containment levees now approximate a landfill's spatial logistics. Use defines the island's interior. These engineering procedures are cast within an island edged by an array of other programs. The island's perimeter juxtaposes not necessarily every other space, as Foucault would have it, but a wide array of unrelated functions. Its plan thematizes and formalizes activities that were previously carried out informally. Once elusive spaces have become concrete. From recreation to institution, from preservation to education, these sites support touring, running, snorkeling, learning, and conserving. Amidst this diversity, the island is a countersite for suspended disbelief, where pirates dressed as Jack Sparrow can board a replica of the eighteenth-century British fleet to rub elbows with Captain William Bligh, where weekend revelers can party on the shores of a National Register historic site, and where beachcombers can stroll through a presidential fallout shelter and pretend to place a call on the red phone. This island floats in a unique zone between the cultural and useful, as a place of deposit that sets up a landscape at once productive and culturally resonant.

In fact, its status as a "beer can island" indicates that neither its social nor its natural context has ever fully accepted its placement and continued deposition of material. "Beer can island" is a common moniker of spoil islands that host social events on their sand outcroppings. "Beer can" points toward the party—drinking beer—and its refuse—discarding the cans. Invoking other metaphors, Peanut rests like an uncorked plug to the inlet's human-made cut. The island effectively makes work for itself by deflecting the shoaling that historically occurred on its present site. Displaced sediment and sand now come to rest in the channel and inlet, where this material awaits re-dredging or settles on the island's northern sandbar extension, the scene of southern Florida's liveliest Memorial Day gatherings.

Though imperceptible to partiers and most scientists, Peanut Island is moving. Over the past decades, it has shifted one hundred feet westward. The "beer can" typology feeds off such transience. Amidst this instability, cycles of dredging alternately traded sandy outcroppings of new dredge spoil for the rapid growth of non-native Australian pines and Brazil pepper trees. From above, during one of its treed cycles, before the start of its recent restoration project, the island seemed an impossible expanse of lush—though exotic and invasive—vegetation lodged in the middle of Palm and Riviera Beaches' urbanization. It stood not merely as an island geographically separated, but also as a psychogeographic conceit, a place of desire. No one in particular would have it, but everyone wanted it.

SHOAL

On a calm May afternoon in 1925, Riviera and Palm Beach residents, journalists, and photographers left their picnics and tea parties to await the Danish cargo ship *Ella F.* The tide had come in, and expectations for the Port of Palm Beach's future ran high with the arrival of its first foreign vessel. But halfway through the inlet, *Ella F.* lodged on a sand bar—like so many that had plagued the harbor—and its disappointed voyagers were ferried ashore.[3] Peanut Island began as a similar shoal. Lake Worth was a lake until mid-nineteenth-century efforts to cut inlets introduced the sea's dynamic currents. In the late 1860s, August Lang excavated the thin but persistent barrier of sand and shell. Lang's inlet quickly filled but was reopened at least twice in the next decade as work continued to establish a permanent inlet.[4] Shoals created by the dissolution of these failed projects remained below the lake's newly brackish waters. One such shoal at the present site of Peanut Island was prominently marked on an 1884 U.S. Coast and Geodetic Survey map. By 1893, when Henry Flagler invested in the cut's enlargement, the Lake

Worth ports were already shipping more winter vegetables than any other part of the state. But the lake continued to reject the inlet, and twelve years later Flagler had to provide funds for re-dredging so that Palm Beach's harbor might still link the state's burgeoning agricultural production—and Flagler's railroad infrastructure—to Cuba and to the northeastern United States. Shown as "Lake Worth Inlet" on the Currie Map of 1907, this cut between sea and lake was one mile north of what was then called "Old Inlet," the approximate location of Lang's failed endeavors .

Two of the early-twentieth-century's most well-known engineers, Isham Randolph and George Goethals, would return to the site of these earlier experiments. In 1915, the Florida legislature chartered the Lake Worth Inlet District to create a port for Palm Beach. One year later on August 15th, voters approved opening a new inlet between Lake Worth and the Atlantic Ocean. The newly formed Inlet District had already approached engineer Isham Randolph to survey the lake and locate the channel and inlet. Famed as the engineer who reversed the course of the Chicago River, Randolph had also filed reports the previous year in Miami about possibilities for its deep water harbor. Randolph located the present-day Lake Worth Inlet, and throughout the fall of 1916 dredgers labored to cut the channel with a low-tide depth of four feet. One year later, small boats made the first entry through this inlet; and in the next two years, dredging increased its width to two hundred feet and its depth to ten feet.[5] Dredgers placed material on the shoal that had formed from previous excavation activities, augmenting a spoil bank that would first be called Hood's Island "in honor of Dr. H. C. Hood, who was the first chairman of the Port Commission." By 1919, this nascent version of Peanut Island had reached ten acres, enough area to support the residence of a Palm Beach Port employee for whom the island was temporarily dubbed "Parker's Island."[6] From his well-placed position, Parker might have seen the first bootleggers skip over the shoals after Prohibition acts were passed in 1919.

Shoaling persisted while dredging continued with greater urgency. Shoals, like the one that supported the new spoil island, would hinder one kind of commercial traffic as they facilitated another. After Prohibition, shallow-draft smuggling boats began to enter the port, outmaneuvering larger police vessels, their pursuit slowed by low tide's sandbars. Opening the inlet had also opened the area to smuggling. Bootleggers, carrying prohibited goods like rum, came in from the Caribbean as cruise ship passengers went out to the Bahamas, and the smuggling boats fared much better on the shoals. In 1921, the dredging vessel *Hercules* set out in earnest to deepen the channel to sixteen feet and enlarge the port's turning basin. The USACE also built jetties over the next five years, readying the port for its inaugural sea-going commerce. Work advanced rapidly, and the spoil area was now called "Inlet Island," officially recognizing its island status, its progenitor, and its new owner. The Inlet District had purchased the site from the state in 1921 for one dollar. Two years later, its area had increased to 47.34 acres, so that the District asked the state for a new title and again purchased the land for one hundred dollars per acre. This hundred-fold increase in price indicated the island's real estate potential, and the four-fold increase in size measured the waterscape's natural resistance to the inlet that required continued maintenance.

Meanwhile, some ships made it through the inlet, while others like the *Ella F.* did not. What some considered the port's first truly ocean-going cargo ship *The Lake Chelan* slid over the inlet's shoals and entered port in December 1925, followed just over a year later—amidst celebration in the beach communities—by the port's first passenger ship *The Mary Weems*, linking Palm Beach to New York and later that year to Nassau and the Bahamas.[7] Other smaller vessels continued to make surreptitious passage from the Caribbean alongside cruise ships on their way into Lake Worth, and the smuggling trade became so rampant that officials requested Congress to provide a Coast Guard station at the inlet to "plug one of the most gaping holes" for the flow of liquor from the Caribbean and the Bahamas.[8] Maintenance of inlet and channel continued.

In 1926, the Port of Palm Beach hired General George Goethals to study the port's expansion and the construction of a twenty-four-foot-deep harbor. Goethals had gained his reputation as the nation's eminent civil engineer with his supervision of the Panama Canal, and the Lake Worth Inlet project would be his last before his death in January 1928. Soon after dredging work began, the land boom's crash halted the $3.5 million project. Only the Merchants and Miners Transportation Company continued its weekly freight service, and the port became primarily a harbor for pleasure boats, its infrastructure damaged by the hurricane of September 1928. In 1935, the federal government took over channel maintenance and deepened the inlet to twenty feet, even though its agencies looked to other ports for the transport of war goods during World War II. The channel was again deepened in 1948 to accommodate oil tankers, and its present depth of thirty-five feet was reached in 1967.[9]

By this time, Peanut Island had become the centerpiece of a wholly artificial, aqueous landscape. What was once a shoal—a ship hazard, but otherwise an understated by-product of technical process—was now part of a much larger territory, mechanized to the extent that was possible amidst the coast's hydrologic forces. A 1946 study by the Corps of Engineers found that the erosion of Palm Beach's coastal sand could be attributed to the inlet and its jetties that extended into the Atlantic Ocean. As the repository for material that might have replenished these beaches, Peanut Island was complicit in this engineered interruption of the coast's natural littoral drift from north to south. The Corps estimated that there was nearly a mile of discrepancy in coastal width as a result of human intervention. North of the inlet, the coastline gained three thousand feet, while as much as one thousand feet had been lost south of the cut.[10]

Peanut Island also moved, and the Corps noted the island's one hundred foot lateral shift westward from 1968 and 1993.[11] Its eastern shoreline retreated, while its fluctuating western edge advanced. Engineers responded to Lake Worth Inlet's kinetic landscapes with mechanical systems. Between 1957 and 1958, the Corps of Engineers installed one of the nation's first sand transfer plants to relocate sand trapped above the inlet south to Palm Beach's sand-starved coast. And in the following decades, shorelines extending far north and south of Peanut Island would lose any remaining natural characteristics as they were edged with rocks, bulkheads, and rip-rap revetment.[12] Reflecting the paradoxes common to spoil islands, the report on Peanut Island in 2000 documented its edge as the only remaining natural shoreline in the northern vicinity of Lake Worth. A ring of exotic vegetation—mostly Australian pine—followed this edge and formed an inner ring that further complicated Peanut's natural status.

Among the *Ella F.*'s crew during its fateful stranding was Gus Jordahn, who had ordered the vessel's cargo of flagpoles. Jordahn served in the Danish army before lifeguarding on Coney Island, and he later founded Palm Beach's swimming facility known as Gus' Baths. As a county commissioner, Jordahn would later contend—legendarily for the island's naming—that the inlet's dredge spoil island was not worth "peanuts." Other theories find alternative origins for the island's name, but the essence of Jordahn's opinion affirmed perceptions that the spoil island lacked direct economic value, whether the precise statement is historically accurate or not. Jordahn knew Lake Worth. An avid swimmer, he often swam between Palm Beach and the mainland. A year before the *Ella F.* fiasco, he had founded the Cowboys of the Sea to save swimmers along the Atlantic beaches. With his firsthand experience of the inlet's silting, it seems that Jordahn might have valued the inlet island's visible inventory of a deeper port, even if he declaimed its real estate potential. In its transformation from shoal to artificial landmass, Peanut Island measures the changing needs of the port and the evolution of the use value of an engineered nature, as it also records the ebb and flow of development proposals for this new-found land. As a waste site, it was particularly open and vulnerable to speculation and to widely divergent valuations that ranged from spoil to paradise and from deposition to development.

PEANUT

Not worth peanuts. Jordahn's legendary statement allocates even less value to the island than a more typical application of the slang phrase. To say something is "worth peanuts" is to associate it with impecunious and marginal existence linked to peanut crops in early American agricultural life. The island's name has two other sources. In 1923, the State of Florida took over the island, which it later leased to W. H. Brown Company.[13] Its subsidiary, Shawano Plantations, had begun farming peanuts in the Belle Glade area of the Everglades, and the island was at one time the proposed site for a peanut processing plant.[14] Plans fell through, and Shawano sold the island back to the Port of Palm Beach for fifty-five thousand dollars in 1945. Others have said the island's name comes from its shape. No longer having the double curve of a peanut's shell, pinched in the middle, this morphology relates more closely to the cracked shell's individual nut, with a slight tapering curve at each end.

In 1958, rumors circulated that the Port Commission of Palm Beach was contemplating an exchange of Peanut Island for Layton Park, adjacent to the port's northern edge. For many years, the Commission had considered options for expanding the port's grounds and facilities, including the relocation of operations to the island itself. Layton Park stood as the sole opportunity for the port's landward expansion. On paper, the proposed exchange appeared reasonable, but the Port had only recently recalculated its island's value from the fifty-five thousand dollar purchase price from Brown Company a decade earlier to the new assessment of one million dollars after Palm Beach Shores and Riviera Beach both expressed interest in purchasing Peanut Island.

Operating from his de facto office in the Colonnade Hotel's coffee shop, no more than a half mile away, Layton Park's owner, John D. MacArthur, set the record straight by saying that he might sell or lease his mainland property but that he did not wish to own Peanut Island under any circumstances. In the 1950s, from his penthouse accommodations at the hotel that he owned, MacArthur had a privileged view of the island with its waving canopy of Australian pines interrupted only by a four-acre clearing on the island's southeast corner. Having methodically increased his wealth with Florida land, MacArthur certainly knew real estate, and it is telling that he had no interest in trading eighteen-acre Layton Park, one of Florida's oldest trailer parks, for the sixty-acre spoil island. A transaction would have to wait more than a decade when on November 1, 1971, Layton Park was sold to the Port of Palm Beach, but Peanut Island was not a part of the deal.

Three decades earlier, even though a hurricane hit and the land boom was reaching its end, it seemed Peanut Island had found an industrial function to complement its use for dredging operations. By 1928, the dredged island was considered a buildable site, one that could accommodate Shawano's peanut plant. Meanwhile, local officials planned to seek the federal government's help to improve both land-based and water-borne infrastructure. A "Notice to Contractors" printed in the *Palm Beach Post* on May 26, 1929, requested bids for dredging the inlet and its channel to a depth of 16 feet and placing the accumulated sand on Inlet Island. To prepare for these activities, the Lake Worth Inlet District purchased 11.48 acres of submerged land adjacent to Peanut Island in 1931. In April of that year, Palm Beach's planning commission considered proposals for the development of Inlet Island. These discussions made official the more informal discussions among developers and business interests over the past two years. The proposed project generated samples of the resistance that would consistently discourage changes at Peanut in years to come. Some opposed the plans "to construct an elaborate system of bridges and causeways at the inlet" because they feared such projects would require the creation of a new taxation district.[15]

In spite of this opposition, the coordinating committee of Palm Beach's civic club approved a Lake Worth harbor development proposal two years later. Presented by New York developer Orrin Randolph and the chairman of Lake Worth Inlet District Harry Corwin, the project meshed federal plans to deepen the inlet with the Lake Worth District's objectives to enlarge bridge and road facilities with aid from the Federal Works Program. The proposal included a controlling depth of twenty-four feet, and the resulting dredge material formed a new island adjacent to the old Inlet Island spoil. In the plan, new spoil material also created a causeway connecting Palm Beach across the old spoil island to an east-west causeway that replaced the Singer Island bridge, which had been destroyed in the hurricane of 1928.

The plans submitted to the Federal Works Program in 1933 treat Inlet Island as a pre-existing condition and propose the creation of another island—an area the proposal referred to as "made land"—to accommodate two wharfs and new road system infrastructure.[16] In the schematic plans, Inlet Island's terrain and shoreline are left as natural contours, while the made land is lined, filleted, and chamfered. Touched only by the north-south causeway, Inlet Island forms the northern edge of the harbor channel. It remains perhaps as a ground for deposition or an outsized foundation for the drawbridge's supports, but more likely it was viewed as the outmoded residue of previous work, now replaced by the newly engineered land to the south. A twelve-foot-deep basin for small craft is carved into the east side of the made land, and its other side forms a passenger and yacht wharf to complement the mainland's larger freight port across the turning basin. The roadways are also designated as "made land." The north-south causeway bisects Lake Worth and runs side-by-side with the ICW to create dedicated transportation corridors for water and land, leaving landward areas unencumbered by infrastructure and open for future real estate speculation. The machined refinement of the new system of roads and harbors also contrasts with the un-ruled shoreline edges of "Inlet Island," an afterthought in this new scheme. This old spoil is the residue of an outdated engineering process, from another time.

Overall, the plan demonstrated the complexity of connecting Palm Beach with the mainland. The transportation system required two drawbridges, a series of drainage openings attempting to maintain tidal flow, and a lengthy causeway running north along the shoal area that bisects Lake Worth and today forms the horseshoe sandbar. It is Peanut Island's characteristic insularity that has frustrated most of its development projects. Access is limited to those with a boat. Most projects have argued the need for access to all as a way of justifying expenses for bridges and causeways. But advocates of these proposals soon found that the public did not object to the limitations of the island's boat access. They in fact preferred it. Nonetheless, limited access has prevented large-scale development.

In 1946, after Shawano sold back the island, the Port of Palm Beach commissioned plans from the New York engineering firm of Theodore Knappen to make Peanut Island the central feature of a new deep-water port. Dramatic renderings of the project show the port terminal taking up the majority of the island. In the birds-eye view, airplanes rush into the frame while steaming ships cut the waters below. The sweeping lines of a viaduct for both automobile and railcar connect the island terminal to the mainland.[17] Despite its economic promise and futuristic renderings, the plan met with forceful opposition from Riviera Beach and Palm Beach residents. In the following years, picnics and boating parties continued on the island, while port officials and engineers could only speculate about the island's potential for development. In 1957, as rumors of the sale with MacArthur circulated, the Port priced the island at one million dollars, twenty times its sale value one decade earlier. No buyers materialized, and with the failure of its deep-water terminal plan, the Port's focus on expanding its mainland holdings left room for other proposals for the island. The U.S. Coast Guard Station continued a ninety-nine-year lease it had begun in 1938, and in 1961 a bunker for President John F. Kennedy was added to this southeast corner of the island.

When Riviera Beach sought to annex the island, conversation centered on the island's value as a recreational area, while other plans fell through because of the lack of vehicular access. In 1967, the Environmental Science and Service Administration considered Peanut Island for a federal oceanographic laboratory but chose Miami for its headquarters. Soon afterwards, Palm Beach County's withdrawal of its offer to fund a bridge also spoiled Ocean Sciences and Engineering's plans to lease a portion of the island. In 1968, another oceanographic research center anchored a much more ambitious proposal by Palm Beach developer Adolph Lewson, who planned to make Peanut the "world center for the ocean science industry."[18] Led by Lewson on October 16, 1968, Peanut Island Properties (PIP) signed a twenty million dollar, ninety-nine-year lease with the Port of Palm Beach for the development of the island.[19] The lease terms required a four million dollar investment by PIP over the following five years. Lewson planned to retain a public beach, but the rest of the island, augmented by filling additional submerged land, would play host to private enterprise—submarine laboratories, a saltwater aquarium, and a hydrofoil engineering plant. But many county officials wanted more than the proposed one thousand feet of beach to be available to the public and countered Lewson's plans with recommendations that more than half of the island be dedicated to recreational use. The broader public argued against the project from an ecological position that the island was not only a recreational but also a natural reserve.

Peanut Island became a pivotal node in the discussions of coastal ecology in the 1960s and 1970s. The group "Save Peanut Island" continued to fight plans for the oceanographic institute, claiming that "ecological balance" would be "adversely affected."[20] Led by Norris Norman, who owned the Buccaneer Yacht Club and apartments overlooking Peanut Island, the group had a vested interest in the island's development as real estate. Saving Peanut Island meant limiting all forms of industry. Norman had once said in one of the many commission meetings that he would not be surprised if a sulfur plant were installed on the island. An editorial published on June 21, 1970 places Peanut Island at the hinge of controversy between "idyllic past" and "new industry," and the Area Planning Board recommended leaving it as open space.[21] But Save Peanut Island's greatest fears—despite Lewson's assurances that only oceanographic industry would be created—were that the island would be developed with high-rise buildings and commercial functions. In December 1974, PIP's lease was cancelled, but Save Peanut Island continued its fight to make Peanut Island an official public park. Almost a decade later, Palm Beach County officials still did not want the spoil island. On August 5, 1983, county commissioner Bill Bailey responded to the Port of Palm Beach's request for the county to buy Peanut Island: "I move we tell them thanks, but no thanks."[22]

Palm Beach County loosely maintained the island as a passive park until speculation about Peanut Island's future began again with the impending departure of the Coast Guard. Due to leave for its new station north of the Blue Heron Bridge in Riviera Beach in 1991, the Coast Guard stayed until 1995, but a flood of proposals for the island began in the late 1980s, including a cruise ship terminal, a luxury hotel, and a National Aquarium of the Tropics. In 1990, entrepreneur Michele Alexander offered $3 million to buy the island for development, but this proposal fell short of its appraised value of $4.7 million.[23] To complicate the situation, 1990 was an election year for the Port of Palm Beach Commission. The incumbent chairman favored a sale to the Florida Inland Navigation District (FIND) for continued use as a spoil deposit site. Another candidate believed the island should be preserved with park-like amenities, and the other did not want any changes—even adding picnic tables—to the island. Those who argued for preservation called the island "pristine," and others believed that "the island is wilderness and should stay that way."[24] In 1988, Lisa Tanner, well-known at the time for her line of designer garbage bags demonstrating "waste with taste," circulated a petition to preserve the island and worked with conservationists from Greenpeace, the Sierra Club, the Historical Society of Palm Beach County, and the Audubon Society. Port Commissioner Stanley Weiss countered the concern over the island with a reminder about its spoil provenance: "People are talking about Peanut Island like they are going to find Indian relics on it. People think it is an archeological treasure. We can build them another one if they want."[25]

PARADISE

> Over the past 15 years the County Parks and Recreation Department has pursued numerous methods of acquiring Peanut Island due to its recognized potential for providing unique public boating and related passive recreational opportunities. . . . The specific purpose for the Lease is to provide an area for passive outdoor recreation and boat use and to promote the protection and enhancement of the public use of local water resources. The terms and conditions of the Lease would also allow both the Port and FIND to maintain their respective use of interior areas of Peanut Island as a needed spoil deposition site for material generated from future maintenance dredging operations.[26]

> Its spectacular location and water conditions make Peanut Island one of the premiere boating destinations in southeast Florida. The island is uniquely situated at the mouth of the Lake Worth Inlet (also known as the Palm Beach Inlet) in the Lake Worth Lagoon estuary. At high tide, Gulfstream-fed ocean water enters through the inlet, bathing the island with crystal-clear water that delights swimmers and snorkelers. It gets even better! The island has been enhanced with the establishment of native wetland and upland habitats and recreational amenities to create a complete island adventure. Come take a tour of our local island paradise![27]

On January 4, 1994, FIND found its much-needed site for spoil. With increased environmental regulations, locating intertidal sites for deposition had become increasingly difficult, and a lease agreement between the Port, FIND, and Palm Beach County ensured that nearly fifty acres would be available over the next fifty-five years. It could be said that this agreement did what conservationists struggled to do, although this "unified approach" represented a compromise. With the removal of topography and replacement of vegetation, the project's implementation was also a reminder that the island is artificial land.[28] Spoil preserved the island. Its continued use as a waste site would keep developers away. And with the management and master plans required by the lease, its outer edge (known as the Palm Beach-Peanut Island, or "PB-PI," zone) would remain open to the public.

Leading up to its final plan in 2000, USACE's recommended scheme for environmental restoration records the complex divisions of the island. This plan also attests to the rich juxtapositions that have made Peanut such a problematic space.[29] The Intracoastal Waterway passes within one hundred feet of the island's western edge, and the Lake Worth Inlet is less than two hundred feet from the southeastern corner. A utility easement for Florida Power and Light runs east-west and separates the two major components of the plan: fifty acres with FIND's spoil disposal area to the north and the island's remaining thirty-six acres with the Port of Palm Beach's disposal area to the south. In the USACE map, a wedge of Palm Beach County land that is the residue of the island's 1941 de-annexation awkwardly cuts into FIND's containment area. In the plan's implementation, FIND's area takes on a pentagonal shape that is reminiscent, particularly from aerial views, of coastal military forts, such as Ft. Clinch's five brick walls built on Florida's coast in the mid-nineteenth century. Deposition of spoil is guarded by sloped containment levees, not unlike the glacis fortification that makes up the artificially sloped earthwork leading up to a fort's walls.[30]

Placed on the southern part of the island and uncannily shaped like a stomach, the Port's spoil area is more organic and curves around the historic Coast Guard property and the chain-linked area of the "old government magazine area." In these central voids can be read the removal of the historic spoil mounds. Between 2001 and 2005, 1.2 million cubic yards of old spoil material were moved to create a series of new spoil islands along the ICW across from the Lake Worth Golf Course. Fifteen hundred barge loads traveled the eleven miles southward to form the new archipelago known as Snook Islands Natural Area. Anchoring the extended territory of these new spoil islands as well as Lake Worth's waterways that must be maintained, Peanut's remaining craters await future deposition that will surely be necessary. These voids also stabilize newly minted public areas that build upon—and in some ways amplify—the recreational opportunities that Peanut has historically provided.

Brochures for Peanut Island promote the island as paradise. Its amenities fill the thin ring of terrain between surrounding water and the levees that define the inlet's spoil deposit areas. The navigation beacon remains in its historic location, high on the eastern shore, aligned with the unrestricted views along the cable easement's east-west corridor. The Maritime Museum occupies the historic Coast Guard station and the Kennedy Bunker, and an observation platform rises above the southernmost point of the island. Human-made snorkeling reefs and a lagoon, adjacent to the station's property on the island's southeastern edge, benefit from the tidal influx of fresh clear water from the Atlantic through the inlet. North of this lagoon, where the FIND disposal area pushes out close to the eastern shore, a maritime hammock anchors the levee's *enceinte* slopes and offers the backdrop to a boater's beach that faces toward the inlet's traditional anchorage area. Docks offer landing areas for boats, fourteen slips on the east and ten on the west. Twenty campsites provide overnight accommodations for boaters, and a fishing pier extends from the northeastern corner. The western side of the island is mainly dedicated to environmental education, with an area that extends the coverage of existing mangroves and a perimeter boardwalk that traces the edge of a tidal pond. This path links to a longer walkway that tracks the circumference of the island. Measured at 1.234 miles, the walk has become the focus of the Peanut Island 24. These "round the island races" test the endurance of runners who must log the most laps during the twenty-four-hour period from 10 a.m. on New Year's Eve to 10 a.m. on New Year's Day.[31]

With such amenities, it is easy to forget that Peanut Island originated as a dump not a recreational paradise. So close to the mainland and barrier islands, but just far enough to require intricate engineering of draw-bridges and access ramps. Closer to the dredged channels it services than the mainland it entices, Peanut has remained indefatigably a spoil island. It was not purpose-built for occupation. It is a place of expedience that has become useful or, more precisely, that has proven useful in other ways. These added, hybridized uses occur in a place where they would not have otherwise been. These mutinous spaces do not abide by mainland rules and constitute another kind of landscape where a waste site might offer a bridge between use and culture. But this link is complicated by a dredging process that implicitly benefits owners of yachts, which require deep channels for their deep drafts. The center of Peanut Island holds the results of this dredging, surrounded by park activities, cultural sites, and a sandbar occupied by a broader public. With such charged spoils, Peanut Island is also a place of contention. A mutiny is a popular disturbance, and populist sentiment that the ICW is for them (yacht owners) and the spoil islands are for us (mainland residents) drives discussions about the continued—sometimes rebellious—occupation of the island's fringe and its surrounding waters.

Throughout the twentieth century, journalists, politicians, and developers bemoaned the failure of the projects proposed for Peanut Island. One writer began his article, which was itself about a current proposal for developing the island, with a string of news headlines about Peanut Island over the past decades: "Peanut Island Sale Delayed," "Port Facilities Planned for Isle," "Riviera Beach Moves to Annex Peanut Island," "Peanut Island as Recreation Center Aired," "Peanut Island Lease Awaits Contract OK," "County May Build Bridge to Island," and "County Now to Review Peanut Island Project."[32] Just as this expectant writer contemplated its futures, Peanut Island has chronically waited for projects. One reason for its lack of development has been the multiple owners and diverse parties who administer the island: Florida Power and Light, Palm Beach County, Port of Palm Beach, as well as the state and federal government. But for the most part, lines of jurisdiction and ownership are clearly drawn, and the realities of both spoil and island are the issue.

Discord and desire, not territorial dispute, fuel mutiny. Want and desire supersede ownership. Understood as paradise from different vantage points, Peanut Island has been both public reserve and wanted landscape. Residents have viewed Peanut Island as a calm place amidst the Palm Beach area's urbanization. It forms a place in reserve. This sense that it is off-limits to development might have factored into John D. MacArthur's lack of interest in owning it. Although administered by multiple entities, it appears beyond ownership, an untenable site for singular interests. Aerial views, particularly those in the 1950s, bear out this idea of Peanut Island as a place out of bounds, reserved for picnics, expeditions, and parties. At Palm Beach's critical, mid-twentieth-century moment of rapid development, Peanut Island stood out as a green zone lodged within a highly urbanized, though not yet completely high-rise, setting. Although covered with Australian pines, exotics considered to be invasive, Peanut Island remained emblematic of nature. Peanut was also a wanted landscape. Progressively more constraining environmental laws made its acreage increasingly valuable to USACE and FIND as a site for the deposit of dredge material—a floating landfill laced by public space.

Refashioning Peanut as an eco-island—at least on its fringe—quelled concerns about private development, although combining practicalities of dredging with exigencies of public access is complex. The cultural and recreational resources sprinkled around its perimeter camouflage the basic concept of the island as spoil, much like the berms hide the actual deposition from view. Formalized nodes of previously informal activities—camping, picnicking, snorkeling—effectively thematize recreation, so that recreation itself becomes the ecological condition. Discrete recreational areas contain program in a similar way that the internal bermed compound contains spoil—as a kind of landfill. Discussions about Lake Worth's and Palm Beach's coastal ecology hinged on maintaining "ecological balance." Groups believed that changing Peanut would affect that balance. The new plan shifts the framework for that equilibrium from natural and human-made to use and culture. Spoil and popular culture form this eco-island. The reserves are being filled with program and dirt, and paradise follows.

Why was Peanut Island not previously developed to what, some would say, is its full potential? Practicalities of access required bridges, causeways, drawbridges. But more significant than logistics of development, residents of Riviera and Palm Beaches displayed an emotional attachment to the island. The public saw it as a remnant of Florida's green space, even though it was human-made. Its placement demonstrated resistance to development pressures from three directions—from Palm Beach eight hundred feet southeast across the main harbor channel, from Riviera Beach six hundred feet across the ICW, and from Palm Beach Shores on Singer Island, less than one thousand feet to the east. Its current development and management plans hold the residue of deposition as well as development. The island's edge archives cultural activities, hybrid programs, and historically open spaces. Picnic areas, raft-ups, camping spots, Coast Guard station, and JFK's bunker all attest to the island's heterotopic, and sometimes mutinous, landscape. Before the current master plan, while developers and engineers discussed other plans for the island, the public went picnicking.

PICNICS AND TEA-PARTIES

Soon after the island was formed, picnickers arrived for afternoon and weekend excursions. But dredging operations continued and "no trespassing" signs were placed on the island to secure this place of deposit and to protect a sometimes wayward picnicking public.[33] Later picnicking expeditions were formalized with clubs and other community groups. In May 1940, the Jay-Cee Band-Parent Association of Riviera Beach organized a "basket picnic supper" for its members. In the late spring of 1956, the Boynton Beach Boat Club held one of its frequent barbecue picnics on twenty-two acres of fresh spoil, a recently deposited sandy upland added to the island by the port's renewed dredging operations.[34] So popular was Peanut Island among picnicking boaters that Coastguardsmen stationed on the island found that the garbage left behind was affecting their quality of life. They spent time each week cleaning up after the weekend's festivities and installed trash barrels across the sandy terrain with signs citing leave-no-trace principles: "If you brought it, take it" and "Leave only your footprints."[35]

Lake Worth's waters also hosted more exclusive gatherings. As the *Hercules* dredged a deeper channel and further augmented the nascent Inlet Island that hosted weekend picnickers, sisters Louise and Agnes Clarke celebrated the first season of their nearby Beaux Arts houseboat. Dubbed *Il Cavalio Marino*, the "sea horse" quickly became a "meeting place for the fashionable" in the middle of Lake Worth. By late February of 1921, this "little floating island of life and color" was not just a space of high society but was also considered a work of art recalling, according to one writer, the "famous Doges Palace in Venice." Cerulean blues, mitis green, and brick red provided backgrounds for fourteenth-century candelabras, sixteenth-century grilled doors, blue and white Delft tiles around the fireplace, and Louis XVI furniture. But this scenography dressed the mundane shell of a houseboat, and the *Miami News* commentator would ask, "whoever heard of a tea room on a houseboat?"[36] The baroque construct mixed the sophisticated with the commonplace, and the traditional with the unexpected. This "bit of Venice" took up residence in the shiftiness of the unmoored house, and despite these paradoxes, the houseboat was at home in Lake Worth. Less than three months after its grand opening, *Il Cavalio Marino* sank when a seacock was left open. The daughters of John Clarke, New York businessman and Palm Beach baron, vowed to raise the boat and renovate in time for the next season.

Protected by its barrier land, and the wealthy resources of its residents, Lake Worth has been a historically calm, relatively shallow anchorage for other houseboats.[37] Henry Flagler frequently docked his houseboat on Singerf Island, along Lake Worth's eastern shore.[38] In 1907, George Graham Currie, a land developer, poet, and philosopher who had settled in West Palm Beach thirteen years earlier, composed a map of the Palm Beaches' points of interest and in the process created, an early version of a chamber of commerce travel guide. At the present-day site of Peanut Island, the Currie Map of 1907 shows a "House Boat." Its location suggests that the shoaling area generated from that time period's Lake Worth Inlet was becoming an established sandbar that would later prove to be a useful site for augmentation and the creation of Peanut Island. Hotel ferries, in particular the launches from Royal Poinciana, brought guests out to this "wondrous tea-house restaurant" raised on pilings and surrounded by shimmering pools of fish attracted by the restaurant's table scraps.[39] The House Boat was at the center of a community of islands. To the north, James Munroe Munyon, homeopathic inventor of Paw Paw Pills, settled on what became Munyon Island, where he also developed Hygeia Hotel surrounded by five thousand rose bushes. Across the lake, the Coniff Hermitage was the dilapidated residence of a noted hermit.

With their limited mobility—not unlike that of a mobile home—and their vulnerability to rough seas, houseboats work as human-made, floating islands. Like spoil islands in general and Peanut in particular, they offer escape from normative domestic life on the mainland. Houseboat living in Riviera Beach has continued, and debates over one floating home redefined this dwelling type's legal status. In January 2013, the U.S. Supreme Court ruled that the city of Riviera Beach could not regulate Fane Lozman's residence as a maritime vessel. Soon after Lozman docked his two-story floating home at the city's marina, officials announced plans for a $2.4 billion private redevelopment project that required his eviction through eminent domain, and the city seized and destroyed the houseboat, which did not have engines, steering lights, or bilge pumps. Lozman initiated the legal proceedings that would end up in Supreme Court in a case significant to state definitions of floating homes and federal definitions of vessels. In one respect, houseboats, which do not or cannot move, approximate artificial islands and are subject to taxation. In another respect, if they are "practically capable of transportation over water," they constitute vessels and are subject not to taxation but to seizure.[40] Lake Worth's waterscape also supports another artificial island typology. Not so much house-as-boat but more like boat as temporary house-party, "raft-ups" build on the spoil island's shoaled, shallow-water landscape to form an ephemeral archipelago of revelry.

RAFT-UP

> We have always been amazed at the host of pleasure boats continually anchored in the charted patch of deep water east of Peanut Island. Entry into this haven calls for a tricky cruise between unmarked shoals, and the anchorage is open to rough water and strong tides from the nearby inlet. Nevertheless, many pleasure craft seem to be willing to put up with these inconveniences to anchor close by reliable Lake Worth Inlet.[41]

At low tide, you can walk from Peanut Island's north shore for nearly a half mile up to the viaduct-shaded beaches under Blue Heron Bridge. Along the way, Conchy Bowe's Bait Barge offers breakfast sandwiches and fresh bait out of a yellow houseboat with painted palm trees, a Carolina Skiff delivers ice cream, and Domino's sometimes sells pizza from a Boston Whaler. This submerged spoil area—locally described as the "horseshoe sandbar"—supports swimmers, snorkelers, sunbathers, tubers, grillers, and fishermen. Appearing to fill the inlet, Peanut Island hosts a festive space, a waterborne party that extends outward from the island's circumference. Boaters also anchor off the west and south beaches for quieter—a relative term on weekends—places to enjoy Lake Worth's waters. Peanut Island is known for its "raft-ups." This technique attaches boats together to form an interconnected floating raft. The term describes social gathering as much as the technical methods of tying; and on any given Saturday or Sunday, rafted vessels extend the island's surface northward to create a buoyant—sometimes Bacchanalian—landscape.[42] Double-row raft-ups create alleyways for veritable aquatic street parties. On Memorial Day, the density is not unlike a late spring day on Central Park's Great Lawn.

At the southern end of Lake Worth, another spoil island began accumulating sand and side-cast dredge material when its inlet was opened. Between 1925 and 1927, dredgers cut Boynton Inlet, also called South Lake Worth Inlet, to purge the intertidal waters, already polluted by increased population and poor circulation. Locals and administrative documents officially referred to the dredge site, which doubled in size in 1955, as Beer Can Island. The island became a popular stop-over and rest site for boaters passing between Palm Beach to the north and Boca Raton to the south. In 1994, its name was changed to Bird Island for a planned bird rookery. Unlike Peanut Island, this smaller seven-acre island was privately held. Its owner William Ziff, who received hate letters from party-goers for its change of use, placed the land in the Bird Island Trust and, with the help of the National Audubon Society, planned a restoration project to make the island an attractive nesting site for least terns. Replacing Australian pines with native vegetation and raising part of the island to a more protected elevation, the "biological wasteland" became an alternative to the terns' urbanized nesting pattern on Florida's white-gravel roofs.[43] In April 1997, island access was restricted, and boaters found that they lacked a public docking area. The South Lake Worth Inlet's advisory committee investigated the construction of a new spoil island to replace Bird Island, and in 2006 the county opened Ocean Ridge Natural Area, a compromise of sorts, providing a dock with nine boat slips nested within a restored wetland habitat of mangroves and other maritime hammock vegetation.

Peanut Island is a permutation of this beer can island typology. Its traditional use as a party destination extends back to early picnickers who dodged the first loads of deposited spoil in the late 1910s. Since these times, Peanut Island's recreational use has been formalized, and campers, beachgoers, naturalists, and snorkelers each have a designated place. But the beer-can moniker retains this direct meaning with the island's continued reputation for revelry. In 2011, at least three websites were dedicated to helping organize the party.[44] An extended meaning of "beer can" designates a waste space of informal use juxtaposed with upscale boating and housing. Indicating an accessible site for populist gatherings, the island's name reinforces this connotation. Discarded, recyclable, and otherwise rife with associations, Peanut Island does seem to rest like a partially crushed aluminum can thrown aside an entryway. John D. MacArthur did not want the island, but thousands of revelers attest to its amenities and propitious location. The politics of dredging found in the Indian River Lagoon are also at work here. The terms of exchange remain tacit, unlike further north where Ft. Pierce residents followed a widely expressed logic of appropriation and entitlement. The Corps of Engineers' official mandate is to dredge for commercial vessels, but new generations of deep-draft super yachts and mega yachts have spurred the maintenance and deepening of waterways and ports like Palm Beach where private vessels with drafts of nearly twenty feet are not uncommon. A new generation of Vanderbilts, Astors, and Carnegies have yachts and mainland mansions, and everyone else has weekend raft-ups.

DOLLHOUSE IN A BRIAR PATCH

> Much of the architectural impact of the property derives from the dramatic placement of two strong, simple buildings in the sloping grassy field, precisely framed by dark forest, with a sparkling waterway filled with boats spread at its foot. From any viewpoint, one is impressed by the serenity of the island setting in the midst of a bustling urban waterscape.[45]

The United States Coast Guard Station on Peanut Island has a basement. Built in 1936, the station followed a federal template as part of a network of naval facilities along the nation's coasts.[46] Almost ten years had passed since the first requests for a station to monitor smuggling operations. In spite of economic hardship, bootleggers as well as cruise ship passengers were still riding the Gulf Stream to its closest point on the U.S. coast, where the inlet opened to Lake Worth. For ten months, builders followed plans for a "standard dwelling" that included the basement, a rarity in Florida because of high water tables and the frequency of storms and flooding. With only minor modifications, the Coast Guard's civil engineering office superimposed the station's turn-of-the-century architecture on the region's landscape. Except for a standardized porch, its Neo-Georgian design does not respond to the Florida coast's climatic imperatives. Overhangs do not extend to shade windows, and the heavily grounded foundation disallows the cooling air-flow of a raised platform. More broadly, planners modeled the "standard dwelling" on the design of a private house. Implementation of the plans to house twelve coastguardsmen results in a disconnect between the scale of the domestic model and the institution it accommodates. The modestly detailed station packs in an unexpected twelve thousand square feet. A journalist writing in 1970 captured the surreal mix of the station's design and its ruderal context when he wrote that it had been "planted like a dollhouse in a briar patch."[47] It could be so rooted and its stock plan could be followed so closely because of the artificial ground that made room for the basement.

But aesthetically it works, as the author of the property's historic survey suggests in his description of its "dramatic placement." The Lake Worth Inlet Station, nested in its domestic shell, rises up from a gently sloping hill. The station's silhouette and the knoll's sweeping curve offer striking contours against southern Florida's tropical sky. The approach from the dock gradually moves up the twelve feet of elevation that was the island's highest point in the 1930s so that this "dramatic placement" was originally a function of the spoil island's location and construction. This southeastern quadrant was the most readily accessible area for the port's dredgers working on the inlet channel, and the high point afforded views up and down the ICW and eastward through the inlet's cut out to sea. The station's octagonal cupola, also at odds with the Georgian architecture, further extended this unobstructed view of smugglers, German U-boats, and stranded yachtsmen. Australian pines growing on newly deposited sand above the station would later obscure this view but would also vividly highlight the station's position on the gentle rise. This prospect would have pleased William Gilpin, raconteur of the English picturesque, who might have cited the "airy pines" as the borrowed scenery of landscape design. Restoration efforts in the 1990s removed this "dark forest" of invasive Australian pines. In its present state with more spare vegetation and established grasses, Gilpin might have noted the topography's "elegant lines" and found repose in the contrast of the "bare knoll" with the uniformity of the station's lined architecture.[48] The careful gaze of the aesthete substitutes for the strategic sight lines of the lookout, workaday topography conveys sublimity, a sandy dumping ground becomes a grassy lawn, and the "dollhouse in the briar patch" becomes landfill picturesque.

The Coast Guard decommissioned the Lake Worth Inlet Station in 1991, and two years later the Palm Beach Maritime Museum began preparations to maintain the historic structures—including the main building and the boathouse—for tours and educational programs. The museum leases the four-acre property from the Port of Palm Beach and has sought to expand its programs. In 2011, when the museum proposed helicopter rides, a bed and breakfast, and a restaurant, Palm Beach commissioners cited the county's comprehensive plan and zoning codes that restrict commercial land use on the island. Museum officials have argued that the port has jurisdiction, and the 1915 legislation creating the port states that neither county nor municipality can regulate the use of port property. The Museum's development plans are a microcosm of the broader schemes for the entire island that resurfaced in the late 1980s and early 1990s when the Coast Guard left the island and developers revisited their plans.

BUNKER

> Tomorrow, I am requesting of the Congress new funds for the following immediate objectives: to identify and mark space in existing structures public and private that could be used for fall-out shelters in case of attack; to stock those shelters with food, water, first-aid kits and other minimum essentials for survival . . . the time to start is now. In the coming months, I hope to let every citizen know what steps he can take without delay to protect his family in case of attack. I know that you will want to do no less.[49]

> In simple terms, the goal is to reach for fallout protection for every American as rapidly as possible.[50]

Just returned from a meeting with Soviet Premier Nikita Khrushchev, President John F. Kennedy announced plans for an extensive network of fallout shelters in a nationally televised speech on July 25, 1961. Less than one month later, the Berlin Wall was in place, followed by Soviet nuclear testing after a three-year freeze. By October, Kennedy advocated, with increasing urgency, access to shelters "for every American as rapidly as possible." Meanwhile, military planners "marked space" on Peanut Island's existing earthwork structure, and before the end of the year, Navy Seabees—the U.S. Navy Construction Battalion—built Kennedy's vacation fall-out shelter on Peanut Island.

The island offered an ideal site for the Kennedy Bunker. Less than five minutes by helicopter from surrounding areas, it was also a quick ride on the presidential yacht *Honey Fitz* across the harbor channel from the Winter White House, the Kennedy family's vacation home known as "La Guerida" in Palm Beach. Just as the Coast Guard Station had been sited for its strategic location and its panoramic views of the coastal region, the island's defensibility created a secure military site for the fallout shelter and command center. Largely undeveloped, and very overgrown, Peanut Island was also a readymade earthwork that allowed for easy excavation—a rarity in Florida's flat terrain and high water table. The Coast Guard station's basement and the bunker both exploited the depth of this artificial ground.

Augmented by continued dredging in 1948, the Port of Palm Beach's spoil disposal area rose to a prominent elevation that exceeded the inlet's controlling depth of twenty-seven feet. With the port's consent to what were cryptically called "improvements" for "the benefit of navigation and commerce," SeaBees burrowed into this mound from the northwest corner of the Coast Guard property.[51] Shown in plan, this angular, kite-shaped annex uneasily collides with the rational rectilinearity of the Coast Guard's four acres. The bunker plot's geometric awkwardness hints at a clandestine use that restoration studies and management plans seem to conceal steadfastly. Planning maps in USACE's 2000 report maintain the site's cover as a munitions depot, labeling the fenced grounds "old government magazine area," even though the report itself discusses the bunker's history. In 1973, when it officially identified the facility's existence and purpose, the U.S. government acknowledged what picnickers and boaters already surmised from their visits to the shelter, unlocked and abandoned since its decommissioning nine years earlier.[52]

The plot's angularity does reveal its function. SeaBees placed the bunker's entry on axis with the Coast Guard's docks, located five hundred feet away and newly renovated at the time with concrete planks. Once inside, the access passage makes a sharp turn to protect the main room from the shock effects of an atomic blast. A camouflaged portal begins this sequence. During an attack, Kennedy would have been escorted through its exterior door flanked by the tunnel's adjustable air valves, two prominent "buttons" partially obscured by overgrowth. Lined with lead and concrete like the rest of the complex for radiation protection, a corrugated steel pipe—seven feet in diameter—leads through a decontamination area with its rinsing shower and then connects to the shelter and command center. This main room was the "Detachment Hotel" described in the Seabee's construction documents. Buried under more than ten feet of spoil material, it could accommodate thirty people for thirty days on fifteen metal bunk beds. As tensions with Russia increased, Kennedy visited the bunker twice during the eleven months leading up to the Cuban Missile Crisis in October 1962. After Kennedy's assassination, the bunker was decommissioned in 1964.[53]

The Palm Beach Maritime Museum maintains the bunker, which opened for public tours in 1999. The Museum's restoration includes apt and timely, if historically inexact, artifacts: a rocking chair—called for by Kennedy's bad back, a desk with red telephone—not part of Cold War military command until after the Cuban Missile Crisis, and a U.S. Presidential Seal painted on the floor—an embellishment not originally a part of the bunker's spartan furnishings. Such historical informality matches the site's improbability, as a special case warranted by the Kennedy family's Palm Beach compound. Now such patrons as paddleboarders in swimsuits can imagine calling the Kremlin, engage in desperate conversation over the Taylor transmitter ham radio, or simply contemplate being underground in Florida. The bunker's short lifespan occurred during a lull in dredging operations and discussions of Peanut Island's development plans. In December 1963, less than one month after the president's assassination, city officials and local groups considered renaming Peanut Island for the President. At the time, the Coast Guard station's docks still harbored the presidential yacht *Honey Fitz*—a visible reminder of Kennedy's significant presence in the area—while the spoil deposits masked a hidden legacy, soon unearthed by the community.

CULTURAL LANDSCAPE

As a site of layered historical significance, Peanut Island extends and reframes the definition of cultural landscape. Paralleling established definitions of cultural landscape as a part of historic preservation practice, modifications of spoil islands combine the work of humans and nature. Cultural landscapes typically entail the human occupation of natural landforms through time. For historians, the study of cultural landscapes "situates the construction of places in complex historical contexts," including design, production, economic development, and the politics of land use.[54] Spoil islands expand these frameworks. Peanut Island begins with the augmentation of a natural landform, but the island's size and prominence result in a wholly manufactured site. Though still susceptible to natural forces, such as those indicated by its one hundred foot shift westward, the historical context for the making of Peanut Island's places is itself human-made. Peanut's first permutation as a useful site for spoil deposition becomes the scene of other culturally—and politically—charged constructions.[55]

This twice-built island mixes the useful with the cultural in a landscape that is productive not just as a continually evolving experiment in waste deposit and place-making, but also as a reminder of the variable ways landscape itself can be defined. Peanut Island continues as both stage and infrastructure for the daily activities—quotidian and festive—of southern Florida's residents. It recalls J. B. Jackson's definition of landscape as "a composition of man-made and man-modified spaces to serve as infrastructure or background for our collective existence."[56] At Peanut, this collective existence is often mutinous—either in the sometimes riotous voices of its disparate parts or in the popular disturbances of public opinion.

Maintaining such a landscape is complicated. Shoal, bunker, picturesque, dollhouse, raft-up and developers' dream collide on the site. Too close for some—Palm Beach residents fear large-scale development, just out of reach for others—real estate requires access, but just right for most—throngs of boaters flock there on weekends. Preserving its diverse components poses greater challenges, but considering the island within the framework of cultural landscape preservation opens up possibilities for Peanut Island in particular and spoil islands in general. It also offers a framework for rethinking the contemporary understanding of cultural landscape. The National Parks Service (NPS) and UNESCO each provide frameworks for how spoil islands might be preserved. NPS's definition of the "historic vernacular landscape" outlines the functional activities that link people to land. The resultant shaping of the landscape captures one aspect of spoil island construction. Although the definition notes "industrial complexes," it privileges the productive landscapes of agriculture.[57] Given NPS's narrow definition, UNESCO's guidelines provide a more apposite starting point for the cultural landscapes of spoil islands.

UNESCO's first category of cultural landscapes is a "clearly defined landscape designed and created intentionally by man." As islands, spoil areas are clearly delimited, but identifying intentionality poses a problem with spoil islands because many were not initially purpose-built. As the by-product of another intentioned process, these places are residual manufactured landscapes, and purposefulness has come about over time. UNESCO's second category also tracks along the spoil island's changeability as an "organically evolved landscape." This category presumes that human intervention is grounded in a natural setting. Just as spoil islands—as wholly human-made sites—extend the first category to its limit, they also expand this second category in what might be meant by the landscape's "association with and response to its natural environment."[58] With spoil islands, this response is one of displacement. Dredged materials are removed from their natural context and deposited in another location that is then modified by natural forces. The discordance of this mutinous space conflicts with what appears to be the UNESCO guideline's interest in traditional, perhaps more harmonic, relations between society and nature. A sub-category in this designation holds greater potential for spoils. The "continuing landscape" retains its active role in society, and spoil islands exhibit what this designation calls the "significant material evidence of [their] evolution over time." With UNESCO's third category, spoil islands provide a uniquely "associative cultural landscape."[59] With their multiple associations, places such as Peanut Island demonstrate the potency of landscape as a work of the mind. While they do not have the overt religious significance of UNESCO's world heritage sites, spoil islands like Peanut connote a diverse set of meanings for those who live near them, ranging from paradise to industry and from nature to wildness.

In strict preservation practice, cultural landscapes fail tests of integrity because of their frequently dynamic contexts. Cultural landscape historians have noted that many preservation criteria are derived from architectural models, and some argue that preparing the site for its interpretation—its interactivity with the public—achieves higher value than more static conditions of the site's faithfulness to past form or use.[60] Places such as Peanut Island are chronically changing historic landscapes. Like buildings, these islands have been created "from the ground up." As a contiguous site, Peanut Island includes a collection of historic artifacts, which themselves have historic, or at least cultural, significance. Both spoil island, as the vessel, and human interventions, as the contained elements, have significance.

Whether framed architecturally or in terms of landscape, Peanut Island fundamentally fails the determination of integrity. Except for the Coast Guard station's relative constancy, the island's grounds have been consistently transformed. Because the island is built out of waste, conventional methods of historical and archeological analysis are altered. A jumble of stones and sand combined with erosion and continued dumping prohibit such interpretation. Historical surveyors of the Coast Guard site on Peanut Island came to similar conclusions as archeologists studying Indian shell middens. Rules of stratigraphy do not apply. But as a cultural landscape, the island holds significant potential as an interpretive site. Continued use of the island for dredging maintains the waterways while it also sustains a procedure that has been an integral part of the nation's coastal landscape for more than a century. Its persistent usefulness calls attention to the continually evolving balance between environmental, industrial, and cultural concerns. The collections of diverse activities and programs it has attracted attest to the waste site's openness to interpretation. Waste areas gauge public perception like no other places. Spoil islands such as Peanut are litmus sites, absorbing and resisting societal change.

Given the complexity of Peanut Island's formation and history, the term "cultural landscape" falls short of defining the place as a whole. The roots of the term "landscape" itself come closer to the spoil island's mix of use and culture. On these islands, landscape is not painterly background. It is instead the productive subject and the constructed site. Simon Schama found similar places when he made the connection between the Dutch term *landschap* and that region's highly engineered landscapes.[61] Such landscapes inextricably combine design with use. Taking landscape's etymological origins back further reinforces this idea: *Land-schap* is "shaped land." Even more fundamentally, it is "created land."[62] Spoil islands are made land, and their landscapes are the work of human intervention. They are artifacts in the most basic sense. There still may be ambiguity about the degree of human intervention, but these islands are truly born of human agency. The landscapes are not just altered; they are established, founded, and initiated by humans. Used in historic preservation documents to describe the work of human culture, "substantial modification," is the foundational term for their development. And this agency occurs within a dynamic natural context.

Wildness is growth, work, seasonal change, transition. When he writes that "in Wildness is the preservation of the World," Thoreau posits wildness in opposition to city, as a remedy to urban and industrial difficulties.[63] But landscape is shaped by the mind as much as it is molded by hand and by nature. Wildness preserves the world, but perception makes the place. As Schama points out, Thoreau's wilds are products of cultural frameworks. They are "imagined gardens."[64] Spoil islands also measure decisions we make about changing the landscape—not just of the island itself, but the territory it services. Spoil heaps might be paradise, and dollhouses might grow in briar patches. Use and culture collide on tides of island-borne wildness. What is called "unspoiled" on Peanut Island, on Indian River's spoil islands, or on Miami's picnic islands—what is perceived as wild has grown out of the work that Thoreau dismisses. Peanut Island preserves a wildness that we did not know we had. Its cultural and useful landscape is messy, its dredging does spoil the water with sediment, its master plan is prescriptive, and its party does not appeal to everyone. But it maintains something that many of us do not want to lose. It remains open.

NOTES

1. Robert K. Oglesby, "Peanut Island," *Palm Beach Post*, 11 August 1968, G1.
2. Michel Foucault, "Of Other Spaces," *Lotus International*, nos. 48–49 (1986): 9–17.
3. For an account of this event, see V. A. Keil, "Inlet Hearing Before Army Engineers at Washington Monday Means Much to Port," *Palm Beach Post*, 5 May 1935, 15–16.
4. U.S. Army Corps of Engineers (USACE), *Peanut Island, Palm Beach County, Florida: Final Environmental Restoration Report and Final Environmental Assessment*, Environmental Restoration, Section 1135 (USACE, Jacksonville District, March 2000), 6.
5. "Dredging Operations Progressing Rapidly at Lake Worth Inlet," *Palm Beach Post*, 27 June 1918. The dredger *Standard* started work in Lake Worth in the middle of June 1918.
6. USACE, 53.
7. "Ship Enters Harbor," *Palm Beach Post*, 3 January 1927, 1.
8. *Palm Beach Post*, 8 January 1928, cited in USACE, 9.
9. USACE, 52.
10. USACE, 10.
11. USACE, 12.
12. USACE, 23.

13. "Sale to Brown Company Seems Agreeable Now," *Palm Beach Post*, 25 October 1923, 1. In 1935, the IIF Board authorized a one-year lease to the Lake Worth Inlet District and the United States government to dump spoil material "between the banks of that certain island known as Peanut Island." See April 10, 1935, *Minutes of the Trustees of the Internal Improvement Fund* 20 (Tallahassee, FL: 1936): 42. For a summary of the complex land transactions related to Peanut Island (then "Inlet Island") in the 1920s, see also October 23, 1962, Minutes 34 (1964): 74–5.

14. For the popular media's presentation of the economic context for this agricultural production, see the full-page advertisement titled "Believe It or Not," authored by Southern States Land and Timber Company," *Palm Beach Post*, 10 December 1929, section 4, page 3.

15. "Announce Plans for Development of Harbor, Inlet," *Palm Beach Post*, 22 April 1931, 1.

16. "Announce Plans for Development of Harbor, Inlet," 1. See also "Civic Club Committee Approves Huge Inlet Development," *Palm Beach Post*, 15 July 1933, 1.

17. Oglesby, G1. The Port of Palm Beach provided the rendering for the newspaper article.

18. Lewson cites this goal at multiple times between 1968 and 1970. See, for example, Lewson's quotes in the article "Peanut Island Lessee Readies Land Use Plan," *Palm Beach Daily News*, 25 March 1970, 1.

19. Todd Wright, "Port Authority Signs $20-million Isle Pact," *Palm Beach Post*, 17 October 1968, 1.

20. Jim Quinlan, "Peanut Island: Oceanography or Recreation," *Palm Beach Post*, 21 June 1970, 1.

21. Quinlan, 1.

22. Gary Blankenship, "Commissioners Reject Chance to Buy Peanut Island," *Palm Beach Post*, 5 August 1983, 1.

23. A year later, in 1991, the Palm Beach County property appraiser assessed the island's value at twelve million dollars.

24. Don Van Natta, Jr., "Island Needs Protection, Official Says," *Miami Herald*, 12 April 1988, 4.

25. Van Natta, 4. Tanner made the following comment: "I'm sick of all this development, all the concrete. The county cannot continue to bring as many tourists as they bring in here and promote this area without having some green space and places for people to go an enjoy."

26. See Appendix F of USACE's *Peanut Island, Palm Beach County, Florida: Final Environmental Restoration Report and Final Environmental Assessment* (2000). The report also noted the following: "More recently through the cooperative efforts of both the Florida Inland Navigation District (FIND) and the Port of Palm Beach District, a unified approach has been developed that would grant the county a long term lease of Peanut Island and thus allow public use of the Island's most valuable shoreline areas for boating and passive recreational purposes."

27. "Peanut Island," pamphlet produced by Palm Beach County's Environmental Resources Management (ERM), also available on ERM's website, accessed May 17, 2012, http://www.pbcgov.com/erm/downloads/pdf/brochures/PeanutIslandBrochure.pdf.

28. See Appendix F of USACE's *Peanut Island, Palm Beach County, Florida: Final Environmental Restoration Report and Final Environmental Assessment* (2000).

29. The "Environmental Restoration Recommended Plan," dated November 1999, is included on pages 3–4 of USACE's *Peanut Island, Palm Beach County, Florida: Final Environmental Restoration Report and Final Environmental Assessment* (2000).

30. Influenced by eighteenth-century French military design, pentagonal forts are common along the U.S. east coast, including Ft. Clinch, Fort Pulaski near Savannah, Georgia, and Fort Delaware on Pea Patch Island.

31. See "Peanut Island 24: 'Round-the-Island Races," accessed June 5, 2012, http://peanutisland24.com. In the 2011–2012 race, Stephanie Miller broke the course record, running ninety-five laps for more than 117 miles.

32. Oglesby, G1.

33. USACE, 53.

34. "Band-Parents Group Plans Picnic Party," *Palm Beach Post*, 30 May 1940, 12. See also "Boat Club to Meet," *Palm Beach Post*, 9 May 1956, 15.

35. Susu Jeffrey, "Guests Abuse Island: Coast Guard Must Clean Up After Litterers," *Palm Beach Post*, 16 May 1972, C3.

36. "Tea House Boat is Remarkable Feature of Palm Beach Resort," *The Miami News*, 24 February 1921, 3.

37. Claiborne S. Young has noted the water's calmness: "In fair weather this just may be one of the most painless inlets to run that you ever encounter. Many aids to navigation in the channel are charted, a sure sign of a stable channel." See *Cruising Guide to Eastern Florida* (Gretna, Louisiana: Pelican Publishing, 1996), 388.

38. While his houseboat was being repaired on Singer Island, Henry Flagler stayed at "Riviera," the residence converted from the Riviera Hotel to a private winter residence by Iowan Charles N. Newcomb in 1901. A scientist and inventor as well as an industrialist, Newcomb devised his own dredger to fill the swampy shore in front of his estate. By 1910, Newcomb had purchased more land to the west and planned a resort community that included a bridge crossing Lake Worth to Singer Island where Peanut Island now rests. Newcomb's "Riviera" became a popular lodging for other Palm Beach area magnates, hosting the Vanderbilts, the Astors, Andrew Carnegie, and Henry Flagler, who stayed at the house while his "houseboat" was being repaired. The dock for his "Riviera" house was also on axis with the area that would be cut for the new inlet in 1918. See Lynn Brink, ed., *A History of Riviera Beach, Florida* (City of Riviera Beach: Bicentennial Commission of Riviera Beach, 1976), 15.

39. James R. Knott, "The Currie Map of 1907," *Palm Beach Post*, 4 January 1981, supplement B.

40. With the majority, Justice Stephen Breyer determined that Lozman's residence was not a vessel, and he wrote that maritime law only applies if the structure looks like practical water transport—something that a "reasonable observer, looking to the home's physical characteristics and activities, would consider . . . designed to a practical degree for carrying people or things over water." See *Lozman v. City of Riviera Beach, Florida*, 568 U.S. 5 (2013). See also Curt Anderson, "Clash over floating home in Riviera Beach reaches U.S. Supreme Court," *The Miami Herald*, 31 March 2012, and Michael Doyle, "South Florida man wins houseboat case in Supreme Court," *The Miami Herald*, 15 January 2013.

41. Young, 389.

42. Such "party barges" have entered the popular imagination in songs like country music singer Craig Morgan's "Red Neck Yacht Club."

43. Neil Santaniello, "Cleaning Up a Dirty Act," *Sun-Sentinel*, 27 December 1994, 5B. Since 1934, the National Audubon Society of Florida has managed a system of islands, many of them created from spoil, in its Florida Coastal Islands Sanctuary program. See http://fl.audubon.org/florida-coastal-islands-sanctuaries.

44. Websites related to Peanut Island's party include: PNUTPARTY.com; peanutisland.fateback.com; and www.reocities.com/palmbeachboating/peanut_island.html, accessed March 4, 2012.

45. "Peanut Island U.S. Coast Guard Station," Historical Structure Master Site File, PB 6291, Florida State Archives, 15 March 1989, n.p.

46. "Peanut Island U.S. Coast Guard Station," n.p. See in particular the section titled "Historic and Present Physical Appearance."

47. Mike Abrams, "Beverly Hillbilly Returns Home," *Palm Beach Post*, 10 January 1970, A21.

48. William Gilpin, *Three Essays on Picturesque Beauty, on Picturesque Travel, and on Sketching Landscape* (London: R. Blamire, 1792), 10. See also Gilpin's description of his approach to Inveraray Castle in Argyll, Scotland, in *Observations, relative chiefly to picturesque beauty* (London: R. Blamire, 1792). Coast Guardsmen also found unexpected beauty in Peanut Island's landscape. In 1972, seaman apprentice Steve Kalamajka noted: "It's so fantastic here. . . . It's covered with flowers and it's so green. . . . It's unexplainable. You have to see it." See Susu Jeffrey, "Guests Abuse Island: Coast Guard Must Clean Up After Litterers," *Palm Beach Post*, 16 May 1972.

49. John F. Kennedy, "Radio and Television Report to the American People on the Berlin Crisis," 25 July 1961, online by Gerhard Peters and John T. Woolley, *The American Presidency Project*, accessed February 15, 2012, http://www.presidency.ucsb.edu/ws/?pid=8259.

50. John F. Kennedy, "Letter to the Members of the Committee on Civil Defense of the Governors' Conference," 6 October 1961, online by Gerhard Peters and John T. Woolley, *The American Presidency Project*, accessed July 3, 2012, http://www.presidency.ucsb.edu/ws/?pid=8379.

51. "Peanut Island U.S. Coast Guard Station," 13.

52. USACE, 4, 9.

53. Tom Vanderbilt's *Survival City: Adventures Among the Ruins of Atomic America* (New York: Princeton Architectural Press, 2002) places the Kennedy Bunker within the context of other Cold War landscapes.

54. Dolores Hayden, "Foreword," *Preserving Cultural Landscapes in America* (Baltimore: Johns Hopkins University Press, 2000), vii.

55. Even though it was purpose-built in 1937 for the Golden Gate International Exposition (1939), the artificially constructed Treasure Island parallels Peanut Island's hybrid programs and its cultural landscape. Its position in San Francisco Bay afforded Treasure Island autonomy from political conflicts on the mainland. The plasticity of its landscape could also be molded to the exposition's needs. The island's supporters believed that its position allowed for "civic values" unattainable in the city's fabric and consequently transported fair-goers from an "ordinary life" to "enchanted isles, distant arcadias, and synthetic utopias." See Eugen Neuhaus, *The Art of Treasure Island* (Berkeley: University of California Press, 1939), 2.

56. John Brinckerhoff Jackson, *Discovering the Vernacular Landscape* (New Haven: Yale University Press, 1984), 8.

57. See Charles Birnbaum, *Preservation Brief 36: Protecting Cultural Landscapes: Planning, Treatment and Management of Historic Landscapes* (Washington, D.C.: National Park Service, Preservation Assistance Division, 1994).

58. United Nations Educational, Scientific and Cultural Organization (UNESCO), *Operational Guidelines for the Implementation of the World Heritage Convention* (Paris: UNESCO World Heritage Centre, 1999), 9.

59. UNESCO, 10.

60. See, for example, Catherine Howett's essay "Integrity as a Value in Cultural Landscape Preservation" in *Preserving Cultural Landscapes in America* (Baltimore: Johns Hopkins University Press, 2000). Howett writes that the "quality and importance of any preservation project is determined not by the integrity of the site, but by the quality of what is made of the site through interpretation of its history. That is the added value that can turn even a precious few evocative fragments, transformed by intelligence and imagination, into significant history; without it, even historic sites of exceptional integrity will remain sterile." (207)

61. Simon Schama, *Landscape and Memory* (New York: Alfred A. Knopf, 1995), 10.

62. Schama, 10. Schama references J. B. Jackson, who discusses *schaft* as unit of human occupation in *Discovering the Vernacular Landscape*, 3–8.

63. Henry David Thoreau, *Walking* (New York: Harper Collins, 1994), 19.

64. Schama, 6.

Singers Island
Palm Beach Shores
Peanut Island

Camp

In the center of the peninsula, fifty miles west of the ICW and Brevard County's dredged islands, a court case centered on a small spoil island. Its outcome would influence land use law not just in Florida but in other locations where water and land meet. At issue was the legality of a barbed wire fence that land-owner Roger Macnamara placed around the dredged island adjacent to his property along the Kissimmee River canal. With the support of the Sierra Club's legal team, Kissimmee River Valley Sportsman's Association brought the case against Macnamara in 1993. Their local goal was to maintain the right to camp on the spoil, and their broader objective focused on the rights of public access to Florida's water bodies. The Second District court ruled in favor of the Association, and appeals went to the Florida Supreme Court, which upheld the ruling in October 1994. The case focused on legal definitions of the ordinary high water mark to locate the spoil island within the indeterminate zone between water and land. But it was the site's favorable conditions for camping that initiated and also, in the end, won the case. Less than three decades in the making, ten acres of spoil strengthened the state's claim to one million acres of waterfront land.[1]

The canal that straightened the Kissimmee River is part of the larger project that impounded Lake Okeechobee, cut canals from the lake's eastern edge to Lake Worth, and dredged the trans-Florida waterway that connects the southern terminus of Indian River with the Gulf of Mexico. Historically, the waters of the Kissimmee River Basin formed a wide and slow-moving system of surface sheet flow from present-day Orlando to Lake Okeechobee and on to the Everglades. Changes to the river began in the 1890s to improve navigation and to control flooding. Expanded flood control projects in the middle of the twentieth century channelized the swampy, meandering course of the Kissimmee River, and in 1966 the Corps of Engineers dredged the C-

37 canal between two lakes—Hatchineha to the north and Kissimmee to the south.[2] Dredging produced a spoil island, 150 feet wide and 3,000 feet long, on the western edge of the canal where it meets Lake Hatchineha. In some areas of the island, spoil material rises twenty feet, and over the years the site has become a favored place to camp. In 1991, Macnamara strung a mile of barbed wire fence along the eastern edge of the spoil island to deter boaters and campers whose activities, he felt, spoiled the site.

More than half of Florida was originally wetlands. This wet ground makes for a unique landscape, and it is not surprising that the peninsula has been called the "major testing ground" for land use law.[3] The ground is so flat that slight vertical fluctuations—water level changes of inches—can influence miles of horizontal land. Porosities of sand and limestone, cycles of tides, and bursts of rainfall add to the landscape in flux. In this landscape of water, the extent of its often fluid boundaries and its navigability determine sovereignty and public access.[4] Spoil islands in navigable water are public lands held in trust by the state for "public use and enjoyment."[5] The Florida Constitution supports this claim: "title to lands under navigable waters . . . is held by the state, by virtue of its sovereignty, in trust for all people. Sale of such lands may be authorized . . . only when not contrary to the public interest."[6] As the plaintiffs, the Sportsman's Association and the IIF's Board of Trustees also cited the Florida Statutes, which are more specific about the topographical changes created by dredging: "the title to all sovereignty tidal and submerged bottom lands, including all islands, sandbars, shallow banks, and small islands made by the process of dredging any channel by the U.S. Government and similar other islands, sandbars, and shallow banks located in . . . navigable waters. . . ."[7]

The IIF Board of Trustees originally deeded Macnamara's property to Ingram Fletcher in 1876. Surveyors at the time based the deed's eastern boundaries on what was known as the "meander line" of the river. This boundary line approximated the sinuous river's curves with faceted segments, but surveying manuals have historically been vague about which water line—high, low, or something in between—determines a waterway's course. Because of difficulties defining fresh water boundaries like the Kissimmee River, courts have instead relied on the "ordinary high water boundary," which is described as the "ordinary or normal reach of water during the high water season."[8] But locating this boundary adds to the confusion. The *Macnamara* court document hints at sarcasm when it describes what would occur with the application of federal methods to identify the high water boundary in Florida's sodden context. This federally defined "agriculture test" excludes areas that are prone to livestock grazing from submerged land sovereignty, and the Florida court document notes: ". . . it suffices to say that this federal test would produce absurd results if it were applied to locate the legal boundary on Florida flatland lakes and rivers."[9] A previous court case had already determined that Lake Hatchineha's high water boundary is 52.5 feet, and the fence was "waterward," rather than "landward," of this line.

In the late nineteenth and early twentieth centuries, steamboats passed across Lake Hatchineha and along the Kissimmee River, where they docked at Shell Hammock landing just south of the spoil island site. This history of steamboat traffic established the navigability of lake and river, along with the fishing expeditions of local sportsmen. The spoil island came to occupy the historic area of vegetated lake bottom, where a line of cypress trees grew from a submerged sandbar known as an excellent fishing ground.[10] Court documents are not quite lyrical but they are poetically specific in their description of the recreational landscape before and after the spoil island's placement. Story-telling and local knowledge enter into the discussion: "the area around the base of these cypress trees [was] a favorite fishing site for sport fishermen . . . boaters who didn't know of the sand bar would get stuck on it with their kicker [motor] boats." Today channel catfish populate the C-37's sedimented channel, and the lake's prized bass hide under invasive blooms of hydrilla.

Camping on the spoil island caused the controversy, and the spoil island's suitability as a camp settled it. As plaintiff, the Sportsman's Association needed to prove that it had "standing to assert the right to use" the spoil island. "Standing" includes usefulness to the public, and the public wanted to be there because it was a nice place to pitch a tent, light a fire, and spend the night. In the end, the island's auspicious camping grounds and its unique—even if foreign—topographic conditions contributed to the court decision: "In the instant case, a substantial number of the Association members use the spoil island for fishing, swimming, and camping. The juxtaposition of deep water and high land in a lake with shallow vegetated margins makes the spoil island a unique and highly desirable site for public recreation. For these reasons, the Association is directly affected by the exclusion of the public from this area, and has standing to bring this action."[11] The human-made island offered what the surrounding landscape could not. The spoil island was a test case for camping and for defining boundaries—between land and water and between public and private.[12] It was not supposed to be there, but once it was introduced, it served its public purpose.

NOTES

1. The case had broad implications for attempts to redefine the ordinary high water boundary. Phosphate companies sought boundaries that would define narrower courses for waterways like Peace River, where they had mined phosphate and might incur multimillion-dollar liabilities for its extraction. The companies lobbied for changes and filed a 1986 case that argued for "terrestrial vegetation" as the new boundary, which would occur waterward of cypress trees and result in a much narrower waterway. Monica K. Reimer summarizes court

cases that have addressed public lands, definitions of boundaries, and navigability determinations in her article "The Public Trust Doctrine: Historic Protection for Florida's Navigable Rivers and Lakes," *The Florida Bar Journal*, vol. LXXV, no. 4 (April 2001): 10–22.

2. The State of Florida and the Corps of Engineers have embarked on a project to restore the Kissimmee River Basin. The court's decision in the *Macnamara* case bolstered these restoration projects and strengthened the state's claim to these lands. Before the case, the state had already agreed to allow the water management district to purchase sixteen thousand acres along the Kissimmee River. Even though the state technically already owned the land, officials believed that the costs to purchase—in the millions of dollars—were less expensive than the costs of legal fees in what was expected to be a lengthy case. The *Macnamara* case, however, showed that decisions could be reached quickly and inexpensively.

3. Luther Carter, "Land Use Law: Florida Is a Major Testing Ground," *Science*, vol. 182, no. 4115 (Nov. 30, 1973): 902.

4. In an extraordinary letter from Florida's Attorney General W. H. Ellis to Governor Park Trammell, Ellis observes that a particular area of land near Biscayne Bay is "always so saturated with water that the imprint of one's shoe will quickly fill." He goes on to determine that this land, because of its "saturated" quality, is the property of the state in its "sovereign capacity." See the letter dated 28 January 1913, included in the *Minutes of the Trustees of the Internal Improvement Fund* 10 (Tallahassee, FL: 1913): 30.

5. *Macnamara v. Kissimmee River Valley Sportsmans' Association*, 648 So. 2d 155 (Fla. 2d DCA 1994). Unless otherwise noted, all quotations in this section have been excerpted from this court document.

6. Florida Constitution, Article X, Section 11.

7. Florida Statutes, Section 253.12(1).

8. This definition works by analogy from tidal boundaries, which are determined by the "mean high water" (MHW) of the daily cycle of tides. The fresh water boundary is based on annual changes.

9. In its presentation, the plaintiff included an exhibit (6M) showing a cow grazing on vegetation "at least a hundred yards" within the boundaries of Lake Hatchineha.

10. Hatchineha comes from a Seminole-Creek word "achinaho" that means cypress tree. See Alma Hetherington, *The River of the Long Water* (Chuluota, FL: Mickler House Publishers, 1980), 87.

11. *Macnamara v. Kissimmee River Valley Sportsmans' Association*, 648 So. 2d 155 (Fla. 2d DCA 1994).

12. For further discussion of how camps define settlement patterns and offer alternative modes of place-making, see my books on the subject: *Camps: A Guide to 21st-Century Space* (MIT Press, 2009) and *Campsite: Architectures of Duration and Place* (LSU Press, 2008).

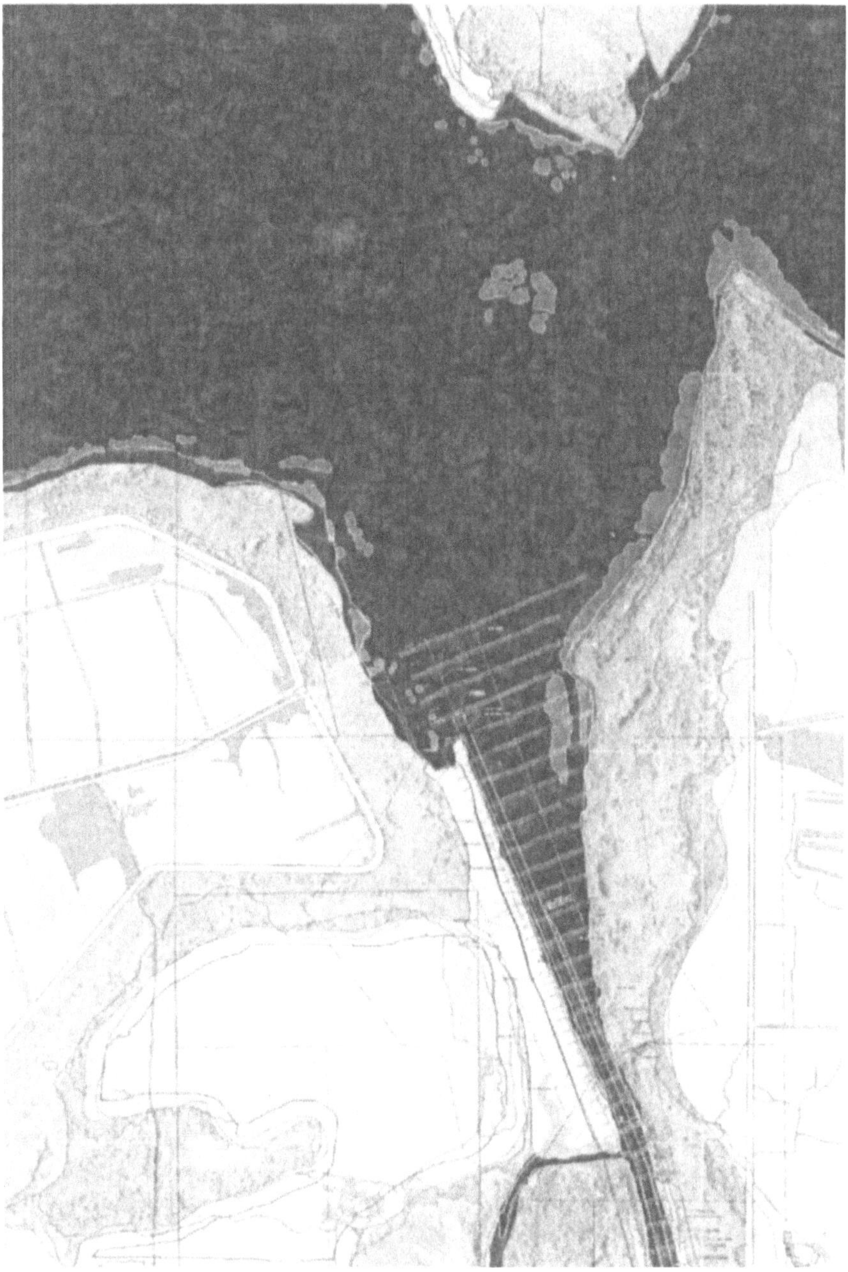

Chapter Five

Rational and Irrational

Developing Biscayne Bay's Lagoon

> Since the war they have paved a wilderness. They have made the spectacular commonplace.[1]

> Spoil islands are not permanent features, they both move and erode completely away.[2]

In the first part of the twentieth century, Miami Beach's mangroves were building land too slowly. Rationales of speculative real estate calculated wasted land where mangroves edged Biscayne Bay's sandy barrier. These unstable barrier islands might become permanent with man's intervention, and developers proceeded to bury cleared vegetation with dredged spoil from the bay's waters and to create new islands over its submerged lands. In the process, they transformed the bay's subtropical paradise into a readily accessible—even "commonplace"—landscape of development. What Baedeker described in 1909 as a "large sheet of clear salt water" became, in the two decades that followed, a turbid solution out of which new islands were raised.[3] According to this logic, development transformed the wasteland of wilderness into useful real estate property.

At about the same time, another dredging project yielded a series of disused islands. When the Corps of Engineers finally extended the Intracoastal Waterway to Miami in the 1920s, dredgers cast aside spoil along the bay's western edge. On each side of the channel and sometimes within one hundred yards of the mainland, the islands were left as by-products of the newly dredged channel. The spoil islands came to be located at the crossing of the bay's interwoven infrastructure, along two axes—the north-south waterway of boaters and the bay's six east-west causeways for the city's motorists. While the developers' purpose-built islands gained permanence, the spoil islands shifted with wind and tide. While the developers' islands privatized the bay bottom, these spoil islands remained resolutely public.

The ICW's walking islands introduced an illogic to the bay's increasingly manufactured landscape. Visually, they play like discordant notes across the causeways' musical staff. They also demonstrate the limits of human control over the bay's natural systems and expose an underlying irrationality in the hubris of land-boom development. Filling and bulkheading might stabilize new islands and shorelines but might also undermine the paradisiacal context that made this a desirable place. As this development became the rule, having normalized a wasted wilderness, the ICW's spoil islands remained marginal sites, though more geographically central to the bay's burgeoning growth. Their waste reinstalled wildness. In an urbanized bay, they became sites for what one resident observer called "green relief." Their susceptibility to natural forces and openness to public appraisal were their strengths. Their irrational presence became a reprieve from the everyday. Biscayne Bay's artificial islands tell two stories of dredged lands that are also stories of a manipulated environment lodged between competing, but not altogether distinct, rationales of conservation and development and less quantifiable and sometimes less reasoned, but equally potent, activities ranging from art installations to picnic excursions. In this dynamic setting, spoil islands emerge as environmentally charged afterimages that mix the rational with the irrational.

MIAMI DREDGE

> And today the dredge is the national emblem of Florida.[4]

> Miami, with its wonderful growth and development, situated on Biscayne Bay with its beautiful shore line and unique climatic conditions, is at present a veritable Mecca for an immense fleet of dredges; great machines which by the skillful manipulation of man, are transforming the dormant and waste places into active scenes of beauty and usefulness.[5]

Dredging improved navigation but it also made land. Nowhere is the story of this dual function more visible than in the waters of Biscayne Bay. In the summer of 1913, dredges filled the bay and already epitomized what Will Rogers would describe as Florida's "national emblem." In his timely article for the *Miami Herald*, civil engineer J. I. Conklin used the dredging activities to orient his readers to the bay's general layout, development, and growth. This dredge geography also served to describe—and promote—the bay's productive future, one that was mostly funded by private interests. Miami's portion of the lagoon has been the focus of development, transforming the northern third of Biscayne Bay, which extends southward to Card Sound. With Miami on the mainland and a chain of outer keys that include Miami Beach to the east, the bay constricts from a three-mile width to a half-mile wide apex where Haulover Inlet joins the Atlantic and the ICW shares a much narrower waterway that continues to the north.

Conklin begins his article breathlessly looking at Ocean Beach Realty Company's work to fill Miami Beach's bay shore. Three dredges—*Davis*, *Florida*, and *Biscayne*—work on the southern part of the peninsula, north of where the main shipping channel passes through Government Cut. Contracted for three million cubic yards, this filled land marks where the County Causeway—later renamed for General Douglas MacArthur—would link mainland to key in 1920. Just to the north, the dredge *Davis* places twenty thousand yards of material in just twenty-four hours at the foot of John Collins' viaduct, which was dedicated only weeks earlier in June 1913 and was—for the moment—claimed as the world's longest short-span bridge. It would soon be remade as a causeway linking Carl Fisher's purpose-built Venetian Islands. Further north, dredges fill land along present-day Alton Road and on top of the tidal mangroves where another causeway would later connect 79th Street with North Bay Village's artificial islands and Miami Beach's north shore. Conklin looks even further north to a dipper dredge with a seventy-five-horsepower engine and sixty-five-foot boom that extends a canal along Indian Creek and augments bay-shore land at Bay Harbor and Bal Harbor.

Conklin then turns his attention to the mainland's shore above and below the Miami River's confluence with the bay. To the north, two dredges build 150 thousand cubic yards of land near Miramar; and to the south at Point View, 200 thousand yards reward its owners with increased land value from $1,600 to $85,000. "Can you take this all in?" Conklin pauses to ask his reader, and then continues his tour with an elevated birds-eye view of the bay's activities. He asks his readers to imagine themselves on level with the sixth-floor salon of Henry Flagler's famed hotel on the Miami River's north bank, in a sense joining its tycoons, developers, and Conklin himself:

> Let your fancy take you now to a point opposite the Royal Palm hotel and high enough to look down on all the dredging activities at once. First look over to the Casino and see the *Davis* pouring in its twenty-inch stream of sand. Follow along to the north and notice the constant swing of the *Florida*. Now look up the canal and note the heavy cutting of the No.2 Highleyman dredge and just beyond the *Gopher* . . . rapidly making homes along Indian Creek. Swing around now and look to the south. . . .

Here, Conklin introduces us to the bay's burgeoning port activities, to the Florida East Coast Railroad's deepening of the main channels to accommodate merchant steam ships like the *S.S. Miami*, and to Virginia Key and Key

Biscayne where Rickenbacker Causeway will effectively section off northern Biscayne Bay, making it an artificial lagoon. With "cross the bay again," Conklin extends his vision southward to the nascent canals of Coconut Grove's "new Venice" before sweeping back along the mainland shore, Point View, and the dredging maintenance of the Miami River. The engineer Conklin's rationalized vision becomes our own "clear" perspective defined exclusively by what the dredges and their private enterprises are creating: "What magnificent work! All these operations carried on within a radius of a few miles can mean but one thing. That Miami's friends with abounding faith and clear vision are spending millions to make her the most beautiful and salubrious spot on the continent."[6] This is the developer's vision that would fill the bay's largely empty waters with islands.

DEVELOPER'S VISION

Dredging is an event. In the first part of the century, the swing of the dredger's boom measured progress, and the opening of Government Cut can be said to have initiated the rapid transformations of Biscayne Bay in the three decades that followed. In 1902, the Committee on Rivers and Harbors approved the dredging of a nine hundred foot wide channel to expand the Port of Miami. The channel's new cut would sever Miami Beach's southern end to create Rat Island, also known at the time as Peninsula Island and later as Fisher Island. On March 14, 1905, a rapt audience—reportedly 3,500 strong—watched the dipper dredge excavating Government Cut seize up, its bucket wedged deep in the sand. Miami's Mayor John Sewell had called for a citywide holiday and invited citizens to witness the final stages of making the channel cut, but the broken machine left undredged a ten foot wide section of beach between bay and ocean. A photograph documenting the event shows the slackened arm and cables supporting its bucket, immersed in the bay's waters, which are significantly lower than the ocean's surf line to the right. Engineers' stakes line the unexcavated isthmus, and a crowd gathers along the sandy cliff. This scene is the moment before Sewell—undaunted and determined—takes off his coat, grabs a shovel, and proceeds with the dredger's work by hand.[7] While Sewell—not quite the Corps' vision of the sailor with a shovel, but more of a politician with a spade—was opening the cut, other parts of the bay were already being closed with dredging operations. Within thirty minutes, he completed a small but effective trench that would widen and signal the starting point of Miami's modern era of sea commerce and Biscayne Bay's process of human transformation.[8]

Henry Flagler's railroad line had arrived in 1896, carrying with it hotel and port development. Flagler opened the Royal Palm Hotel one year later and dredged a channel from Miami's port around Cape Florida to the Atlantic Ocean. For him, dredged material was the superfluous residue of a development process. For his successor Carl Fisher, spoils were a valuable commodity. Spoil islands deposited during Flagler's projects to deepen the Miami River's channel and the city's harbor entrance would wait fifty years before becoming Brickell Key, and the rest of the disused material would wait another twenty to be repurposed as part of the Dodge Island port.[9] Fisher, on the other hand, knew how to make land out of water. If Flagler constructed railroads and hotels, then Fisher built islands. To illustrate how the dredge became Florida's "emblem," Will Rogers told the following story: "Carl discovered that sand could hold up a Real Estate sign, and that was all he wanted it for. Carl rowed the customers out in the ocean and let them pick out some nice smooth water where they would like to build, and then he would replace the water with an island, and you would be a little Robinson Crueso [sic] of your own."[10] Rogers' folksy story presents a cautionary tale embedded within the telling of an American success story, foreshadowing this ephemeral real estate's rise and fall. In Rogers' manner of speaking, something cannot come from nothing, paradise islands are often desert islands—if not Robinson Crusoe's Island of Despair, prospective buyers might become castaways, and developers might also be pirates.

With the state's readiness to sell, submerged land had quickly become a valuable component in bay real estate ventures. Fisher's aspirations for island-building coincided with a 1913 Act that authorized Florida's Internal Improvement Fund (IIF) to protect sovereignty submerged lands under navigable waters. IIF already facilitated drainage and land reclamation projects on the mainland, and the Act extended its authority to what had always been subject to the Public Trust Doctrine. This state law moved away from previous concepts that these areas were exclusively in the public trust to an idea that the material could be bought and sold on a case by case basis with decisions by the Trustees of the IIF. The 1913 Act effectively facilitated the transfer of submerged lands for filling new land.[11] Although not directly affecting dedicated waterways and channels, this shift complicated the idea that these areas should remain free of impediments for unhindered navigation.[12] Fisher and other developers like John Collins, Thomas Pancoast, and John Lummus, could purchase bay bottom, apply for dredge permits, set wood piles to claim their territory, and proceed to pump material to make land.

Miami's residents largely saw the dredging fleet that filled Biscayne Bay as an indicator of progress. In the summer of 1913, the dredges—many of them a part of Carl Fisher's enterprise—provided an "index of Miami's prosperity."[13] Fisher's promotional advertisement for his development on Miami Beach's western bay shore conjures a magical approach that glosses the expense and labor of dredging and construction: "Transformed from a wilderness into a park, almost in an instant as though by the waving of a magic wand, Alton Beach today seems a fairy land, and the story of its development during the past twelve months reads like a romance, to which, however, 'finis' has by no means been written, for each day sees some new beauty added to it."[14] With this announcement printed in the *Miami Herald*'s December 10, 1913 edition, Fisher also links earth-moving and island-building with the process of beautification. In southern Miami Beach, workers cut down mangroves as dredgers cast spoil material over the stumps and tangled roots. Dredging substituted one waste area for another. To many, the eastern islands' natural accumulation of mangroves was an impediment to development—a "waste place" that could only be woken from its unproductive dormancy by "skillful manipulation." For Fisher, transformation to usefulness would save these areas, but the initial result was another kind of wasteland. Dredged sand, mud, and marl—workmen called it "soup"—reeked with decaying sea life and vegetation. Photographs from the time show deserts of sand extending to a horizon active with the smoke and steam of the dredge fleet. This dredging project was mostly confined to the augmentation of land already begun by the mangroves. But material left by the dredger *Davis* spawned one of the bay's first spoil islands that would soon become its first dredge and fill island.

Biscayne Bay has a long history of the consolidation of spoil material to create purpose-built islands. Fisher began the first such project with Star Island in March 1917, when the Secretary of War issued a permit to John H. Levi, project engineer and president of Ocean View Company. A month earlier, Dade County had begun construction of the County Causeway that would eventually provide access to the island, which today is a public neighborhood despite appearing to be a limited access development—an image strengthened by the simulated guard house installed along the entrance road. The spoil that went to form Star Island had come up for auction during the previous summer's meetings of the IIF Board as a partially submerged tract of land with a "general circular form."[15] Discarded spoil was the raw material for the bay's first wholly artificial purpose-built island. This practice of consolidating previously unused spoil continued with Lummus and Dodge Islands, which had emerged from dredging Government Cut and the main shipping canal.

> A tourist asks, "What's the name of this spoil island here?"
> "Has no name," answers a bureaucrat.
> "Why not name it Dodge Island," suggested Dodge.
> "OK."[16]

When Robert Waldron, onetime director of Miami's port, recounted the origin of Dodge Island's name, he touched on both the anonymity of spoil islands and the way their informality at the same time attracts multiple names. Before it was built and opened as Miami's new port facility, Dodge Island was a string of spoil islands deposited with the dredging of the main ship channel that passes through Government Cut. In a more specific version of the story, Miami Planning Board's director Frank Stearns named the island for his friend Ray Dodge, who was visiting from Wisconsin and touring the city's infrastructure in 1950.[17] In another account, the name's history extends back to one of the original Government Cut spoil areas named for Dr. R. L. Dodge, a member of the Port Authority.[18] Consolidation of this site along with the other spoil islands yielded the world's largest cruise ship terminal, so that designating the once nameless island for a tourist—the everyman tourist who needs to register the place with a name—makes sense.

Critics initially saw plans for the new island as a form of planned obsolescence that would be outdated in five years and as a potential eyesore further congesting this already crowded section of the bay. Supporters of the project believed that the port's shift from the city's downtown bay front would improve the skyline and that this plan was the most economical solution.[19] The final plan followed John Orr's 1925 idea for a massive island project founded on spoils from the main channel and the turning basin but requiring additional, extensive dredging.[20] In spite of the plan's support, another forty years transpired before Dodge Island Seaport was officially opened on October 4, 1964. Within twelve years, more than one million passengers were traveling through the port. This traffic accounted for one-third of the world's cruise ships by 1982. Now, more than four million visitors arrive along the rectilinear edges of these resolutely purpose-built islands without any ambiguities to their name, status, or purpose.[21] No ragged or unformed edges, unless an inquisitive visitor happens to look north past the Coast Guard's rectangular Causeway Island, between South Miami Beach's faceted edge and Star Island's rounded corner, to glimpse the eroded mass of Flagler Memorial Island.

FLAGLER ISLAND

After seven years of digging and filling and not wanting to waste any of this valuable residue, Fisher prepared his monument to Flagler. The northern industrialist had died the day after Fisher's first dredge arrived in the bay, and the inheritance was not lost on Fisher.[22] Surplus dredge material—presumably privately owned—became the foundation for a public monument funded by an individual donor. Perhaps Fisher saw himself returning some of the spoils of development for public consumption, infused with didactic messages of progress. The memorial island to Flagler triangulated visual connections within Fisher's real estate empire—Flamingo Hotel to the east and the strong vertical feature of Star Island's water tower to the southeast. Fisher exhausted his funds dredging the bay, removing mangroves, and filling wetlands and mangrove stands, to develop bay-front land and to speculate with projects like the recently completed hotel. But, in his desire to venerate Flagler, he also underestimated the power of the bay's waters. Despite the strength of its planning, Flagler Island, memorial to the indomitable industrialist and symbol of a city's progress and affluence, was built without sea walls. Without funds for the solid bulkheading enjoyed by Fisher's other developments like Star Island, Flagler Memorial Island's construction would soon return to its previously amorphous form, sculpted by the bay's currents.

The island, originally a perfect circle, is an allegory of development. News reports praised Fisher's monument after its completion in 1921, when the bay's watery expanse was already becoming a manufactured landscape. The memorial could be seen from Collins Bridge—soon to be known as the Venetian Causeway—and from Fisher's Flamingo Hotel. As one news article noted, the obelisk commanded "an imposing view of the bay."[23] From the water, it anchored the bay's increasingly active recreational space. Its placement served as a useful navigational reference for boaters, particularly during Fisher's well-attended sailing regatta, an event that also benefited from the deeper bay created by the developer's dredging activities. News articles about the memorial island's construction and inauguration read like engineering reports, exhaustively describing the monument's dimensions and building materials. Quantification precisely tracks commemoration: three-acre island, four hundred feet in diameter, $125,000 budget, 110-foot high obelisk resting on a slab thirty-five feet square, five and one-half feet thick reinforced concrete with white cement finish, shaft to be twenty-eight feet by twenty-eight feet at the base and two by eight feet at the top, 253 piles driven to support the base, and sculpted figures measuring eighteen feet and weighing twenty tons.[24]

After dimensioning and celebrating the monument's form, news reports unpacked the meaning of the four figures sculpted by Ettore Pellegatta and H. P. Peterson and built by John B. Orr.[25] Each figure is oriented toward a cardinal direction and alludes to an aspect of Flagler's legacy. Their gaze links symbolic meanings to the geographical locations of events that influenced the bay's development, which was reaching a highpoint during the memorial's construction. The sculpture also genders its axes. Analog to the vertical obelisk itself, the north-south axis is male. Industry and Pioneer anchor the perceived strength of commerce and development made possible by Flagler's coastal rail system. This direction has supplanted the feminized east-west vector, no longer one of frontier expansion but now Education and Prosperity.

On the north side, the figure of Industry looks resolutely northward along Flagler's railroad toward New York and the northeast's storied traditions of manufacturing and production. Mapping the extent of this progress, a scroll in the sculpture's right hand designates the east coast of Florida as the current region for the project's continued expansion and—by expanding the map to include Key West—captures the continental edge to suggest that values of diligence can extend beyond the mainland, decisively island-hopping just as Flagler's railway traveled across the keys. The figure holds a model of this technological success—the overseas railway—in his left hand, but also grasps an engineering level and a deep-sea measuring chain. These tools of construction serve builders and engineers on land as well as across water, and they are reminders that dredging displaces earth as much as it deepens channels.

If this northern gaze marks limitless industrial progress, then the southerly figure engages the mythic theme of the Pioneer, shading his eyes from the sun and casting an expansive gaze over the bay's growing artificial landscape and its newly dredged archipelago, with its first—Star Island—in the foreground, and the widening waters of the southern bay in the background. "Flagler" is etched under the pioneer's form, and here the tools do not reflect conventions of the pioneer myth. Dressed as a mariner, the figure does not shoulder a rifle but instead hoists a shovel. So, it is the sailor with a shovel who will continue Flagler's project of southerly progress. The sailor navigates aqueous terrain rather than the solid expanse of the frontiersman's American interior. This landward frontier had effectively closed more than two decades previously with the U.S. Census' official 1890 statement about the land frontier's disintegration, but it remained an important American idea illustrated by Frederick Jackson Turner in his 1893 speech "The Significance of the Frontier in American History."

On the waterborne frontier, land must be made. With the help of sailors with shovels, developers of this time might have imagined the bay's bottom and the seabed itself as unclaimed territories rivaling the expansiveness—and mined richness—of the American West. This artificial island memorializes Flagler and his dream of infrastructural connectivity, but it also celebrates the developers' foresight and makes tangible Fisher's own vision for making new land out of what is not visible. If America's western pioneers expanded territory simply by seeing and surveying, the nation's developers, particularly those in Florida, sought lucrative territories in the unseen. Submerged lands were the new frontier, and dredged islands were gold.

The figure of Education looks east, not back toward Miami but to the newly dredged grounds and recently platted lots of Alton Beach and Miami Beach. Here, the sculpture identifies the plans of Carl Fisher with what early commentators called Flagler's legacy of "intelligent development" amidst "Florida's rapid growth."[26] This figure faces the Flamingo Hotel, but it also looks further afield, across the Atlantic to European traditions of art and city-building and to Venice for lessons on island-building. The sculpture welcomes visitors arriving through Government Cut with a sophisticated visage enlightened by the Book of Knowledge held by the figure who solidly treads on the book of ignorance. To the west, the figure of Prosperity beckons to Miami's residents and tourists arriving on Flagler's railcars with a sculpted cornucopia representing the boundless potential of a nature now overflowing and made accessible by human progress. Spoils from dredging had found purpose as the spoils of development, and the memorial celebrated the time's profligate wealth and its artificial bounty.

The four figures stand at the base of the central obelisk, now dwarfed by Miami's towers and engulfed by the port's development. When it was completed, this vertical element was prominently visible from causeways, hotels, and boats navigating the channels. From the air, the circle of land looked like a lily pad, resting just above the buoying tension of the water's surface. The artist Christo would later form his own lily pads with the fabric fringes surrounding the bay's spoil islands. When Carl Fisher announced plans for the Flagler Memorial Island on June 17, 1919, construction of the Venetian Islands was proceeding rapidly, and augmentation had transformed the tidal flats of Bull's Island into a thirty-acre landmass already partially platted as Belle Isle. The nearly homonymic transfer from "Bull" to "Belle" is indicative of the way Fisher recast the bay's landscape both topographically and toponymically. With geomorphic wordplay, the developer reigned in unoccupiable wilds, seemingly overrun with mosquitoes and mangroves, and he gave a name to the "improvement" of this irrational environment that was deemed inadequate for Miami's future and its new residents.[27]

The Venetian Islands carried forward the geometric island-building project Fisher had begun with Star Island. Between 1918 and 1925, five islands were strung along what had previously been Collins' pioneering viaduct. A 1925 advertisement called the islands the "gems of America's Mediterranean" and "man's masterpiece in development projects," set within a bay that was "nature's masterpiece."[28] Other developers and landowners shared this vision of Biscayne Bay as a Venetian lagoon, and James Deering's channel to his Vizcaya Villa is reminiscent of the Grand Canal. With such ongoing projects as a backdrop, Fisher completed the Flagler memorial before the Venetian Islands, and it is the earlier project that has more in common with Venice's visions for its own lagoon.

VENICE (1920)

> Tra-la-la-la-la-la-laire—nil nisi divinum stabile est; caetera fumus—the gondola stopped, the old palace was there, how charming its grey and pink—goats and monkeys, with such hair too!—so the countess passed on until she came through the little park, where Niobe presented her with a cabinet, and so departed.[29]

> I made one swift summer visit to Europe without Carl . . . and from Venice I shipped him a red, black and gold gondola, complete with fringed pillows and the costume for a gondolier.[30]

Some islands are sinking; others are artificially rising. In 1920, as T. S. Eliot was completing his poem of Venice and Fisher concluded his homage to Flagler, Jane Fisher returned with a gondola as a souvenir for her developer husband. Italy's Venice crumbles—Eliot's "rats underneath the piles" of "protozoic slime"—while Florida's Venice emerges from millennia of sand and marl. It was not uncommon for Florida's developers to invoke Venetian images to describe their real estate projects. The city of Venice on the state's west coast, the intricately canalized areas of Ft. Lauderdale, and the unearthing of Venetian islands from Biscayne Bay all referenced the Italian place by name, sometimes by morphology, and at other times with thematized accessories that made for developments like theme parks. Fisher, Collins, and Pancoast knew the romantic allure of Venice. Advertisements and news articles cited the "Modern Venice," "Venice of America," and "New Venice of the World" to describe Biscayne Bay's artificial islands. This focus on developing the bay in the 1920s translated its waters into a grand urban space without precedent in Florida's built environment. Developers made the link with Venice for its perceived sophistication, its cultural traditions, and its identifiable image base. They might have known of Venice's origins in the human-made augmentation of natural islets, but their work also unwittingly touched on debates about Venice's Renaissance development. Those contentious discussions still serve to frame the relation between technology, nature, and aesthetics in the Venetian lagoon today.

TWO PERSPECTIVES

On Valentine's Day in 1922, the Bay Biscayne Improvement Company devoted a full-page in the *Miami Daily Metropolis* to advertise Venetian Islands real estate. Two headings present the project. "Venetian Islands in Biscayne Bay" stretches in capitalized letters above the redundant but tersely direct line announcing "The Venice of America—The New Venice of the World." Symmetrical blocks of text describe the location, prices, and lot sizes with a heraldic tone. In the center of the page is a plan view of Di Lido Island that shows the platting and property numbers of the lots for sale. At the top of the page, two gilded images flank a central, vertically formatted text block that seems to undermine the inclusion of these renderings but still serves to identify the project and its context with Venetian artistic traditions: "No painting or reproduction can ever hope to picture the glorious setting and incomparable beauty of the completed Venetian Islands, as they dot the surface of the sparkling waters of Biscayne Bay." That said, the advertisement still attempts to represent this vainglorious beauty. On the left is a birds-eye view of the four islands "as they will appear when completed in every detail," and to the right is the depiction of a typical home "set in the heart of Biscayne Bay." This pairing extends the developer's view, presented in the Di Lido plan, to illustrate the grand urban space and the ornate extravagance of the private villas—both representative of an increasingly human-made environment.

Both renderings have Venetian precedents. The view on the left of the advertisement can be compared to Jacopo de' Barbari's famous birds-eye map of Venice. Completed in 1500, Barbari's multi-block woodcut combined map-making and perspectival construction techniques to create a remarkable image of the city. The document was accurate and complete in its fine detailing, but Barbari also manipulated the image to present specific ideas about the city at the height of its Renaissance.[31] With multiple vanishing points, he emphasized spaces—lagoon, piazza, canal—rather than just locations and particular buildings. In particular, he was also able to increase the scenic value of the Grand Canal. Perspectival distortion straightens its banks for a clearer view along its course and its entry into the Bacino San Marco. Barbari has similarly smoothed and compressed the city's western shoreline. The manipulation of the image parallels revisions to the urban form during the Renaissance and indicates a consciousness of the role of civic space and an understanding of the city's malleability to match human goals.

City and image can be similarly molded. Not long before Barbari created his map, engineers straightened the Grand Canal, lined it with new facades, and smoothed awkward corners; and soon after the map was published, land reclamation expanded Venice's northern edge. Like Barbari's constructed image, Bay Biscayne Improvement Company's birds-eye map exhibits the developer's vision, and it also narrates human intervention by placing human-made forms within a naturalized context. This setting is compliant—it can be framed, composed, and ultimately sold. In the rendering, unruly vegetation masks Miami's urban center and provides a textured horizon to complement the bay's scalloped water that floods all but one-tenth of the image. Its perfect circle cropped by the frame, Flagler Island is overgrown with greenery at the base of its exaggerated obelisk. The Venetian Causeway, its elongated axis vanishing into a distant Miami mainland, crosses Belle Isle in the foreground and Rivo Alto, Di Lido, and San Marino islands in the composition's center. Biscayne Bay is a civic space of water and made land.

The image on the right of the Improvement Company's advertisement is reminiscent of Antonio Canaletto's eighteenth-century views of Venice. Canaletto was known for stitching pieces of the city together to give a sense of the place and its atmosphere, and the Company's artists have rendered a similarly hybrid composition. To the left, silhouetted figures stand on shoreline docks while others occupy small boats in a lagoon bounded by sea walls in the foreground, stately cypress trunks on the right, and Venetian palazzos along the center of the painting. Not only has land been made, but an image of Venetian life—on the water itself—has also been fabricated. Canaletto's paintings, such as *Veduta del Palazzo Ducale* (1735), combine the "minutest details" with "picturesque freedom," and the artist tunes specific aspects of topography, atmosphere, and edifice to the realities expected by his audience.[32] Artists for the real estate company have adapted a "boats-eye" view so that prospective buyers can visualize an approach to the island and imagine their arrival as if by gondola at a palazzo's water entrance. In this representation, water is no longer the background, but the substrate for viewing the islands and accessing the houses.

The boater's perspective also blurs distinctions between what is human-made and what is natural, or at least what appears not to be manipulated by humans. Captain Claiborne Young captures these indefinite readings in his piloting guide to the ICW:

> Boaters studying the various channels south of Broad Causeway will quickly note five causeways crossing the bay between the mainland and the eastern islands. Three of these spans are partially comprised of a series of heavily developed, artificial islands. It's easy to see from the chart that these landmasses are unnatural, but from the water they appear as small islands that have every right to be there. Actually, these isles are some of the most valued real-estate property in greater Miami. As chart 11467 indicates, deep water surrounds most of the various land bodies, and careful boaters are free to take a closer inspection. Please don't abuse the privilege by trespassing![33]

From above, most of the islands in Biscayne Bay's archipelago are quickly understood to be artificial, but on the water they read as natural horizons. What the map-maker—and the map-reader—might rationalize as unnatural, the boater still might view as the work of nature, with "every right to be there." The blurring of human-made and naturally-occurring insular form becomes even murkier with Biscayne Bay's spoil islands. Some of these have been augmented for purpose-built landmasses, but others—particularly ones that cannot be developed—have been considered only for more ephemeral, and sometimes fleeting, purposes. Most of these islands are irregular in shape and have been sculpted by currents and wave action.

From the aerial perspective and its angle of vision that might be used by a developer to plan a real estate venture, two-dimensional landforms can be rationally planned, measured absolutely, and quantified with confidence. Unnaturally straight edges and geometric configurations—such as the Venetian Islands' rounded corners—help distinguish what has been artificially modified from what has been left to the forces of waves, tides, and currents. Despite water-level correspondences between artificial and natural keys, reading differences between types of artificial islands is made easier by degrees of development. From a boater's perspective, dredge spoil islands typically offer undeveloped, sometimes overgrown, areas with open access.[34] Recreational enthusiasts consider the spoil areas to be anchorages for fishing, sunbathing or swimming. As Captain Claiborne's account in this section's epigraph suggests, these islands also signify public and private, as well as a tourism of celebrity. In contrast to the dredge spoil islands in the public domain, many of Biscayne Bay's purpose-built artificial islands accommodate residences that on the mainland might normally occur within the wealthiest of high-end gated communities. In the bay, they rely instead on the isolation and limited access provided by their aqueous context.

The bay forms a field condition where multiple viewpoints are at work. Drawing and land-making share a procedure in which islands are tools of representation and agents of planning. Artificial islands, spoil islands in particular, play an active role in this construction of civic space, influencing the image of the city and creating the setting for its recreational and cultural use. The plan's overview favors the real estate venture's landforms that developers can quantify with confidence—two-dimensionally platted, rationally planned, and measured absolutely. The birds-eye view complicates this rationality by turning the bay's water surface to meet its viewer and consequently casting island development's rationale within the broader environmental context, still controlled by rules of perspective but transformed by other factors of light, air, water, and vegetation. The boaters-eye view further obscures the human role in controlling the environment, allowing the boater as viewer to embark on a more personal exploration of the now communal space, where differences between what is human-made and what is natural matter less than the overall ecological functioning of the bay. Christo's project will pick up on these birds-eye and boaters-eye views to establish the centrality of the spoil islands that run along the bay's north-south axis, but a contrasting pair of sixteenth-century planning projects for Venice's lagoon sets the background for spoil islands as technological and cultural agents in choreographing art and nature.

VENICE (SIXTEENTH CENTURY)

More than three hundred years before Flagler and Fisher arrived in Biscayne Bay, two Venetians proposed alternative plans for their lagoon. Both men respected Venice's many histories and, like other Venetians, understood their city's tenuous position between earth and water. In the sixteenth century, amidst these debates about the city's development, the humanist Daniele Barbaro noted the temporal influences on the Venetian lagoon and its human-made works: "Time, which brings every advantage and disadvantage, works in concert with two elements to wage war against us and do us great harm; I am speaking of the sea and the earth, one of which seems to want to recede and the other to occupy the space of this lagoon."[35] Earth retreats, and water intrudes. Entropy resides in Venice, where equilibrium has long been disrupted by human occupation. In the thirteenth century, islands had already been lost under the sea. Venice's foundations are hidden by—and resolved on—water.[36] Buildings rest on submerged pilings that extend through layers of sand and mud toward more compressed clay, but even this substrate relies on water for support, and Venice's twentieth-century subsidence has been attributed to the lower-level aquifer's depletion. In Venice, hydrogeography and politics are unavoidably intertwined, and its commissioners—not unlike Florida's IIF Trustees—have been appointed to manage the protection and reclamation of land since the ninth century.[37]

In the middle of the sixteenth century, Cristoforo Sabbadino and Alvise Cornaro both understood the politics of water and earth, but they proposed divergent plans for maintaining Venice's lagoon and its canals. A Minister of Waterways and hydraulic engineer, Sabbadino proposed in 1557 the use of dredged material to reclaim land and augment the city's form. Three years later, Cornaro, a patron of the arts with an interest in hydraulics and reclaiming wetlands for farming, proposed to limit urban development and to use dredge spoil for altogether new urban forms, away from the city's existing land. Planning to extend the city, Sabbadino believed that Venice's urban landscape could accommodate both "naturalness" and intensive human-made production of land. Maintaining a fixed urban form, Cornaro sought a newer model that privileged technology's future over Sabbadino's prudence. If Sabbadino relied on nature's time-honored narratives, Cornaro wanted to invent new myths. Cornaro played with spoils, and Sabbadino wanted to put them to productive use.[38]

With his plan, Cornaro took a mainland perspective of the lagoon in contrast to Sabbadino's relative attunement to the lagoon's systems. Viewing the lagoon from the city's land, Cornaro operated from a humanistic, patently Renaissance, perspective with man at the center. His 1560 plan firmly placed *homo faber*—as the victor over nature—at the "center of an ideal cosmography."[39] Seeking to understand the lagoon's systems, Sabbadino worked from a more traditional—and pragmatic—lagoon perspective. He cited the earliest Venetians as having preserved nature through art, and he believed that preservation could be renewal. Both men accepted technology as a restorative agent against the vagaries of time and sea, but they had different views of how technology might work between man and nature and how it might influence the image of the city. For Sabbadino, technology could "restore corrupted naturalness," and for Cornaro, technology "did not possess its own language" but could be directed by men to produce artistic urban forms.[40]

Technology, here, emphasized the work of nature. For Cornaro, nature followed the terms established by artistic production.

Sabbadino's 1557 lagoon plan follows a logic that prudent intervention is restoration and that work to maintain the lagoon for human use can at the same time conserve its natural context. Sabbadino took a broad approach to the project of land augmentation, viewing the process of dredging as part of an urban system, a network of infrastructure much like the canals and channels themselves. In this model and its related technological processes of clearing channels, new land emerges from cleaning the lagoon. Opposing dirty to clean, Sabbadino's plan relies on rigid boundaries between earth and water and between waste and nature; and Manfredo Tafuri has noted that this strategy anticipates modern instruments of urban planning that separate different programs.[41] The plan also foreshadows modern attitudes—such as those formalized in Corps of Engineers projects—that dredging might serve not only to maintain transportation networks but also to create natural environments. The plan parallels recent conservation strategies working from the assumption that intervention can return waste sites like spoil islands to a natural state without exotic vegetation. Restoration and enhancement projects like a 1999 plan to protect Biscayne Bay's spoil islands seek to stabilize shorelines and install native vegetation.[42] Sabbadino's productive work contrasts with the more playful and synthetic approach of Cornaro, who was not satisfied with nature as the only programmatic context. He wanted entertainment.

In contrast to Sabbadino's systemic approach, Cornaro focused on a localized, more theatrical, place within view of the city's grand piazza. Two spoil banks were the targets of Cornaro's attention and the cosmographic nexus of his new mythology. Augmented with centuries of dredged waste from the lagoon, one shoal was only five hundred feet southeast of the Doge's Palace, off axis from Piazzetta San Marco and spanning an area of the Bacino between the Rio Palazzo and the Rio del Vin. A second shoal lay a quarter of a mile to the west, in the center of the canal between Santa Maria Della Salute and La Giudecca island. Cornaro's plan transformed these spoils—invisible but just below the surface—into two highly visible islands, "spectacular objects," that could emerge out of the residue from Sabbadino's more visually neutral, productive process. Cornaro referred to the one closer to San Marco as a "shapeless little hill," and the other was to be a hemicycle amphitheater.[43] A fountain placed between the piazzetta's two columns completed the triangular relationship and tied the islands to the main urban space of the piazza. Both islands were to serve as sites for training citizens in virtue—one through the experience of theatrical productions and the other, at a distance, as an object of contemplation seen from the piazzetta and as a site to look back to the city. The theater island took the form of a hemicycle for performing plays. With a loggia at its highest central elevation, the "shapeless little hill"—sized to match the piazza's open space—was a highly contoured, irregularly shaped mound that could be seen from the piazzetta, as an extension of Venice's central urban space, and from the main port passage, as a didactic vanishing point that would inspire mariners, tourists, traders, and citizens.

With his islands, Cornaro gave shape to the technical operations of dredging. He resolved one of the logistical and technical issues of dredging—what to do with the leftover material that was an impediment to transportation—by using it to create an object for "aesthetic contemplation."[44] Cornaro's plan went beyond mere removal—the core objective of Sabbadino's restoration efforts—and became a spectacular focal point, however mundane its origins might have been. Calling the hill a "hill" accepts the residual and everyday qualities of this earthen residue. Calling it "shapeless" acknowledges the forces of nature that can shape it as well as the cast-off nature of the material. It is unformed because it is waste. "Little" though it is, it remains visible, and its scale belies its purpose to capture a visual effect. Cornaro saw the island as incomparably beautiful because it was a work of art made by human skill and because it was carefully placed in the lagoon.[45]

Echoing a developer's zeal, Cornaro could be selling water-laden lots in Biscayne Bay, alongside Carl Fisher, when he describes his project:

> And this will be done with little expense, and very conveniently with the dredged soil, and mud that will be taken out of the Canals, and the hill will be planted with trees, and its streets will be made beautiful, and it will be a very beautiful place to go for amusement, and at its summit there will be a loggia open on every side and protected from the sun; and one will see this hill standing in the piazza, where there will also be a fountain of fresh flowing water; and thus in one glance one will see fountain, hill, and theater, and between them many large ships, which will then be able to enter port; and this will be a spectacle, and more beautiful, more charming . . . different from any other that has ever been seen, or that one will be able to see in the future in the whole world.[46]

Views from the "shapeless little hill" convert the lagoon to a stage for the "many large ships" and the city to a theater set. This "charming," unparalleled spectacle didactically places the human subject—the viewer—at the center of an orchestrated landscape. The island, fountain, and theater triangulate a larger theater where human agency plays out. For Cornaro, only human intervention and its resultant island, which is no longer waste but itself the focus, can make this happen. Fisher similarly places the mythic figure of Flagler, as *homo faber*, at the center of a nation's burgeoning development and positions the memorial island in the center of the bay that was rapidly becoming the pivotal zone of development, industry, and transportation.

Just as Cornaro's artificial islands provided focal points for the Venetian lagoon, Fisher's plan included what Tafuri has called—in the Venetian context—a "caput ad triangulum." In Venice, the geometric triangular form was complemented by a triad of wonders: natural elements—the fresh water and wooded area of the hill, artificial elements—the islands themselves, and an antiquarian element—the theater.[47] Working in a bay that human intervention converted into an artificial lagoon for an upstart Venice, Fisher triangulated his own didacticism and theatricality between Flagler Memorial Island with its sculptures and obelisk, the artificial Star Island with its monumental column serving as a water tower, and the Flamingo Hotel on Miami Beach's larger body of natural land. These connections demonstrate that the bay is no longer empty space and that it has unlimited use value for private development and for civic identity. Although the artificial islands have been placed in unnatural locations, they offer specific connotations for the metropolitan area of Miami and Miami Beach. The islands are not just engineering marvels—they are the wonders of Biscayne Bay itself.[48]

Cornaro's project offered public spectacle. The theater was to be a setting for dramatic entertainment, but its island form and placement created another spectacle on the stage of the lagoon: ". . . it would itself become a spectacular object, while its placement in the lagoon would have endowed it with phantasmagorical and estranging characteristics."[49] Christo's project to surround Miami's spoil islands will have a similarly defamiliarizing effect on Biscayne Bay's aqueous public space. Also, there is an aesthetic impulse associated with the augmentation of something that is pre-existing but not completely visible. Cornaro amplified a well-placed shoal to create one vertex in his triangulated spectacle, and Christo will expand the size and shape of Biscayne Bay's spoil islands with pink fabric. Like Christo's, Cornaro's project extended the urban planning of the spectacle beyond a singular object to include an entire system, an archipelagic composition of diversely formed elements made possible by shapeless little islands.

AFTERIMAGES

A few decades before Cornaro submitted his plans for the "shapeless little hill," Venetian painters had already placed imaginary, often wooded, islands in the Bacino San Marco. These painted islands transform the lagoon into an ideal garden and extend the storied mixing of land and water in Venice's urban identity.[50] In 1516, Vittore Carpaccio painted *Leone di San Marco* for the State Treasurers. Carpaccio knew Venice and he knew the lagoon, having been born on a small island in its waters.[51] In the painting, the Lion of St. Marks, symbol of the city of Venice, stands with two back feet in—actually on—the lagoon and two front feet over land—one resting on an opened book and the other on the island's sandy shore. This imagined island occupies the lower left corner of the composition, and dark tree limbs strengthen the left edge of the frame and contrast with Piazza San Marco's campanile in the distance. To the right of the lion, ships rest comfortably on the water's horizon; some are anchored and others have full sails. The island represents land, and the lion's position—appearing to move from water to land—further suggests that this is emergent land, having recently been made. It is also important that the lion's position is our own position as viewer of the painting and viewer of Venice. We are on the island with the symbol of the city. The artificial, imagined island is the idea of a city that could not exist without human agency, without the art or technology. In 1512, Sebastiono del Piombo also painted an island into the Bacino. In *Morte di Adone*, the island is larger, and its land and the events of Adonis' death occupy the majority of the composition. In the background, across the lagoon's polished waters, the Doge's Palace and Piazza San Marco's partially complete campanile float above Adonis' prone body. A cluster of six figures mourn their loss next to dark trunks under the shade of a mature tree canopy. Less about the politics and commerce of land-building, here, artificial land provides a distinct, mythological setting to narrate the event. This distance suggests the escape to another nature afforded by human-made, painterly landforms.

Biscayne Bay's spoil islands also offer escape from the city. They allow island-goers a necessary distance to look back at the city of Miami. Even without Cornaro's deliberate theatricality, Carpaccio's didacticism, or del Piombo's mythology, these islands—which will be the focus of the next chapter—offer antidotes to purpose-built islands' private and commercial uses. They also capture the romance of Venice. The story of the *Leona Pearl* links the boater's perspective with the sense of escape afforded by the spoil islands. In the early 1990s, this 36-foot, Venetian gondola disembarked from Bayside across from Dodge Island, dodged the wake of cruise ships, and carried couples over the bay's moonlit waters. With taped music described as Barry White's version of *Volare* playing in the background, gondolier Karl Ruhnke steered couples along the ICW to candlelit picnics on one of the bay's spoil islands.[52] With champagne, wine, Italian cheeses, stone crabs, lobster, and Swiss chocolates on a blanket around a campfire in the glow of Miami's skyline a half mile to the west, this was a popular excursion for tourists and residents alike. In at least one case, the couple returned to be married on the island. Festooned with pink carnations, the ebony gondola found Venice not in the sculpted canals of Coconut Grove or along the rounded bulkheads of Rivo Alto but on the ragged shores of the picnic islands.

NOTES

1. Anne O'Hare McCormick, "Miracle Men on Florida's Gold Coast," *New York Times*, 8 March 1925, SM3.

2. Peter Wayne Harlem, *Aerial Photographic Interpretation of the Historical Changes in Northern Biscayne Bay, Florida: 1925 to 1976*, Sea Grant Technical Bulletin 40 (University of Miami, 1979), 106.

3. Karl Baedeker, *The United States: Handbook for Travelers* (New York: Charles Scribner's Sons, 1904), 453. Baedeker continues with a note about dredging: "The bay is now being dredged and improved by the U.S. Government, and it is expected that vessels drawing 17 ft. of water will be able to enter the new harbor."

4. Will Rogers, "Carl Took Florida from the Alligators and Gave it to the Indianians," *Will Rogers' Weekly Articles*, volume 2 (Stillwater, OK: Oklahoma State University Press, 1979), 88. Rogers' weekly essays appeared in the newspaper *The Tulsa Daily World.*

5. J. I. Conklin, "Dredging Now Going on in Biscayne Bay," *The Miami Herald*, 7 August 1913, 17. The article included a panoramic photograph of the bay with yachts in the foreground and dredgers in the background.

6. Conklin, 17.

7. Arthur Chapman, "Watch the Port of Miami," *Tequesta* 53 (1993): 11. The title of Chapman's article is drawn from Henry Flagler's phrase following his early development of railroad and port facilities in 1896 and 1897. Legend has it that Julia Tuttle sent him a package of orange blossoms to invite him, along with his acumen for development, to Miami. Flagler dredged his own channel with a twelve-foot depth around the Florida Cape to the Miami River.

8. See Howard Kleinberg, *Miami Beach: A History* (Miami: Centennial Press, 1994), 23, for an account of the story and a photograph of the event.

9. The two spoil areas at the mouth of the Miami River were created in 1896 when Flagler's crews dredged the river's nine-foot channel. Toner notes that additional dredging activity in 1923 augmented what was then known as Burlingame Island. See M. Toner, "Bay's first surveyor wouldn't recognize how it looks today," *Miami Herald*, 22 April 1979, 19-A. See also NOAA, *Biscayne Bay: Environmental History and Annotated Bibliography*, Technical Memorandum 145 (Silver Spring, MD: Center for Coastal Monitoring and Assessment, July 2000). In 1943, Edward Claughton bought the islands along with adjacent submerged bay bottom so that he could join the land for what would become Brickell Key. See D. C. Gaby, *Historical guide to the Miami River and its tributaries* (Miami: Historical Association of Southern Florida, 1990). See also the *Minutes of the Trustees of the Internal Improvement Fund* 27 (Tallahassee, FL: 1950), for additional records related to problems with the transfer of land to Claughton and his proposals to place spoil on the original island and expand its size fourfold.

10. Rogers, 88.

11. The Act became effective June 5, 1913, Chapter 6451, 1913 Fla. Laws 122. See *The Compiled Laws 1913 of the State of Florida* (St. Paul: West Publishing Company, 1914). See also Sidney F. Ansbacher and Joe Knetsch, "Negotiating the Maze: Tracing Historical Title Claims in Spanish Land Grants & Swamp and Overflowed Lands Act," *Journal of Land Use and Environmental Law* 17 (Spring 2002): 351–374. Ansbacher and Knetsch note two effects of this Act: first, it "authorized the Trustees to convey islands and other property that was not adjacent to any privately owned riparian lands," and second, it "led, unsurprisingly, to further development of, and off of, barrier islands." (357) The authors have also published another article that provides background on the Public Trust Doctrine: "The Public Trust Doctrine and Sovereignty Lands in Florida: A Legal and Historical Analysis," *Journal of Land Use and Environmental Law* 4 (1989): 337–351.

12. See Ted Steiberg, "Do-It-Yourself Deathscape: The Unnatural History of Natural Disaster in South Florida," *Environmental History and the American South: A Reader*, eds. Paul Sutter and Christopher Manganiello (Athens: University of Georgia Press, 2009), 400–431.

13. Conklin, 17.

14. Fisher's advertisement is cited in Mark Foster, *Castles in the Sand: The Life and Times of Carl Graham Fisher* (Gainesville: University Press of Florida, 2000), 164–165.

15. June 16, 1916, *Minutes of the Trustees of the Internal Improvement Fund* 11 (Tallahassee, FL: 1917), 286.

16. Fredric Tasker, "Dodge Island named after tourist," *Miami Herald*, 28 January 1983, 1E.

17. "The Landmark Name Game," *Miami Times*, 25 July 1973, 1C.

18. Chapman, 18. There were two main spoil areas created by the dredging of Government Cut and the main ship channel. Dodge Island was originally a cluster of the westernmost islands, and Lummus Island was on the eastern end of the main channel before it reached the cut. The two islands have been joined to form a continuous port facility more than two miles long. Chapman drew information from the following sources: Juanita Greene, "New Port Waits at Crossroads," *Miami Herald, 26* April 1956; "What's To Do About Miami Port," *Miami Herald,* 14 October 1956; "City Faces Challenge on Seaport," *The Miami Herald,* 19 April 1959; and Paul Einstein, "Port Battles Slug In Open," *Miami Herald,* 8 May 1959.

19. Juanita Greene, "What's to do about Miami Port?" *Miami Herald*, 14 October 1956, 1B.

20. "Council Ready to Push Harbor under Orr Plan," *Miami News*, 17 November 1925, 1, 23. The article notes that the IIF board offered the sale of bay bottom for one thousand dollars per acre.

21. "Port Statistics," Miami-Dade County and Port of Miami, accessed June 1, 2012, http://www.miamidade.gov/portofmiami. The authorization of the "Deep Dredge" project in 2011 planned for fifty-foot depths for the Port of Miami by 2014 to coincide with the Panama Canal's expansion and deepening. See the Port of Miami website: http://www.miamidade.gov/portofmiami/deep_dredge.asp.

22. See Polly Redford, *Billion-Dollar Sandbar: A Biography of Henry Flagler* (New York: Dutton, 1970). Flagler died on May 20, 1913, after a fall down Whitehall's stairs in Palm Beach.

23. "Flagler Shaft is Imposing Figure Monument to the Memory of Great Builder Who Developed Florida East Coast is Work of Art," *The Miami Herald*, 16 April 1921, 5.

24. The following articles discussed the monument's construction and completion: "Erect Memorial to H. M. Flagler," *Miami Herald*, 18 June 1919, 1; and "Flagler Shaft is Imposing Figure Monument to the Memory of Great Builder Who Developed Florida East Coast is Work of Art," *Miami Herald*, 16 April 1921, 5.

25. Both Pellegatta and Orr previously worked at Vizcaya, James Deering's estate and gardens along the bay's western shoreline. Pellegatta was an Italian sculptor who sculpted the lions guarding the New York Public Library's entrance and Mercury on the front façade of Grand Central Station. In Miami, many consider his masterpiece to be the figures on the bow of Vizcaya's stone barge on the bay's shore. Orr, who had trained as a mason in his native Scotland, was Vizcaya's contractor, where he coordinated the work of its artisans and builders. An expert on stucco and plastering, Orr was also active in the American Concrete Institute, and in 1917 he presented a paper on "Artistic Stucco," at the ACI's 13th Annual Convention. See American Concrete Institute, *Proceedings of the Thirteenth Annual Convention*, volume XIII (Chicago, February 8–10, 1917): 262–274.

26. "Flagler Shaft Being Erected," *Miami Herald*, 8 March 1921, 7.

27. See Foster's account of mosquitoes and mangroves in *Castles in the Sand*, 160.

28. See "Venetian Isles," *Miami Daily News*, 15 November 1925, 13. On the two-mile stretch of causeway, the following sites make up the Venetian Islands (from west to east): Biscayne, San Marco, San Marino, Di Lido, Rivo Alto, and Belle Isle. Eleven more islands were planned. In 1925, the full-page advertisement in the *News*' Sunday edition included a plan that showed a line of additional islands connected by the "Drive of the Campanili" and extending northward from Di Lido up the middle of the Bay. Pilings for one of the planned but never-built Venetian Islands are still visible in the Bay's waters.

29. T. S. Eliot, "Burbank with a Baedeker: Bleistein with a Cigar," *The Waste Land and other poems* (Buffalo, NY: Broadview Press, 2011), 44–5. Eliot opens another 1920 poem "Sweeney Erect" with similar images of coastal waste: "Paint me a cavernous waste shore / Cast in the unstilled Cyclades, / Paint me the bold anfractuous rocks / Faced by the snarled and yelping seas" (46). Two years later, his poem "The Waste Land" will begin with April as the "cruelest month, breeding / lilacs out of the dead land, mixing / Memory and desire . . ." (63).

30. Jane Fisher, *Fabulous Hoosier: A Story of American Achievement* (New York: Robert M. McBride and Co., 1947), 146. Fisher also writes: "Carl was as enchanted as a child with a new toy. He had Edward Purdy, his boatbuilder, make copies of this gondola. Painted gold and circus-red, and carrying musicians with guitars and mandolins and sweet-voiced singers in Venetian costumes, these gay craft drifted though the canals and the lagoons of Miami Beach. The little sailboats for hire on Biscayne Bay were set with red, henna and turquoise sails. Italian troubadours in gay costumes strolled around the Roman pools, singing and playing." (147)

31. See Deborah Howard, "Venice as a Dolphin: Further Investigations into Jacopo de' Barbari's View," *Artibus et Historiae* 18 (1997): 101–111. Howard makes the case for Barbari's manipulation of the birds-eye view to reflect the form of a dolphin.

32. Detlev Baron von Hadelin, "Some Drawings by Canaletto," *The Burlington Magazine for Connoisseurs* 49 (December 1926): 310–317. Details made his paintings deceptively realistic. Katharine Baetjer has noted that a "Canaletto image was a carefully adjusted simulacrum that the viewer read as an accurate transcription." See "Canaletti Painting: On Turner, Canaletto, and Venice," *Metropolitan Museum Journal* 42 (2007): 169.

33. Claiborne S. Young, *Cruising Guide to Eastern Florida* (Gretna: Pelican, 1996), 491.

34. Navigators speak of charted and uncharted spoil banks, pointing toward the constantly shifting nature of these areas.

35. Manfredo Tafuri, *Venice and the Renaissance*, trans. Jessica Levine (Cambridge: MIT Press, 1995), 140. This statement (cited by Tafuri) was a part of Barbaro's commentary on Vitruvius, whom he translated into the Italian.

36. Peter Ackroyd, *Venice: Pure City* (New York: Doubleday, 2010), 8. The lost islands were Ammiana and Constanziaca.

37. Ackroyd, 15.

38. Tafuri, 155.

39. Tafuri, 143.

40. Tafuri, 140, 156.

41. Tafuri, 141.

42. See Gary Milano, "Island Restoration and Enhancement in Biscayne Bay, Florida," Miami-Dade Department of Environmental Resources Management (DERM), Miami, Florida, 1999.

43. Tafuri, 152.

44. Tafuri, 151, 156.

45. For comparison, Christo noted the aesthetically pleasing placement of the spoil islands in Miami's Biscayne Bay, and his treatment of the bay's spoil islands also temporarily made them "spectacular objects."

46. Tafuri, 150.

47. Tafuri, 153.

48. Perhaps realizing the civic quality of this endeavor—or just seeking to divest himself of its maintenance, Fisher turned the Flagler Memorial Island over to the city in 1939.

49. Tafuri, 148.

50. See Tafuri's discussion of the Arcadian landscape that Cornaro sought in the lagoon (153, 157).

51. Frederic Chapin Lane, *Venice: A Maritime Republic* (Baltimore: Johns Hopkins University Press, 1973), 213. Also see Tafuri's brief summary of the island motif in Venetian painting (157).

52. John Barry, "Hedonism, two exotic cruises allow you to indulge yourself," *Miami Herald*, 8 April 1992, 1E.

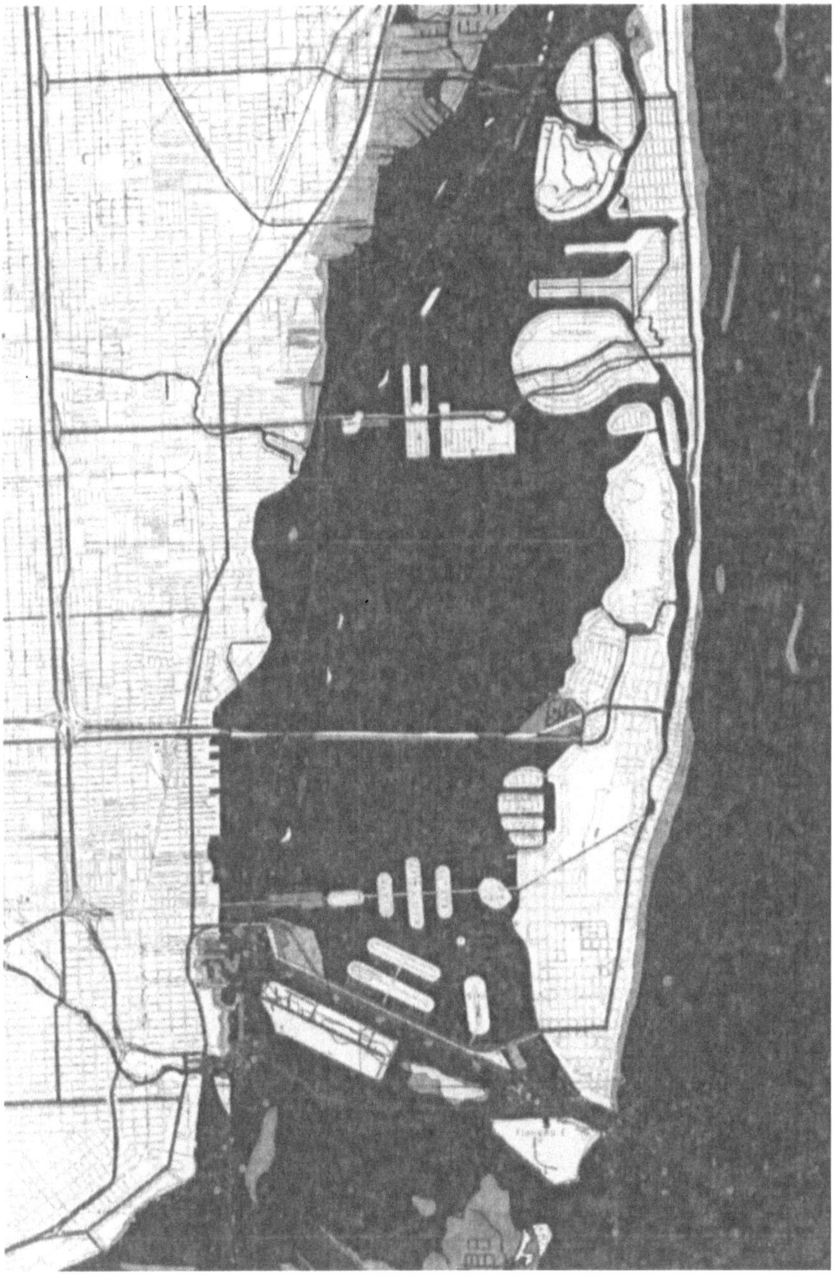

Spit

I wanted to create a whimsical, surreal experience. It's out of the everyday for the boater.[1]

On January 2, 2011, Nicholas Harrington hoisted a charred baby grand piano onto his family's twenty-two-foot Open Fisherman boat. With the help of his brother and neighbors, the sixteen-year-old Miami resident deposited the musical instrument on the highest ground of a spoil area less than a quarter mile east of Miami Shores. Thirty years earlier, Christo wrapped the spoil islands to the north and south of this area in his *Surrounded Islands* project. Hoping to garner attention for his college application to art school, Harrington described his placement of the piano as a movement rather than a prank and cited Christo and Banksy as inspiration.

For more than three weeks, the "piano on the sandbar" garnered attention from international media outlets and remained an enthralling mystery. Suzanne Beard photographed the piano, covered with pelicans, for *National Geographic*. The *Miami Herald* sponsored a caption contest, with entries including "You can tune a piano but you can't tuna fish" and "Play it again sand. . . ." Two hours before it was removed on January 27th, the Miami Mermaids, a local performance troupe, positioned a candelabra on the piano, draped it with pink fabric, and carried out a photo-shoot for their promotional literature. When independent filmmakers William and Anais Yeager claimed credit for the project, Harrington came forward, and the mystery was solved.

The Coast Guard had released an earlier statement that they did not plan to remove the piano unless it became an "environmental threat."[2] Without a single point of responsibility for its removal, the piano lingered. After calling the *Herald* to claim authorship, Harrington was notified by authorities—from the Florida Fish and Wildlife Conservation Commission, Miami-Dade

County marine patrol, and Florida Department of Environmental Protection—that placing the piano on the sand spit constituted a felony. For them, the piano was garbage, and because it exceeded five hundred pounds, dumping it was a criminal offence. For others, the piano spawned not only a mystery but a game of associating the piano's unlikely presence with Miami's "weirdness"—one cartoon showed a manatee playing the piano, with the caption "Forget the piano-playing manatee, how did the piano *get* there?"

Locals have known it as "Billy's Shoal" along the extreme southern end of the spoil area connected to Quayside Island, which was island Number 7 in Christo's project. But those reporting on the event underscored the site's genericism and its ordinariness, perhaps to highlight the piano's dramatic effect. Some journalists located the piano on a "narrow sand bar," others described a "small sand spit," and still others made it into the "piano bar." Neither recognized its mixed provenance in both natural and human forces. The sandbar was a spit, originally connected to a now submerged spoil area, dating back to dredging projects of the 1920s. The island is gone, but the spit lingers. At low tide, it emerges triumphantly. At high tide, the bay's waters swirl playfully above its sand and broken coral.

By definition, spits extend or project—whether as land, horns, or secreted liquid. If they extend too far, as the excess of spittle does, they might exceed boundaries. Deposited on what had long ago been a waste site, Harrington's project did that. More than excretion, it was an excrescence. The piano on the spit burned twice in three days, once on the mainland during a New Year's Eve party and a second time on January 2nd when it was first deposited and Harrington's friends photographed him pretending to play the flaming piano—Harrington had planned to create a video for his college application. When the piano came to the spit, two discarded things came together. A much older waste site hosted new waste. For most authorities, it was garbage. But the Coast Guard's view offered a different frame, suggesting that it was not garbage until it proved disloyal to its context or until it upset the bay's balance of natural and human systems as well as the more subtle balance of rational and irrational forces that, like the sea itself, shaped this zone.

When authorities told Harrington that he had twenty-four hours to get rid of the piano, he began making arrangements for its removal; but another Miami resident hired a salvage company and claimed ownership of the piano. The rules of the sea allow such salvaging of abandoned materials. Although the bay's hydrologic forces already continued slowly transforming the site, media coverage had augmented its visibility—and its imageability. A day after the piano was removed, another project was installed on the island—a small table set for two with wine, red roses, and a chef statue.

NOTES

1. Charles Rabin, "Piano Mystery Solved," *Miami Herald*, 27 January 2011.

2. "For piano, first hot flames then cool sandbar," *Miami Herald*, 26 January 2011. The Coast Guard's spokesman Nick Amin noted: "We will not get involved on this issue unless it becomes a legitimate danger to navigation or environmental threat, which isn't near at this point."

Chapter Six

Real and Surreal

Surrounding Biscayne Bay's Spoil Islands

BEAUTY SPOTS

> It is most unfortunate that Biscayne Bay is a shallow body of water. It has often been said that if there were thirty feet of water there, nothing in the world would exceed it in beauty and usefulness. But we are compelled to take things as they are and make them over to suit our demands.
>
> He [Dr. E. V. Blackman] suggests that the city build an island of a hundred acres or more out in the bay, connecting it with the city by a wide causeway. On the island are to be located the public auditorium for which there is now so much need, the fair grounds, a park, play grounds and other public conveniences. Such a plan, if carried out, would make Miami the Venice of America in a very real sense.
>
> As the years go on, we will probably find that the most feasible way of utilizing the bay is suggested in this idea of making a city park by digging up the bottom of the bay and depositing the spoil where it will make islands that can be beautified and made beauty spots.[1]

In the mid-1920s, as dredgers were completing their work on the Venetian Islands, the Corps of Engineers arrived in the bay's still brilliantly aquamarine waters to dredge an extension of the Intracoastal Waterway. A compliant but still dynamic Biscayne Bay added these projects to its index geography of displaced land and turbid water. From above, dredged channels and borrow areas now disclose these geomorphic histories in a chaotic weave of land-boom island-making, submerged lands bought and sold, disused waterways, and other channels maintained at all costs and at increasingly greater depths. In the 1920s, the Corps' ICW project added a north-south weft to the Venetian Islands' east-west warp, which had already been strengthened with bulkheads and sea walls. Rationales for the east-west strands of purpose-built islands and causeways were clear in their economic logic of real estate, trade, and commerce; and the ICW's underlying principles of transportation were also unambiguous. But this north-south dredging project interwove an irregular pattern of land into the bay's increasingly controlled environment. The resultant chain of spoil islands—more than a dozen in the northern part of the bay—drew the attention of city officials, picnickers, boaters, felons, tourists, and artists. For the artists Christo and Jeanne-Claude in particular, the islands formed an archipelago that was simultaneously real—in its public nature and stubborn lack of development—and surreal—in its unexpected aesthetic composition and the sheer incongruity of its existence. Just as they afforded Christo an artistic medium—*Surrounded Islands* could not have happened if its islands were not spoil, the bay's spoil islands also answered Miami's need for parkland to insure that public islands would form a part of the bay's civic space and the spectacle of development initiated by Carl Fisher.

On March 12, 1917, the *Miami Herald* printed an editorial titled "A Novel Suggestion," supporting the construction of a one hundred acre island in Biscayne Bay. The unattributed essay responded to a previous week's letter whose author proposed an artificial island to host public activities in the particular Venetian spirit of Piazza San Marco and with the broader objective of fashioning an American Venice. In the ensuing decades, challenges posed by the bay's shallowness were overcome; but the purpose-built public island never materialized, and Miami's lack of bayside parks continued. Citing the "popular imagination" of the time, the essay reads like a manifesto of island construction, or at least island conversation. The author notes that "intelligent city planning" must include publicly accessible parks and cultural sites such as a new auditorium, which is not unlike the island theater that Alvise Cornaro included in his sixteenth-century project for Venice's lagoon. This prescient scheme recognizes Miami's bay as an open space whose usefulness included recreation and municipal activities. It also predicts the transformation of spoil to parkland. The city would rely on filling projects in the 1960s to create much-needed parks. The city would also look to the chain of spoil islands as ready-made "beauty spots," useful as both "picnic islands" and "green relief." Unplanned in their placement, overgrown with exotic vegetation, and eroded by the bay's aqueous forces, the spoil islands may seem an unlikely solution. Despite their origins in discarded waste and their fragmented land area, these islands were within two acres of the "Novel Suggestion's" recommendation for one hundred acres of park space.[2] These imperfect spots were not blemishes but opportunities.

PICNIC ISLANDS AND GREEN RELIEF

Dredging operations, like those catalogued by the civil engineer J. I. Conklin in his 1913 article, expanded the bay's shoreline and made its purpose-built islands. Dredging also created a tapestry of waterways and channels that forms a historical context for the ICW's north-south strand and its island by-products. The bay includes the dredged traces of a canal project begun near Jacksonville in the northern part of the state in 1881. The Florida Coast Line Canal (FCLC) company never fully completed its planned route southward, but 350 miles down the coast, Dr. John Westcott oversaw the dredging of Biscayne Bay's section. Westcott, who became Florida's Surveyor-General, believed the bay would soon become a commercial center for shipping. Only a mile and half of the canal was completed in the 1890s, and now a ghostly line points north from Julia Tuttle Causeway toward Sunny Isles in the middle of the bay. Henry Flagler had seen the FCLC's ineffectiveness, which was logged in IIF's minutes as "circuitous routes and awkward elbows," and he cut his own channel around Cape Florida and through the southern end of the bay to link the open Atlantic with his train line and port facilities at the mouth of the Miami River.[3]

In the 1920s, the federal government picked up where these other initiatives had left off, and the ICW finally came to Biscayne Bay. The eight-foot channel provided a two-fold increase in depth for boaters—mostly in recreational vessels—navigating north of Miami's port. Dredging the northern bay's ICW section, the Corps of Engineers worked southward from Haulover Inlet in a phased project. By June 1926, the northernmost segment had reached a seven-foot depth; and, by November 1927, another portion connected with a remnant of the FCLC canal, which had cut across a particularly shallow shoal, just south of Biscayne Point and Normandy Isle. Lack of maintenance meant that shipping interests looked southward first to Flagler's 1897 channel with its reliable eleven-foot depth around Cape Florida and later to the main shipping channel with its even greater depths through Government Cut. Developers of Miami Beach and the Venetian Islands also sought alternative navigation routes and maintained their own north-south channel along the eastern edge of the bay. By 1930, with its first pass complete, the Corps straightened the meandering north-south corridor. Its controlling depth remained no deeper than eight feet and primarily benefited recreational interests, but its spoil islands were now in place and would soon be scrutinized by local as well as national politicians.

> Mr. Bolton: In the case of this canal, who do the islands belong to that are created by this soil thrown out in dredging the canal, the Government or the State of Florida? You see, all those lagoons are very shallow, and the engineers undoubtedly are creating a lot of islands. That has been a bone of conten-

> tion between the Government and the State of Florida for years as to who owns those islands.
> The Chairman: I think both sides will win, but I imagine that Florida may be a little bit more on the job.[4]

This conversation between Chairman Wallace Dempsey and Representative Chester Bolton coincided with completion of the Corps' long-awaited Biscayne Bay segment of the ICW project. In their March 1930 meetings, Rivers and Harbors House Committee members entertained, in passing, the matter of what to do with material left over from the dredging of waterways. Questions of ownership and oversight were also raised. Dempsey suggests that this surplus of land constitutes a "win-win" situation for both state and federal governments. He also recognizes the significance of Florida's IIF Board, which had been monitoring such issues since the end of the previous century. Interests of developers, industrialists, and tourists all converged in the state's exceedingly shallow inland waters; and soon after they were deposited from the bay's turn-of-the-century shipping channels, claims were made for these spoil areas. But there were already precedents—specific to Biscayne Bay—for these discussions.

In 1910, the IIF Board denied a request to purchase a sand bank across from the Royal Palm Hotel. This area, forming only about an acre of land at low tide, was north of the Miami River's confluence with the bay and less than a half mile from Biscayne Drive. The board confirmed that the state maintained sovereignty over the island "by reason of its formation and location." And even though the applicant Mitchell Price had received a letter from the Assistant Secretary of War noting that the U.S. government did not claim the island and only reserved the site as a place of depositing dredged material, the IIF Board maintained that it had "no right to dispose of the island."[5] The island was state property, and the board decided that it could not be sold, rented, or transferred. In deferring the issue, the board's motivations were unclear, but the decision confirmed the island's complicated status in which "formation and location" were only one part of a complex system of ownership and possible use. The board's tentativeness indicates the early stages of a decision-making process that will eventually frame how these irregular types of land are managed and sometimes codified in the public interest. The board may have guessed, and the applicant must have known, that this particular place would soon be a nexus of commercial and recreational activities. Directly across from the mainland site of where Bayfront Park would emerge in 1925 as Miami's central public recreational space, the tiny island formed the western edge of what would become Dodge Island and marked what would be the southern terminus of the ICW's extension through Biscayne Bay.

By the 1950s, Biscayne Bay had essentially reached its capacity—and its tolerance—for islands. Dodge Island's expansion between 1956 and 1964 marked the end of large-scale land-making opportunities in the northern part of Biscayne Bay.[6] Its wayward spoil areas had been accounted for and consolidated, all except for the lingering possibilities associated with the ICW's spoil islands. Private landowners also found it more difficult to extend their property by claiming or purchasing submerged lands. In 1921, the Butler Act reasserted the control of riparian land owners over submerged areas adjacent to their upland properties. Restating the earlier Riparian Act of 1856, this legislation sought to encourage waterfront improvements that supported commercial goals, and upland owners could gain title to these submerged lands by bulkheading, filling, and permanently improving the area between the channel and the upland's high water mark.[7] But, effective June 6, 1957, the Bulkhead Act repealed the Butler Act and authorized local governments to set bulkhead lines as the outer limits of filling projects by private riparian landowners.[8] The vision of a purposeful and productive, if not beautiful, bay was complete.

Though crowded by artificial islands and isolated by bulk-heading and its privatized shorelines, the bay's waters afforded the city a grand public space but did not solve its dearth of parks. The City of Miami and Miami-Dade County have historically lacked public park space. In studies by the Trust for Public Land, Miami consistently ranks near the bottom of U.S. cities with high population density.[9] As more than two hundred acres of the Dodge Island port facility were being built, city commissioners and other local officials formulated their proposals for a system of public parks along the bay's western shore and on its remaining spoil islands. Spurred by the area's continued rapid development, officials, who were once reluctant to commit public funds to parkland, sought "green relief." Attention turned to the ICW islands, often overlooked, wildly residual, but compliant after many years of informal community activities.

When the Corps of Engineers planned another cycle of dredging for the bay's ICW in the mid-1960s, local officials saw an opportunity to address the city's lack of publicly accessible land and formulated a plan for a system of urban parks between the MacArthur and 79th Street causeways. Their proposal included both shoreline sites on the mainland and offshore sites on the ICW's spoil islands, typically less than a half mile out in the bay. On December 10, 1963, representatives from Miami's public works and law departments presented their case to the IIF Board for filling two shoreline areas and reserving two spoil islands for park space. Time was of the essence because the Corps of Engineers was preparing to advertise for dredging contracts the following week. Noting that they had already received approval from FIND, the applicants summarized the project's goals as "making provision for public access to the bay, creating recreation areas, straightening the shoreline where debris-catching pockets existed, and using spoil material for public purposes rather than for private uses or undesirable fills."[10] One aspect of this scheme sought to replace waste-collecting notches in the shoreline with another, more purposeful waste material from the imminent dredging of the ICW. Filling these polluted inlets would create pocket parks. Depositing materials to augment mainland shoreline areas required the IIF Board to authorize alternatives to the spoil easements that had been granted to the U.S. government. Extensions of these easements came into conflict with private landowners on the mainland.

The project's public intent was clear, but its application was complicated by previous dedications associated with the sites selected for land augmentation. The board's divergent viewpoints on this relation between public and private interests framed the next few years of debate about the disposition of spoil islands. One of the sites was located along the shoreline near the Venetian Causeway's mainland terminus. The City of Miami objected to the continued disposal of dredged material in the spoil easement just north of Biscayne Island and requested that spoil be used to augment shorelines east of the Miramar neighborhood, where the Miramar Hotel occupied Bayshore Drive's half-circle extension into the bay. The hotel's owners filed objections at the meeting because they believed the spoil would devalue their prominent location that offered "restricted clientele" access to "beautiful tropical gardens" connecting with the bay's open waters.[11] Immediately to the north of the hotel, the median of Northeast 18th Street continued across Bayshore Drive and extended eighty feet into the bay as a pier with a concrete platform. The city based its claim for "riparian upland ownership" of this narrow strip of land on a 1960 court decree that equated its perpetual dedication to the "free use of all purchasers of lots" with the public access of a city street or sidewalk.[12] Noting that the court case did not name all lot owners as defendants, IIF's director believed the project's filled area would adversely affect private ownership and recommended denial of this part of the application. Plans to fill the area between the shore and bulkhead lines would push out almost four hundred feet into the bay, land-locking the hotel and bayshore residents.

The city's second proposal also met with opposition. Immediately north of the recently completed Julia Tuttle Causeway, a "narrow sliver" of land—320 feet long and 10 feet wide at one end and 25 feet wide at the other—was dedicated to the owners of sixty-one lots in Magnolia Park subdivision. Following the same argument as the Miramar site, city officials wanted to extend this parcel 340 feet into the bay. IIF's staff contended that the project should only proceed after permission was received from Magnolia Park's residents. In spite of the staff's conditions and the director's objection to the two projects, the Governor of Florida Cecil Farris Bryant weighed in and the applications were approved. Bryant summarized his position: "the Board was committed to multiple uses of waterway maintenance spoil areas and . . . the city plan for placement of spoil on alternate areas within the established bulkhead line was desirable." Attorney General Richard Ervin concurred by noting that the IIF Trustees' policy "for years was to maintain the interest of the public, and this was a public project."[13] This decision resulted in the formation of Margaret Pace Park east of Miramar and Magnolia Park, now known as E. Albert Pallot Park, north of the Julia Tuttle Causeway. This latter park's shoreline remains one of the few places on the mainland where Miami residents can readily access and actually "touch" the bay's waters. The city withdrew another similar proposal to deposit spoil along the shoreline property of the Harvey Seeds American Legion Post, which it was concurrently negotiating to purchase as a municipal park.[14] The board deferred the designation of two spoil islands as "picnic parks" until consent from the Corps of Engineers and FIND was granted, but the links between island and mainland were now firmly established.

A legacy of these projects is the association of mainland park with spoil island park. Though not perfectly aligned, there are strong visual and spatial connections across the half-mile stretches of water between the open shoreline areas and the green tufts rising from the bay's bluescape. Park-island combinations include Legion, Morningside, and Pace Park.[15] The spoil islands preceded the mainland park areas. It might be said that they inspired their landward counterparts. The islands had already served informally as places for picnics and boaters' retreats. Because they were not new land like the fill areas on the mainland, their status was more complicated and their official dedication as parkland took more time. But on May 11, 1965, the IIF Board authorized the dedication of three spoil islands to the city of North Miami for "public park and recreational use."[16] These were the areas of "unspoiled" freedom that Bay Harbor's mayor Sheppard Broad sought "as green relief . . . in the area with so many concrete seawalls" and so much development.[17]

The city of Miami also continued pursuing this relief. In March 1965, only two months before ribbon-cutting ceremonies at Dodge Island, the Miami City Commission authorized the city manager to make an application for six spoil islands to become city parks. At the start of the following year, the IIF Board accepted the city's proposal to designate these "permanent maintenance spoil areas" as "picnic island parks for the boating public."[18] By the end of 1966, the palette for Christo's project was in place—its islands had been formed and its public status had been confirmed. In the interim, the spoil islands would provide recreational opportunities but would also support exotic vegetation and harbor the refuse of their enthusiasts. The picnic islands were "green relief" but they were also "beer can islands," and Christo's art would benefit from both legacies of this spoil.

SURROUNDED ISLANDS

> I remember the two of us standing there, looking out at those little green islands in the middle of the bay. I said, "Wouldn't it be beautiful to surround them with pink floating fabric? What do you think?"[19]

In late December 1980, Jeanne-Claude and Christo stood on a Biscayne Bay causeway and formulated their idea for what would become *Surrounded Islands*. They had spent the day crossing the bay and paused to consider these islands as potential sites for a project. The following month, it became clear that their proposal for *The Gates* project in Central Park was unlikely to go forward, and Christo flew back to Miami on February 27th. From the plane, he again saw the watery canvas laid out below him and began formulating a new project. Two years later, air passengers flying to and from Miami's airport would also see eleven islands, now riotously fringed in pink polypropylene fabric.

Biscayne Bay and its spoil islands offered what Central Park could not. The bay provided sites that were spatially isolated but still highly visible. The spoil islands were also resolutely public. After their designation as "picnic islands" nearly two decades earlier, limited oversight left their use open-ended. Many of the bay's residents ignored the islands, some saw them as "beer can" islands that needed improvement, and others rested on their sand beaches or picnicked in the shade of their vegetation. Jeanne-Claude's and Christo's first idea for the project came before they knew these landforms were human-made. But it was this condition and the islands' indefinite status that allowed the project's completion. Like their Central Park proposal, this was an urban project, but exchanging grass for water and anchoring fabric offshore on already spoiled grounds made Christo's vision possible.[20]

Biscayne Bay's spoil islands provided the human-made infrastructure that the artists needed for their project of surrounded islands. Fourteen islands—later winnowed to eleven—initially presented to the artists a painterly line. They were first marks on the bay's landscape, a medium incrementally stretched taught by human occupation in the decades preceding this 1983 intervention. The islands marked and made visible the ICW's relatively narrow channel, and they also revealed the interactions of the bay's communities. Christo summarized this process of disclosure in his artistic statement for the project: "*Surrounded Islands* underlines the various elements and ways in which the people of Miami live, between land and water."[21]

When Christo arrived, Biscayne Bay was already a function of southern Florida's rapidly changing society, rationalized by economies of trade, tourism, and real estate but considered irrational with a reputation of excess, leisure, and boosterism. The spoil islands were products of this setting, for which they also provided antidotes. In spite of this complexity, Christo did not see an irreconcilable context but one that might accommodate his technical vision and the project's experiential logistics.[22] Christo saw a place waiting for a project to reveal and poetically demonstrate a logic behind its absurdity. In the wider scope, he saw "a bay highly manipulated by man, already managed by houses and bridges and causeways . . . a place man organized for himself."[23] Manipulated, yes, but the spoil islands also added a degree of play. For Christo, the islands were "whimsically manipulated" and at the same time "aesthetically well-placed."[24]

Their rational placement followed an assured and pleasing north-south axis, but their arrangement along this line was erratic. The unevenly spaced islands are variably placed on either side of the waterway. At the bay's north end, there are seven on the west, then further south, there are three on the east, then four on the west, then three more on the east. These alternations place islands near the shoreline, close to the middle of the bay, and next to apartment blocks. Christo is quick to point out that the surrounded island Number 4—known as Pelican Island—was within two hundred feet of North Bay Village. Christo tapped urban whimsy in the juxtaposition of publicly viewed art and private real estate. The ICW's north-south corridor that organized boating traffic also helped engage a sometimes unwitting public in the art of the spectacle: "it's unavoidable that all the boats are going north-south very near our islands. All of that is already entirely organized."[25]

The spoil islands also mirrored Christo's broader views about art. Christo found fellow travelers in the spoil islands. These constructs were already on their way to artistic form, and they resonated with his own process of artistic expression, which is itself "so absurd, so irrational, so obsessional."[26] For him, they followed the arc of the artistic gesture, and when he was interviewed during the setup of *Surrounded Islands*, he added "irresponsible" and "poetical" to his description of art's expressiveness.[27] Capriciously shaped, haphazardly placed, yet readily available, these islands were afterthoughts of the dredging process that would yield lasting afterimages in the artistic process.

Christo also saw the bay as a canvas that provided its own natural palette. His vision of human intervention was matched by a painterly appraisal of the bay's rich colors, recalling earlier views of its waters. In 1915, Everglades environmentalist Marjory Stoneman Douglas observed: "Oh, it was beautiful, so clear and clean. The color of the water changed with the nature of the bottom. It was pale green and dark blue and almost purple in spots."[28] Nearly seven decades later, Christo translated this coloration into an artist's palette: "All together it is a work of art. We were blessed that Biscayne Bay has an incredible variety of color from very deep green-blue to the light green in the northern part [with] shallower water."[29] These pigments lay pressed across an exceedingly flat canvas. Christo would later speak of the "incredible flatness of the Miami landscape that is not only horizontal but completely flat and a very fluid situation between earth and water."[30] The shallowness that generated the spoil islands is the painterly flatness that characterizes Christo's project, and the spoil islands were first marks across this nascent painting. Cast between earth and water, they also measured the bay's horizontality and flatness. Their beaches tracked tidal changes, and their vegetation added subtle texture to the smooth blue-green surface of the water. Christo has said that this was the "most painterly" of his and Jeanne-Claude's projects. They considered the bay "like a giant painting," and the spoil islands shaped its contours.[31]

LANDSCAPE PAINTING

Surrounded Islands formed an occupiable landscape painting. The project extended more than seven miles from just north of the Venetian Causeway up to Haulover Inlet. For reference, the artists numbered the islands from south to north. The project's first island—also known as Teacher's Island and Pace Picnic Island—lies a half mile from Biscayne Island, Carl Fisher's last Venetian Isle. Its spoil islands traverse not only the bay's geographies but also the histories of its manipulation. This southernmost part of the project provides visual connection to Miami's active port zone. To the north across the Julia Tuttle, Christo surrounded Morningside and Osprey islands, skipped Legion Island and Bird Key, which is not human-made, then included the next eight spoil islands. Here, the island chain—as if at the end of a brush stroke—tapers at the bay's narrow northern end with the smallest islands, Little Sandspur and Crescent. This geographical gesture also seems to draw the project seaward toward Haulover Inlet and the Atlantic's open water. Along the way, causeways appear to pry open and hold the flanking land apart. And wakes of boats have streaked white across the rolling shadows of wave and bay bottom.

With the water as an active park setting, Christo's landscape painting enlisted what he called "the energy of everyday life" through the wave patterns that are a combination of human activity, boating and bulkheading, and natural forces of wind and current.[32] He also recognized the role of the islands as interruptions that reflect this energy and amplify its textures. With the islands like dripped paint, the artist reacted to the accidents of their placement. At the same time, they offered a regulating axis with fixed points where Christo's fabric could respond to the bay's complex energy. Christo practiced his art in unlikely places and circumstances, and the islands were equally displaced from what he has called the "normal art object."[33] They became tools for occupying the "very complex space" of the bay and the immersive environment of this full-scale landscape painting: "Now we use that space and we have all the consequences of [its] incredible energy."[34]

The project's color and detailing are reminders that the spoil islands mix the natural with the artificial. For Christo and Jeanne-Claude, pink was Miami—associated with retirees' slacks, flamingoes, and the painted reveals of Art Deco hotels. It is also the natural color of sunsets and frangipani. But hues of the latter vary and do not sustain the intensely saturated, monochromatic pink of *Surrounded Islands*. The fabric that surrounded the spoil islands was unmistakably artificial: "My colors are not only the landscape, the sea, the water, but also all these human elements."[35] The pink fabric transformed its context, capturing the shadowed contour of waves and projecting pink reflections on pelicans' wings and low clouds. This use of pink parallels Salvador Dalí's experiment with surrealist painting. Dalí proposed that painting must become a "colored snapshot" in order "to make concrete . . . the most fugitive visions."[36] The pink-fringed islands anchor the ephemeral work of art that also references the transitory nature of human agency.

Engineering the project broke down boundaries between the rational and the irrational, and the real and the surreal. If its shocking color conjured unexpected associations, then the precise detailing of the fabric's anchors and floats brought Dalí's "delirium of obsessive exactitude" to each island's unruly shoreline. Christo and his team experimented extensively with seams, booms, anchors, and buoying systems so that the design of each component could address natural phenomena of tidal fluctuation, surface tension, and wave action. This technical rationalization recalls the dredging technologies that originally produced the islands. The anchoring systems and suction booms of dredgers also contended with the dynamics of currents and the properties of both water and solids. During its assembly, Christo's project recalled the event of dredging. In the first step, its ribbon-like cables were pulled from each island out into the bay. These strands visually and symbolically reconnected the islands to their origin in the bay. Pink fabric would later extend 225 feet between land anchors at each island's vegetation line and floating booms with twelve-inch-diameter octagonal profiles. To hold the booms in place, anchorage points on the bay bottom were located on radial lines that reached exactly 250 feet from each island's high tide mark. But even the logic of this engineered system is countered by makeshift adjuncts. Christo's photographer Wolfgang Volz recalls that land anchors were covered by crushed beer cans, found objects indicative of the islands' informal use and emblematic of their origins as waste.[37]

The floating booms, in particular, call attention to the islands' artificiality and their participation in the production of art. The design team chose octagonal rather than rounded profiles for these elongated floats that marked the outer boundary of the pink fabric. Christo notes that this decision was based on aesthetics, performance, and function: "The booms were an aesthetical decision. The booms have a man-made shape; we [did not want] to have this sausage boom. They are very geometric . . . the boom was aesthetically very important not only that the fabric edging was very precise because, with the light, the boom created shadow. And it was very important that the boom stop any debris coming to the fabric."[38] The booms also magnified the form and what Christo saw as "the very geometric shape" of the islands. In the engineering drawings, a delicate tracery of radial lines extends perpendicular from the geometrically defined chords that regularize each island's curved shoreline. Helkers, the fourth island from the north, required fifty-seven such lines. Its measured drawing presents a view of what appears to be a volcanic island, its radial lines marking a precipitous mountain slope from rim to sea. In their application, each one hundred foot boom adds a facet to a newly drawn outer edge, transforming each curved island into a geometrically defined polygon.

Fabric and booms effectively magnified the size as well as the enigmatic contours of the islands. The fabric increased the islands' area fivefold. 149 acres of fabric, the equivalent of ninety football fields, surrounded about thirty acres of land above the high-water line. For Christo, surrounding the islands not only enlarged them quantifiably but also made "bigger how the islands were shaped by the humans."[39] As a result, minor elements—along with the facts of its origins in human artifice—became major features in the bay's composition. The islands Christo discovered in the making of this project were also molded by entropic forces of wind, wave, and tide. In the bay, they rest like surrealist objects, powerless in the face of time, earthen permutations of Dalí's clocks or analogs to the amorphous objects of desire found in Joan Miro's floating shapes and Max Ernst's unidentifiable forms.[40] Christo's surrealist experiment plays off the imprecision of the spoil islands as well as their indefinite status as found objects. Like surrealism's unexplored "treasure islands," the spoil islands reveal unexpected and new environmental epistemologies that constitute what Dalí calls an "irrational knowledge of the universe."[41]

With its sheer scale and complexity, the island project resists conventional forms of measurement and representation. Engineering plans served a vital purpose to survey, locate, and quantify the project's components and systems. In one of the many photographs that document the project's implementation, Christo stands on a beach and scales an aerial view of island Number 5 with a pen and measuring tape.[42] Like the scaled aerial, plans served a technical purpose but did not convey the atmosphere of *Surrounded Islands*, so Christo used other modes to represent his project. Christo does not produce drawings after the completion of projects. He instead develops collages as prospective—and sometimes speculative—documents that can be sold to fund the work. Christo's vignettes take an impressionistic, almost romantic, approach to present the installation and its conceptual approach. Each collage mixes pastel, charcoal, enamel paint, photographs, tape, crayon, handwritten notations, and swatches of fabric. Maps and photographs of the context join the construction lines of the perspective view and other regulating lines that form a mesh, not unlike the radiating ribbons that extended outward from island to boom. Technical specifications, dimensions, and explanations like "the floating fabric will be attached to a long boom" quantify and detail what to expect in the upcoming project. In places, dimension lines measure the collage itself, suggesting that the document is inherently linked not just to a preliminary process but to the ongoing project itself. Textured shadows, tactile material samples, and heavily brush-stroked surfaces provide more qualitative effects.

Most striking is the viewing angle found in all of the collages. Reminiscent of Jacopo de' Barbari's aerial maps of Venice, the birds-eye, elevated perspectives place the viewer in a privileged position where the chain of islands recedes to either the sea's or the city's horizon line. From this vantage point that will also be the helicopter view for Christo and for those who will tour the project, the surrounding fabric of the nearest island recedes toward this backdrop but also dramatically extends toward the viewer. This exaggerated foreground—a function of the collage's forced perspective—places the viewer on the threshold of the project. The viewer simultaneously experiences the project from the water and above the ground. With these representational techniques, the collages successfully work between the experiences of the map-maker and the boater. Christo expands the artist's role to cartographer, engineer, politician, botanist, helmsman, weaver, and poet. He extends this seemingly impossible mix of responsibilities to the citizen-viewer and suggests that the project elicits an array of experiences that elide categories of viewership as well as simple distinctions between rational and irrational.

At the scale of the bay, it is building that becomes a kind of drawing. Though the final layout conformed to the plans, the process of construction—extensively recorded by Christo, Wolfgang Volz, and others—presents an extraordinary document of a landscape's incremental transformation. Workers first installed the ribbons that would hold individual sections of fabric in place, pulling them from the land anchors outward to the anchored floats. In aerial photographs from the first day, the initial cables are barely visible radial lines, extending more than two hundred feet out from the islands and suggesting that it might be the islands that are now lightly tethered to the bay bottom. The thicker and more visible booms came next, and their deliberate positioning inscribed a new island territory. As the booms faceted this boundary, the water surface between boom and island shore was calmed and became a smooth, polished lagoon. Simultaneously, a protective underlay was installed across the islands' beaches to make sure the rocky coral did not tear the larger swathes of fabric. The resulting inner and outer pink rings dramatically outlined each island's shape. Soon to be painted, these full-scale drawings floated momentarily like desiccated leaves delicately laid within the bay's urbanized context. This subtle tracery was the vein structure of emergent lily pads stepping across causeways and along bulkheaded shorelines.

Christo's lily pads were suspended between sky and water. Like Claude Monet's *Water Lilies*, the unfurled fabric increased the sense of aqueous depth and at the same time intensified the reflective capacity of the bay's surface.[43] A thin lens of water on the fabric presented passing clouds at noontime, tree shadows after sunrise and before sunset, and glistening sunlight reflections throughout the day. Biscayne Bay is a landscape manipulated by human intervention, and it is not surprising to learn that Monet diverted water to form the artificial lake that provided the conditions for his famous painting. Christo built upon these traditions and techniques to paint an ephemeral landscape that celebrated its artificiality while acknowledging its place in a new nature.

MAY DAY

Christo's project also celebrated human labor. A festive weeklong setup event followed up on the project's intensive logic of production. On April 27, 1983, more than four hundred workers registered at Pelican Harbor and began three days of training. The weekend pushed the start of work to 5:30 a.m. on Monday May 2nd. Otherwise, the project would have begun on May Day, a fitting start date for work that had much in common with May Day's demonstrations and marches to observe workers' rights. Known as Workers' Day in his native Bulgaria, May 1st events joined other large-scale events, like commemorations of the October Revolution, to influence Christo's work. His early education in Bulgaria also exposed him to the Russian avant-garde's constructivist architecture that transformed public space, and he read Vladimir Mayakovsky's poems and looked at Vsevolod Meyerhold's theatre productions. In many of his projects, particularly *Surrounded Islands*, Christo is interested in how human labor activates a landscape.[44]

Biscayne Bay provided unique circumstances for this interest because it mixed natural beauty with a history of human manipulation. The bay's dredged islands are visible consequences of work and are part of a larger set of operations that have transformed the east coast of the United States. The islands formed a reciprocal environment and an active background for the artists' own exploration of the limits of human skill and production. In Christo's event and its landscape of labor, parades of workers, many in bathing suits and other Miami springtime attire, exuberantly hoisted a total of seventy-two sections of cocooned fabric attached to the one hundred foot long booms. Photographs show work beginning before day-break and extending after sunset. Choreographed by Christo's rigid schedule, hundreds of laborers stretched out rolls of fabric, strained to pull ribbons taut, and stitched seams while floating on inner tubes and rafts. Even though many would have volunteered, they were paid minimum wage. Photographs narrate a sense of community and easy collaboration, reminiscent of the Works Progress documentary photos like those of Lewis Hines, who presented workers' sense of determination and pride to reveal the extraordinary in the everyday. Christo did not miss this connection. He said at one point, "we are WPA'ing Miami" to describe the large amount of work embodied in the project, particularly at a Hialeah warehouse and Opa Locka's blimp hangar, where forty-five seamstresses worked for four months between December 1982 until late April 1983 to prepare the fabric.[45]

Christo described the assembly process as an "unfurling." Derived from its root word *fardel*, to unfurl is to open a pack or a burden.[46] In *Surrounded Islands*, cocoons attached to the floating booms held the large swathes of rolled fabric. Teams of workers opened these "packs" to begin the onerous task of stretching the fabric along each island's radial lines. Each instance of unfurling was a temporary liberation from the expected. It was an irrational event at the whim of natural forces. Winds caught fabric edges, waves flooded surfaces, and at one point a giant bubble emerged under one island's material. The official unfurling instructions narrate a technical process that makes an unusual set of tools—dinghies, surfmats, cocoons, and grommets—seem routine:

> UNFURLING INSTRUCTIONS: One section at a time, open the cocoon, remove it from around the fabric, find the pulling lines and, using dinghies, bring the cocoon and lines to shore. Slowly pull on the lines until the fabric unfurls and reaches the shore. Secure the edge of the fabric to the land anchors and to the tether lines. Lacers will start as soon as two sections have been unfurled. Lacers will float on surfmats and after having lined up the grommets, starting at the boom, will lace the two sections by running a line through the grommets, along the overlapping hems of the floating fabric.[47]

Unfurling pulled fabric toward the islands, underscoring the fact that this disruption emerged from elsewhere, definitively a product of human intervention. Unfurling toward the islands was technically necessary, but it also dramatized the festive event in which the extraordinary—the cocooned pink fabric—was brought to bear on the ordinary—the yielding archipelago. This was Mikhail Bakhtin's carnival, the "feast of becoming," where there was a "temporary liberation from the prevailing truth and from the established order."[48] Now draped in fabric, the islands seized "the rules of a social apparatus" and revealed mainland truths. According to rules of zoning and environmental regulation, the project needed permits, but these constraints, in the end, strengthened the carnivalesque presence because the ensuing discussions—and controversy—engaged the community. The islands were in fact artificial. They were small but significant components in Miami's growth, and they would remain as emblems of public space.[49] After the two-week spectacle, the islands continued to provide breaks in the everyday that were not sanctioned as a carnival and were partially scripted by increased conservation measures but at least provided for a long-term picnic.

On May 7th at 2 p.m., the project reached "full bloom." At this moment, Christo relished the islands' proximity to human experience: "It was important that these islands were very close to the causeway and close on some occasions to the apartment buildings. People can see it from the causeways, from the high bridges, from the apartment buildings. Of course, the best thing is to go very near to the islands with a little boat."[50] But for many, viewing the surrounded islands was a challenge. For Christo, water creates distance; and Biscayne Bay's northern section—ten miles long and almost three miles wide at its midpoint—provided the necessary expansiveness for his canvas. The islands allowed for perception of that distance, but only in episodic fragments because of their uneven spacing. A totalizing view of *Surrounded Islands* was difficult, if not impossible, and countered many of the project's goals. Christo tells the story that commercial airline pilots added a loop on their approach so that passengers could view the project. From this cartographically distant perspective, the project lacked detail and did not engage the sense of depth Christo had presented in his preparatory collages.

Lower in the sky, views from small planes and helicopters might have come closest to this artistic vision. On a typical flight, the most distant islands faded to the horizon, its slight curve visible at about a mile above the earth. For the two-week period of the project, helicopter rides were offered, but this option was expensive and unavailable to the majority of Miami's citizens who shared in the bay's public status. From the causeways, the project's horizontal layout tilted slightly to include a fringed island in the foreground with the urban edge—and possibly another island—in the background. From boats, the project was a painted line that faintly separated the islands' vegetation from the bay's aquamarine context.[51] From this vantage point, the project quite literally answered the call of the last sentence of Christo's artist's summary—to highlight the diverse ways Miami's residents live "between land and water." While it "underlined" some experiences, it resisted others. From the boater's perspective, distinctions between what is natural and unnatural collapse along the water's surface, and these ambiguities come together at the pink horizon line of Christo's project, just as the turbid sediment of dredging has mixed with the bay's opalescent waters. The line in the water also maintained distance. Boaters, some of whom might normally take breaks or even picnic on the islands, were denied island access by the fabric rings. For the two-week period, these public sites were off limits, and they approximated one model of environmental conservation—

you can see nature from a distance, but coming closer might spoil the view as well as its ecosystem.

TEMPORARY MONUMENT

This temporary impoundment of the islands behind pink fabric is one of the project's environmental legacies, which would transform how Miamians view their bay, their islands, and their city. The spoil islands also became a critical—if largely unrecognized—part of an environmental aesthetics dialogue that would carry over to the start of the twenty-first century. *Surrounded Islands* reminded many of the pitfalls of environmentalism. The visibility, and perhaps the artistic nature, of the project became a target for what some saw as misdirected criticism that ignored core values of environmental consciousness. In a letter to the *Miami Herald*'s editors, Charles Lee of the Florida Audubon Society outlined the environmental ironies of the *Surrounded Islands* project:

> It is unfortunate that the furor created by so many new-found defenders of Biscayne Bay in response to the Christo 'Surrounded Islands' project could not have been channeled against some of the real and long-lasting threats to the Biscayne ecosystem. Where were these good folk when the city of North Miami was dumping garbage for profit at the Munisport dump in the mangroves adjacent to the bay? Christo's colorful plastic will be there two weeks and then be gone. It has more environmental safeguards than most development projects to last for centuries. The ospreys that nest on telephone poles beside U.S. 1 in the roar of traffic on the way to the Keys will really envy their brethren on the Biscayne Bay spoil islands, who will be isolated from the shouts and outboard roar of Sunday boaters by a 250-foot perimeter of floating fabric—even if it is only for a little while.[52]

After the ospreys mentioned by Lee took up residence on island Number 9, a court case nearly halted the project in its final week of preparation. Islands provide a distinctive context for protection and conservation, one that can be more easily delimited than mainland areas, such as the shoreline dumpsite in Lee's letter. It is more difficult to transfer this conservation throughout a coastal region or territory. Spoil islands offer an outlet for conservation when it is not possible elsewhere because of development pressures and large-scale land ownership. As other commentators noted, many of the project's critics had already participated in the bay's degradation, and their disingenuous views failed to see an important lesson of *Surrounded Islands* that the bay is a manipulated landscape that now relies on human intervention: "Folks who live in air-conditioned houses on dredge land beside human-made canals have stood up and screamed that he would destroy [the bay's] 'natural beauty.'"[53] The spoil islands were already spoiled.

Another side in debates over the merits of Christo's project argued that his art was an improvement to a waste-strewn landscape. Miami's city commission chair referred to the islands as "beer can islands" that should be cleaned up. From this position, art provided a different kind of improved nature, in contrast to the real estate improvements made by developers six decades earlier. Jan van der Marck, founding director of Miami's Center for the Fine Arts, noted: "Thanks to Christo, people will now see those once-grubby islands as jewels."[54] This view borrows from a real estate model in which development improves unused, or at least under-utilized, land. In contrast to the bay's profit-driven development projects, *Surrounded Islands* remained a not-for-profit artistic experiment. Biscayne Bay's waters constitute a public space, and the dredge spoil islands created by the dredging of the ICW have for the most part remained an integral part of the bay's public domain. Christo's artistic work applied private funding to transform the bay's public space through its equally public islands. Christo's work is not, however, driven by profit; and project funding cycles back into artistic production. Six decades earlier, when developers created privately funded islands from submerged land owned by the State of Florida and presumably administered in the public interest, the bay's natural landscape was permanently transformed for profit.

As a non-profit enterprise, Christo's temporary repurposing of the islands placed his private vision into the public domain and invited not real estate valuations but assessments based on aesthetics, tourism, and resource management. Christo's "improvement" of the environment was conceptually the reciprocal engagement between lived experience and place, rather than a commodification of landscape as saleable property. At the same time, its iconic image was welcomed by the Chamber of Commerce and provided a boost to the bay's economy. Practically, the improvement of the islands could be quantified in the amount of material that was cleaned up. Christo's workers removed forty tons of garbage from the spoil islands. But, environmentally, this clean-up was matched by the addition of an almost equal amount of fabric—thirty-five tons—to the South Dade Landfill. Recognizing this dilemma, Christo requested that it be recycled but only if it went toward a "non-profit, industrial use." It did not, and assistant chief of solid waste disposal department Tony Sabrino ironically quipped: "This gives new meaning to 'art in public places.'"[55] *Surrounded Islands*' fabric remains in the county landfill.

On the spoil island north of 79th Street causeway, there is a small plaque with Jeanne-Claude's and Christo's handprints pressed into pink concrete. Inscribed with cursive text "Surrounded Islands, 1980–1983," the square slab is an unexpected monument, crypt-like in its construction and closeness to the ground. It rests near the middle of the project's archipelago on Pelican Island, the fourth island from the south and fifth from north. The artists might have chosen this site for its stability—Pelican is one of the larger and more established islands. They also might have chosen the island because it came into closest contact with human dwellings, the private apartments on Biscayne Island, less than two hundred feet from the pink fabric edge. Christo had opposed a permanent monument or memorial to the island installation. Less than a week after the fabric was removed, there were calls for official tributes to the project. Ideas included leaving one island surrounded, adopting pink as the official color of Dade County, and naming the island chain "Christo's Islands." Only the small pink slab remains.

It was Christo's attorney Joseph Z. Fleming, who called *Surrounded Islands* a "temporary monument to the environment."[56] In the Fall of 1982, as criticism increased, Fleming presented this description in a seven-point document, a latter-day version of "frequently asked questions," titled "Answers to questions regarding Christo's *Surrounded Islands* Project." When asked about this assessment, Christo decries the idea of monument as "pompous," but Fleming's reference to monuments sought to answer the question "why is Christo doing the project in Biscayne Bay" and to promote the artist's respect for the bay's "unique beauty."[57] Fleming's commentary not only proposes art as an homage to the bay's beauty but also references a nation's rich heritage of natural monuments, those sublime, often geologic, features that citizens identify with beautiful landscapes. Human-made monuments are productive constructions that purposively emerge out of a cultural context.[58] As artifacts of dredging and now surrounding, the islands are monuments *to* the environment. They engage both the human construction of monuments and the designation of natural features as monuments.[59] In any case, Christo's intervention also monumentalizes the spoil islands, even if just for a short time, because the project was temporary in terms of its duration as well as its participation in the fleeting time of nature and the entropic forces that have shaped the islands.

In an enduring question for the bay in particular and a globalized community in general, Christo's project asks whether humans can occupy a landscape at a large scale without permanently altering it. In the early 1950s, minimalist artist Tony Smith took his Cooper Union students on a drive along the unfinished New Jersey Turnpike. Without highway lighting, signs, or even a moon overhead, there was only "the dark pavement moving through the landscape of the flats."[60] The drive was a "revealing experience" that prefigured an art form that could occupy urban space without filling it in.[61] It changed Smith's views on art: "It seemed that there had been a reality there which had not had any expression in art." With the midnight drive, Smith was translating infrastructure's rational presence into an unlikely, but wholly artistic, experience. This was an art installation temporarily completing something that was previously disused, underutilized, or wasted. Smith summarizes his ideas for art's vastness: "The experience on the road was something mapped out but not socially recognized. I thought to myself, it ought to be clear that's the end of art. Most painting looks pretty pictorial after that. There is no way you can frame it, you just have to experience it. . . . Artificial landscape without cultural precedent began to dawn on me."[62] For Christo, the spoil islands were the last outliers, the last places where control and conformity had not been entirely installed.

Christo's project also creates an "artificial landscape without precedent." Except perhaps for Flagler Memorial Island as an antecedent, there were no clear models for the large-scale use of the spoil islands' artificial landscape. In the absence of cultural precedent, Christo recognized the islands as nodes of informal activity, intersections of natural phenomena, and adjuncts to a formal system of water transport. They were also loaded with meaning as sites of labor, public space, and places of escape from everyday life. For Christo, they were in no way neutral canvases. Instrumentalized as aesthetic tools, the islands revealed what Christo believed was the "unique beauty" of the bay. His two-week project drew attention to the logic of the system and the sometimes irrational heterogeneity of natural and human impacts on the islands. In the process, on the terms established by the spoil islands, he demonstrated ways that a landscape can be occupied without permanent alteration.

Taken from start to finish, *Surrounded Islands* became its own environmental impact statement. This project did not prompt the formal statement outlined by the National Environmental Policy Act (NEPA), but its reports, community debates, permits, and the installation itself demonstrated the project's environmental "effect" as well as the islands' aesthetic and ecological value. More recently, the Bureau of Land Management (BLM) has required Christo to prepare an official Environmental Impact Statement (EIS) for the *Over the River* project and its plan to install six miles of horizontal fabric panels over a forty-two-mile stretch of the Arkansas River. According to Christo, a work of art "for the first time" has joined "oil fields, mining operations, airport construction, highways, bridges, and dams" in the report required by NEPA. For him, the three-and-half-year study "revealed what the project is."[63] Making islands also requires an EIS, and *Surrounded Islands* revealed what that process might be.

AFTERIMAGE: PLATO

VISITOR: I tell you what, Socrates: if nothing else strikes us as obviously suitable, why don't we take weaving? . . . If we take even just this aspect of weaving, I think it will give us the information we want.

YOUNG SOCRATES: Yes, I don't see why it shouldn't.

VISITOR: All right. Now, previously we distinguished all the various subsections by dividing them one after another, so why don't we do the same for weaving?[64]

And the visitor—also called the "stranger" in other translations—proceeds to connect weaving with the practice of governance. Studying *Surrounded Islands* left an afterimage that sent me to one of Plato's dialogues. In the *Statesman*, Plato explores how skills of leadership relate to the organization of political communities. Casting this conversation into the already multivocal discussions emanating from Biscayne Bay and its urban context might seem out of place. But the *Statesman* is not simply about governing skills; it also presents the work of philosophical method. It is an exercise in political thought. *Surrounded Islands* demonstrates an interaction of art—implying a *techne*, or skill, not unlike political discourse—with *polis*, particularly in its root sense of community. For Christo's project, this interaction of art and *polis* is inseparable from, and completely reliant on, its environmental context. As Christo noted, the project is not solely about the islands, but the relation of spoil islands to bay and city. My afterimage of the project is the ghost of these islands, now tethered not just to the waterway from which they emerged, but also to the bay that holds them and the city that grapples with their future. This afterimage oscillates not between the positive and negative of a burned image but between the islands' solidity and fluidity as well as their numerous other dialogical characteristics. It is a woven image, and the islands are the places where crossing and overlaps occur.

Surrounded Islands wove the spoil islands back into the natural and artificial fabric of Biscayne Bay. In this project, weaving operates by way of material, system, and dialogue. As Christo's preferred material, fabric offered buoyancy and flexibility in the bay's waters. It also made possible the broad strokes of color. The islands were surrounded, and the water was effectively wrapped. For five months, Germany's Adolff AG plant produced more than 217,000 miles of pink polypropylene ribbon to weave 6.5 million square feet of fabric—the most woven material ever used by Christo and Jeanne-Claude.[65] Jan van der Marck, who first contacted Christo about participating in the city's 1982 New World Festival of the Arts, has noted the significance of fabric for Christo: "Fabric is to Christo what steel had been to an earlier generation of modernist sculptors. He likes the dual meaning of woven fiber and fabric in the metaphorical way—that which holds our society together. In explaining his work, he often refers to the human interaction, legal fencing, environmental posturing, and technical problem solving as what he calls 'the real fabric' of his work."[66] The project's fabric is not just pink material. People, debates, and the various, often conflicting, systems that the artists found in and around Biscayne Bay are the "real fabric."

Plato's metaphoric treatment of weaving extends the meaning of Christo's fabric. The dispositions of citizens are the "raw materials" in the statesman's process of governing and weaving a state.[67] Christo calls them the "participatory public."[68] In this model, the statesman sees those who "incline towards courage and have a solidity of character" as the warp, and others who "incline towards restraint" are the weft.[69] Weaving produces fabric by interlacing the strong, taught fibers of the warp with the softer threads of the weft, which often lack the warp's tensile strength. It is the statesman's task to weave the weft's gentleness and moderation with the daring and bold spirit of the warp. Debates leading up to *Surrounded Islands*' installation weighed restraint against audacity; and the resolution of these dialogues approaches what Plato, speaking through the visitor, would consider to be an ideal state and what some critics of Christo's work have read as a utopian vision.[70]

Christo is by no means a statesman, but there is an aspect of governance in his projects. His dialogue with Miami's citizens is not unlike the question and answer exchange between Plato's conversants. Like the visitor, also translated as the stranger, Christo is an outsider. His suggestions are sometimes provocations, and these challenges parallel the boldness of spirit that Plato identified with the strong warp of the visitor's commentary. Seeking definition and clarification of the project, interlocutors meanwhile serve as the weft with calls for restraint. Just as Plato's work outlines a method, Christo believed that dialogues with local officials and citizens were as much a part of the project as the final result.

The islands mark another warp-weft intersection. Water meets land. These natural raw materials have already undergone what Plato would consider a process of division, but their distinction then becomes blurred when they are deposited back into the water by the dredging process. The bay's natural context, along with its social and infrastructural setting, prepare for Christo not only a painter's canvas but also a weaver's loom. In one sense, the causeways are the rigidly defined warp, and the waterway and its line of spoil islands constitute the delicate, malleable weft. But Christo's project rethinks these classifications, and essentially "restrings" the bay. Vibrantly colored and figurally augmented, the islands become a solid, if still fragmented, warp. Miami's existing infrastructure becomes a weft that accounts for the islands' presence.

This reversal of warp and weft hinges on the spoil islands. When Christo makes them visible, the bay's systems are transformed. For the moment, islands cannot be accessed, causeways become viewing platforms, and airlines reroute their approaches so their passengers can see the project. Warp meets weft on spoil islands. In my afterimage of the project, distinctions of water and land have disappeared, and spoil islands anchor other imbricated systems, such as bird migrations, boat wakes, wave refractions, and tourist gazes. Christo describes the fabric's response to some of these phenomena:

> Moving with the tides. This is all part of the intricate relation between the tides and the form of the fabric shifting, coming up and down, playing with that special underwater buoy that was acting like a knee, coming up and down.[71]

Surrounded Islands proposes a natural contract. My recollection of Plato's *Statesman* was spurred along by Michel Serres, who summarized his own thoughts about the weaver-king: "In a memorable passage describing the art of governing, Plato sketches the king weaving rational weft yarn onto a warp bearing less reasonable passions. Now we are at the dawn of an era in which the new prince will have to cross the woof of law with a warp born of the physical sciences: political art will follow this weaving."[72] For Serres, the warp and weft are closely aligned with the rational and irrational. Laws of society and laws of production cannot ignore what might seem the unreasonable workings of the physical world. Leading up to this point, Serres argues that distances between rational ideas and real settings result in waste and residue. Certainly, the spoil islands originated from just such a gap. More closely joining the rationalized dredging process with its real context would lessen this residue. Christo's project attempts to bring the rational in contact with the real.[73] His work rationalized the spoil islands' unruliness and reintroduced them to the bay and the city. It also worked with pre-existing urban complexities. After describing how the planes that made extra loops on their approach filled the air above *Surrounded Islands*, Christo noted: "It's part of how the reality of space, upper space, the water, the highway, the causeway, everything was such non-stop human use."[74]

Even though the project's ephemerality means that its materials become waste after its conclusion, its attention to each island's unique topography, the bay's hydrography, and the city's multi-layered regulations interlace constraint with illogic and the technical with the experiential. This weaving the rational and the irrational provides a foundation for what Serres identifies as the natural contract. In his argument for establishing a natural contract, Serres makes the case that reason by itself seeks to defeat an external nature. Serres bases the natural contract on the balanced convergence of these two sides, which he understands as two types of knowledge. Both are essential. Christo's art is not overtly political, but it takes place in the *polis* and is enmeshed in environmental politics. Like Serres' natural contract, *Surrounded Islands* provides a new paradigm in which the natural world is no longer defined by property—whether private or public. Tenets of the natural contract also force the social contract to enter the world of experience. In the new commons—a redefined central park—the natural world instead becomes "our symbiont," where intersections become a series of bonds.[75] I now understand why Christo thought the bay was so beautiful for the two weeks of *Surrounded Islands*. The beauty of the project is not just aesthetics of form, color, and placement but the environmental aesthetics of this natural contract.

Surrounded Islands has left us with an afterimage of Biscayne Bay's spoil islands ringed with meaning. The aesthetic repurposing of the islands has meant that we will never look at these or any spoil areas in the same way. Christo's project differentiated them from the bay's landscape and at the same time established them as an inextricable part of the bay's archipelago, raising a city's—as well as a global art public's—awareness of these sites and moving the public to a deeper understanding of their location and potential. *Surrounded Islands* created an expanded visual field that could then frame a broader environmental aesthetics for the region.

NOTES

1. "A Novel Suggestion," *Miami Herald*, 12 March 1917.
2. Official documents cited in IIF Board minutes calculate the islands' area as 98.48 acres. See January 4, 1966, *Minutes of the Trustees of the Internal Improvement Fund* 35 (Tallahassee, FL: 1962): 401.
3. June 2, 1883, *Minutes of the Trustees of the Internal Improvement Fund* 3 (Tallahassee, FL: 1889): 234–235. See also chapter 3 of Howard Kleinberg's *Miami Beach: A History* (1994) in which he notes the "blundering" activities of the Wescott's Florida Coast Line Canal and Transportation Company.
4. *Hearing before the Committee on Rivers and Harbors, House of Representatives, Seventy-First Congress, Second Session, on the Subject of the Improvement of Miami Harbor, Fla.*, March 1930 (Washington, D.C.: Government Printing Office, 1930), 13–14. "Mr. Bolton" is Chester C. Bolton, state representative from Ohio, and "The Chairman" is S. Wallace Dempsey, representative from New York.
5. May 4, 1910, *Minutes of the Trustees of the Internal Improvement Fund* 8 (Tallahassee, FL: 1910): 387.
6. A 1934 navigational map shows the dredge spoil from the port's channel. Before they went into the formation of Dodge Island, these unconsolidated areas, south of the County Causeway and the channel, contrasted with the rounded, geometric forms of the Venetian Islands to the north and Fisher Island to the east.
7. Edwin A. Steinmeyer, "Dredging up the Butler Act," *The Florida Bar Journal* LXXIII (April 1999): 69.
8. Legislation repealed the Butler Act by implication in 1951 when title for submerged tidal lands was conveyed to the IIF Trustees. Dade County (including Biscayne Bay) and Palm Beach County were not included in this transfer, but the 1957 law—officially identified as Bulkhead Act of 1957 (Chapter 57-362)—included all Florida counties.
9. Center for City Park Excellence and Trust for Public Land, *2011 City Park Facts* (Trust for Public Land, 2011). According to the 2011 report (with data from 2010), Miami provided 2.8 acres of park land for every one thousand residents. The median for this statistical category is 6.8. The previous year (2009), Miami offered 2.3 acres per one thousand residents, and its park land constituted only 4.2 percent of overall land area. In 2010, that figure increased to 5.2 percent but was fourth from last and still half of the median figure for other cities. Neighboring Hialeah provided the least percentage (1.4) for all high-density U.S. cities.
10. December 10, 1963, *Minutes of the Trustees of the Internal Improvement Fund* 34 (Tallahassee, FL: 1964): 363.
11. This text comes from 1940s promotional cards for the Miami Miramar Hotel.
12. December 10, 1963, *Minutes*, 363.
13. December 10, 1963, *Minutes*, 364.

14. The IIF Trustees conveyed thirty-seven acres of submerged land in Biscayne Bay to the American Legion Post, but because it had not developed the land, a reverter clause required the payment of one thousand dollars per acre at the time of its sale. In May 1966, members of the Post began negotiations to convey submerged land east of the bulkhead line back to the Trustees, who would then waive the reverter clause for the three acres remaining between shore and bulkhead lines. Later, the city would develop this filled area as a three hundred foot extension of Legion Park. See May 31, 1966, *Minutes of the Trustees of the Internal Improvement Fund* 35 (Tallahassee, FL: 1966): 504–505.

15. Magnolia Park is at the western end of the Julia Tuttle Causeway, and its corresponding island park would be associated with the dredged material that supports the causeway. On August 19, 1964, the IIF Board approved the City of Miami's application to extend Morningside Park shoreline eastward by filling 3.75 acres of bay bottom land. See August 18, 1964, *Minutes of the Trustees of the Internal Improvement Fund* 35 (Tallahassee, FL: 1966): 39–40.

16. May 11, 1965, *Minutes of the Trustees of the Internal Improvement Fund* 35 (Tallahassee, FL: 1966): 226. The islands were "within a maintenance spoil area granted by the Trustees to the United States" (in sections 27 and 33, Township 52 South, Range 42 East), and they would revert in the event of "non-use or conversion to other uses." The three islands are Quayside (Christo's island Number 7), Heikers Islands (Christo's island Number 8), and Crescent (Christo's islands Number 9 and 10, and the USACE's spoil island No.1). Its two islands were evidently considered to be one at this particular time. North Miami had approved the project on April 13, 1965. By 1971, the cities became more amenable to partnerships, as long as leases involved a public purpose. The company Interama received approval from the IIF Board to lease Sandspur Island as a dedicated preserve: "The purpose of the lease is to preserve the spoil island in its natural state to provide a windbreak and wave barrier for a proposed marina installation." As a part of the lease, Interama was to be responsible for "policing and maintaining the area free of debris, trash and refuse and will allow public use." See the lead-up discussions on December 15, 1970, *Minutes of the Trustees of the Internal Improvement Fund* 38 (Tallahassee, FL: 1973): 115.

17. November 17, 1970, *Minutes of the Trustees of the Internal Improvement Fund* 38 (Tallahassee, FL: 1973): 125. In 1970, the Bay Harbor Mayor sought to maintain the islands after their 1965 dedication to the city had expired. See the discussion in chapter 5's section "Green Relief."

18. January 4, 1966, *Minutes of the Trustees of the Internal Improvement Fund* 35 (Tallahassee, FL: 1966): 401. Because the islands were in the "permanent maintenance spoil areas granted by the IIF Trustees" to the U.S. government, their use was also approved by the U.S. Army District Engineer and FIND. This dedication followed through with the park projects initiated in 1963 and 1964. It included what would be known as Teacher's Island—later Pace Picnic Island—to the south, Osprey and Morningside Islands east of the ICW, and Mangrove and Legion Islands to its west. IIF Board minutes locate the islands in the following sections of Township 53S and Range 42E: 17, 18, 20, 29, 30, and 31.

19. Burt Chernow and Wolfgang Volz, *Christo and Jeanne-Claude: A Biography* (New York: St. Martin's Press, 2002), 295. Jeanne-Claude continues with the recollection of this incipient idea for the project: "It was my idea, but if I'd waited ten seconds, he might have opened his mouth to say the same thing. I was thinking two or three islands. In the end, he chose eleven. The choice of islands was entirely his."

20. Christo has noted: "It was an urban project, not a rural project, like *Over the River*, *Running Fence*, and the *Umbrellas*." (Christo, interview by author, tape recording, Gainesville, FL, 21 September 2011). See also Grace Glueck, "Christo Drapes Miami Isles in Pink," *New York Times*, 5 May 1983, C19. In the article, Christo pointed out that Miamians "use Biscayne Bay as a water, rather than a grass, park."

21. The project summary is included in *Christo: Surrounded Islands*, trans. Stephen Reader (New York: Harry Abrams, 1985), 29.

22. See Christoph Gerozissis, *The Florida Landscape: Revisited* (Lakeland, FL: Polk Museum of Art, 1992).

23. Cathy Lynn Grossman, "Christo: It's a Wrap," *Miami Herald*, 1 May 1983, 1L.

24. Christo, interview.

25. Christo, interview.
26. Grossman, 1L.
27. Grossman, 1L. Christo also noted: "The essence of art is that it belongs to man. It is an irresponsible, irrational, poetical gesture."
28. In her book *The Long Frontier* (New York: Harper and Row, 1967), Douglas continues this description: "the great bay shimmering and streaked with pale jade and azure, holding the creamy reflections of clouds in its fine polish" (4).
29. Christo, interview. Christo continues: "And you should see that the color of the water was also a principle element playing with that work of art. It would be very boring if the water were all the same color. And this is the incredible amount of light green and bluish green in the northern part, and the southern part has deeper water, more algae—or more vegetation—in the water."
30. Chernow and Wolfgang Volz, 295.
31. Christo, interview.
32. Christo, interview.
33. Christo cites *Wrapped Coast* (1969) and *Ocean Front* (1974) as precedents for this project, but the bay's urban context and scale made *Surrounded Islands* unique. In a bay nine miles north of Sydney, the artists wrapped eighty-foot-high cliffs and a portion of the beach for *Wrapped Coast*. In Newport, Rhode Island, *Ocean Front* also explored where sea meets coast: "There is a continuous play between water and the solid earth in our projects." (Christo, interview)
34. Christo, interview.
35. *Christo: Surrounded Islands*, 14.
36. Salvador Dalí, cited in Wayne Andrews, *The Surrealist Parade* (New York: New Directions, 1990), 95.
37. In his "Chronology" for *Christo: Surrounded Islands* (1985), Wolfgang Volz notes that the anchors on the islands were hidden with crushed beer cans (96). Interviewed in 2011, Christo does not remember this detail, but the story suggests an interesting mix of spoil island and garbage products, which in the case of beer cans have become associated with the islands.
38. Christo, interview.
39. Christo, interview.
40. The surrealism found first in the islands and then in Christo's project relates to the "subconscious of the landscape" that Werner Spies finds in Christo's work (*Christo: Surrounded Islands*, 11).
41. Andrews, 95.
42. See *Christo: Surrounded Islands*, 39.
43. Christo has also referenced Monet's *Water Lilies*: "these are my [inverted] water lilies," and he compares the cables and the seams of fabric to veins: "In the seams of the fabric we put five inches high foam, so that the seam was acting like a vein in your hand that was sticking vertically from the flat surface of the fabric and creating incredibly beautiful, almost a red line when you are close to the island in a boat. You are very close to the fabric, and it created hundreds of these like a bloodline vertical." (Christo, interview) In early photographs, Fisher's memorial island for Flagler is a perfectly formed lily pad, delicately fringed with what will prove to be inadequate bulkheading.
44. See Werner Spies, "Christo," *Christo: Surrounded Islands*, 10–11.
45. *Christo: Surrounded Islands*, 18.
46. See the entry for "unfurl" in the *Oxford English Dictionary*, 2nd ed. (New York: Oxford University Press, 1989).
47. *Christo: Surrounded Islands*, 114.
48. Mikhail Bakhtin, *Rabelais and His World*, trans. Hélène Iswolsky (Bloomington: Indiana University Press, 1984), 10.
49. In this sense, *Surrounded Islands* refutes Terry Eagleton's critique of Bakhtin's argument. Eagleton wrote: "Carnival, after all, is a licensed affair in every sense, a permissible rupture of hegemony, a contained blow-off as disturbing and relatively ineffectual as a revolutionary work of art." It is true that Christo's project needed permits as a "licensed affair," but its artistic event came out of the pre-existing whimsy and public nature of the islands that would

remain after the project. See Eagleton, *Walter Benjamin; or, Towards a Revolutionary Criticism* (London: Verso, 1981), 148.

50. Christo, interview.

51. Christo described his impression of the fabric that created an "incredibly beautiful" condition, "almost a red line when you are close to the island in a boat. You are very close to the fabric, and it created hundreds of these like a bloodline vertical." (Christo, interview)

52. Charles Lee's *Miami Herald* editorial, cited in Spies, "Christo," *Christo: Surrounded Islands*, 20. Lee incorrectly refers to the fabric as plastic—a distinction Christo was very careful to maintain.

53. Grossman, 1L.

54. Glueck, C19.

55. Dary Owens, "35 tons of Christo Pink Wrap Down in the Dumps," *Miami Herald*, 1 July 1983, 2D.

56. *Christo: Surrounded Islands*, 16.

57. Christo, interview.

58. Also see chapter 3's discussion of Alois Riegl's definition of monument.

59. It is interesting to note that the United States' first national wildlife refuge preceded its first national monument. President Theodore Roosevelt dedicated Pelican Island—not Biscayne Bay's spoil island, but the natural island 130 miles to the north—in 1903, three years before he made Devil's Tower the first national monument with the Antiquities Act. See chapter 3's discussion of Pelican Island Refuge's origins.

60. Tony Smith, "Talking with Tony Smith," interview by Samual Wagstaff, Jr., *Artforum* (December 1966): 18–19.

61. See Anna C. Chave, "Minimalism and the Rhetoric of Power," *Arts Magazine* (January 1990): 53. See also John Powers' commentary, "Indicator Spaces," accessed January 28, 2012, http://www.johnpowers.us/indicatorspaces.

62. Smith, 18.

63. Christo, interview.

64. Plato, *Statesman*, trans. Robin Waterfield, eds. J. Annas and R. Waterfield (Cambridge: Cambridge University Press, 1995), 35.

65. See *Christo: Surrounded Islands*, 51.

66. Chernow and Volz, 298.

67. Ruby Blondell, *The Play of Character in Plato's Dialogues* (New York: Cambridge University Press, 2004), 335.

68. Christo, interview.

69. Plato, 82 (306a-311c).

70. Spies, "Christo," *Christo: Surrounded Islands*, 16–17.

71. Christo, interview.

72. Michel Serres, *The Natural Contract*, trans. Elizabeth MacArthur and William Paulson (Ann Arbor: University of Michigan Press, 1995), 44.

73. Werner Spies discusses this in terms of the "irrational grounded in realism" that is found in Christo's work (9).

74. Christo, interview.

75. Serres, 44.

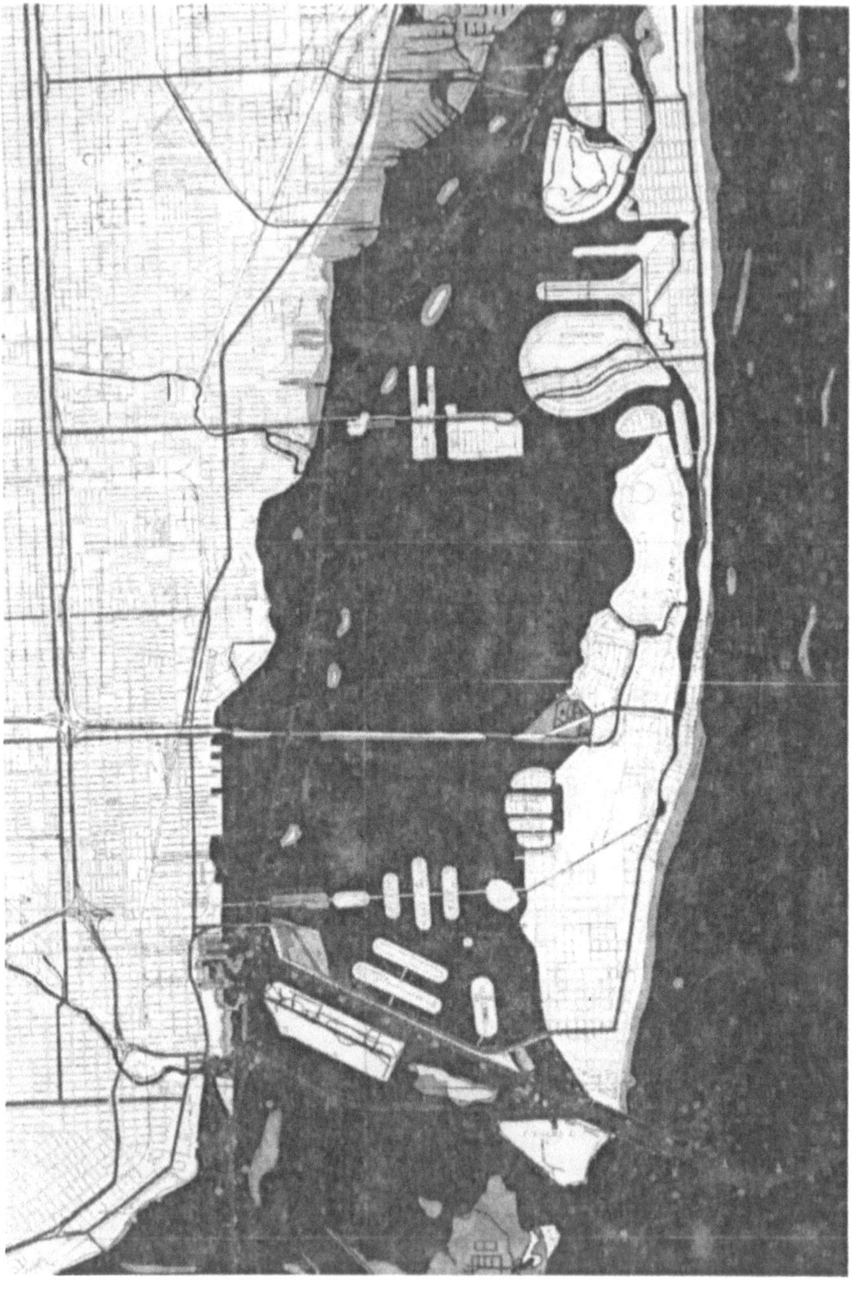

Sand

Lightning flashes, heavy clouds darken, and the air horn sounds. Lifeguards call swimmers in as the late afternoon squall signals an end to the weekend and sends beachgoers to their cars and back to the mainland. The storm is a reminder that this is an unlikely place for a beach, particularly vulnerable to changes in weather like a boat on open water or the board sailors who favor its winds. The island's position also allows for a perspective on the storm, a 360-degree view of water and wind. As the clouds approach, changes in color, light, shadow, and texture are measured by an admixture of concrete, asphalt, palm trees, sea oats, rip-rap, and sand.

Its sand, white like paper stretched between steel jetties, Fred Howard Beach floats one mile offshore in the Gulf of Mexico. A gently curved causeway tethers the beach island to Tarpon Springs' mainland like a kite born on water rather than air. On its west side is the concave beachfront, and to the east is the parking lot formed with billowing curves at its landward corners. The twelve-acre island is half sand and half asphalt—a smaller version of Robert Moses' paradigmatic beach infrastructure on Long Island's south shores and the Bronx's crescent-formed Orchard Beach. Like those, this Florida beach is artificial and it certainly privileges the car. Unlike those, Fred Howard is unmoored. Without a mainland's shoreline, it is quite simply sand, with a lifeline of automobility that allows access to two million annual visitors. Named for Tarpon Springs' four-term mayor, Fred Howard feels like a city park bringing diverse uses to the middle of Saint Joseph Sound. Before the storm, a church group gathered to pray during its Sunday "open waters" service in the Gulf of Mexico.

It is a pocket beach without headlands. Only the remnant stitching of two reinforced jetties keeps the sand in place. With the tops of their twenty-two-inch-wide concrete beams sometimes revealed under a dusting of sand, the jetties bracket a wholly constructed landscape that borrows its details from road-building. Like the precast construction used for retaining walls, sound barriers and temporary dividers along highways, a post-and-panel concrete fence separates parking lot from beach and catches windblown sand to form bermed dunes. A sidewalk approach to the island's northern edge doubles as a sluiceway to drain the parking lot's impervious surface. Along the causeway's length, the road itself becomes a double-sided beachfront for cars and people. On both sides of its exaggerated convexity, the road slopes into the water for fishermen, swimmers, and people-watchers. Angled parking spaces line the southern edge, with bumper stops to prevent them from rolling into the water. On the other side, a wide shoulder makes room for parallel parking and sunbathing.

From the air, the island's molded sand educes formal associations. With undulating causeway tail, it is a stingray gliding out to sea, or a cobra head ready to strike one of the outer keys to the west. Its utilitarian organic form and biomorphic shape also recall the supple Art Nouveau engineering of Hector Guimard. This aerial view is a diagram that disengages beach from land and suggests a planning model for beaches that is eccentric and justifiable, heavy-handed and sensitive, disruptive and balanced. Making the beach an island altered currents and damaged underwater lands, but moving the beach offshore left the mangrove shoreline largely intact. The mangrove habitat remains integral to the mainland park's upland hammock of live oaks, longleaf pines, and sea grapes. Although the beach island might seem to be an outlandish and unique approach, it was not the area's first when it opened in 1966. One mile south, Sunset Beach also presses out into the Gulf waters. Smaller and older, this beach island began as a private dredging project by a local landowner sometime before 1941 when it first appeared on aerial views. Now, the five-acre sand outcropping is a public park that follows a beach island typology.

Sand is a limited commodity here in Pinellas County's offshore areas. Very little new sand enters what has been described since the 1970s as a relatively closed system. After its six-thousand-year eastern migration, the sand is now reserved to dunes, berms, near-shore bars, and inlets. Sand blowing across the asphalt of Fred Howard's causeway and parking lots underscores the sand's transient presence on the limestone bed of this coast's submerged lands. With this topography, the emergence of sand bars along the Pinellas coast was not uncommon, and mid-century concerns were raised about the status of these embryonic islets. So, it is coincidental but not unexpected that Fred Howard's history would be intertwined with its naturally-formed counterparts. At its December 31, 1963 meeting, the Internal Improvement Fund (IIF) board approved the dedication of submerged land for the Fred Howard project, and, at the same meeting, it heard a report from Florida's Attorney General that "emerged" lands officially fell under state sovereignty. The day before IIF's meeting, Secretary of Interior Stewart Udall's ruling disapproved an earlier interpretation of the Submerged Lands Act, strengthening the state's claim to these offshore lands, both natural and human-made.[1] Though bolstered by this decision, the project's endorsement ran a protracted course, slowed by the dedication's transfer from city to county and by the governor's concern that the project might be privatized. In 1966, the IIF board granted final approval for what the project's planners claimed would "improve" the "very shallow water" for public recreational "beach purposes" and would also facilitate access to the Intracoastal Waterway.[2]

The beach was renourished in the mid-1980s and again in 2002. In this most recent augmentation, sand came from land instead of sea. More than four hundred truckloads of beach-compatible sand were harvested from an upland area of the state's interior. The umbilical causeway provided access for almost seven thousand cubic yards of material. Beach nourishment is common along Florida's nearly seven hundred miles of beaches, and the Fred Howard project falls under the broad heading of human enhancement of natural features—technically termed "anthropogenic economic activity." But putting a beach out where one did not previously exist also means something else. Enhancement assumes an existing value structure, which can be changed by degrees and which, it is assumed, is temporary and will have to be repeated. And places like Fred Howard represent another kind of place altogether. Its sand posits a new typology—a working hypothesis that leads to a radical experiment in alternatives for both beach and island. It also cuts a very clear diagram of what spoil islands do. This four-sided beach of sand and asphalt is a reminder that spoiling one view might afford many others and that not spoiling a place will have implications for its alternatives.

NOTES

1. See December 31, 1963, *Minutes of the Trustees of the Internal Improvement Fund* 34 (Tallahassee, FL: 1964): 383–384.
2. For the initial dedication of more than one hundred acres of submerged land, see July 16, 1963, *Minutes of the Trustees of the Internal Improvement Fund* 34 (Tallahassee, FL: 1964): 257. For the later re-dedication to the county of almost two hundred acres, see June 7, 1966, *Minutes of the Trustees of the Internal Improvement Fund* 35 (Tallahassee, FL: 1966): 508.

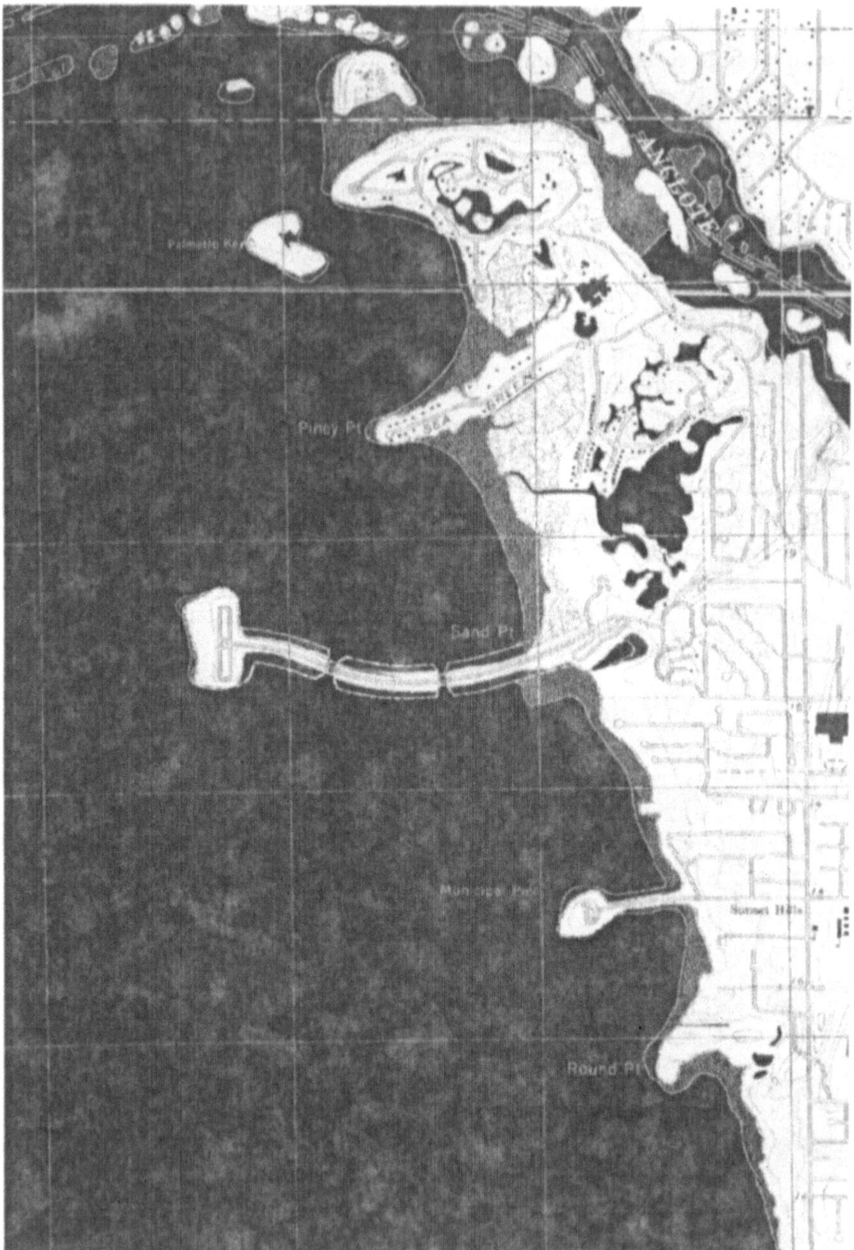
Palmetto Key
Piney Pt
SEA
Sand Pt
Round Pt
Sunset Hills

Chapter Seven

Order and Disorder

Navigating Key West's Western Margin

SUNSET

If it was only the dark voice of the sea
That rose, or even colored by many waves;
If it was only the outer voice of the sky
And cloud, of the sunken coral water-walled,
However clear, it would have been deep air,
The heaving speech of air, a summer sound
Repeated in a summer without end
And sound along. But it was more than that,
More even than her voice, and ours, among
The meaningless plunging of water and the wind,
Theatrical distances, bronze shadows heaped
On high horizons, mountainous atmospheres
Of sky and sea. [1]

Sunset at Key West. On its western shoreline, Mallory Square's performers are out. Refrigerators balanced on foreheads, flaming batons juggled, and Rondini the escape artist shackled and suspended. At this daily sunset celebration, Dominique the Cat Man takes his stage, a prominent location above Westin Pier. Behind him the sun dips lower, and orange sky polishes rolling water. But there is another backdrop, breaking the western horizon line and borrowing from both sea and sky. Backlit by the coloring atmosphere, Sunset Key is a "bronze shadow" against which Dominique performs. It is not a key but a human-made island and now a Westin Resort. If you were to rent one of the cottages, typically costing more than one thousand dollars per night, you would board a ferry at the resort's eponymous pier, here below the Cat Man, for the fifteen hundred foot ride across the harbor channel. This "theatrical distance" extends Mallory Square's performance space, while the spoil island reminds us that there is a sunset to be seen further west, another island available to some, inaccessible to most.

Here, more than one thousand miles from the official start of the Intracoastal in Norfolk is the terminus of the east coast's waterways. Here, also at land's end, mile zero of Route 1 begins a nation's roadway infrastructure. Here, fifteen hundred miles by water from Belmont Island where this book's narrative began, a spoil island interrupts the horizon and upstages Key West's westerly status. Actually, two such islands stretch the continental limits. Sunset Key pairs with Wisteria Island to spoil Key West's liminal position and—some would say—spoil the sunset. Development of the former suggests one kind of order, and eight hundred feet northward riotously swaying pines of the latter signal a disorder and another kind of ordering.

Wisteria came first. It continued to emerge and expand as Wallace Stevens was composing his poem "The Idea of Order in Key West." Key West's main channel and harbors had recently been dredged to a depth of eighteen feet, so that by late 1933 the warships from its Navy base were anchoring comfortably just offshore as a nation pondered how to address Fulgencio Batista's military coup in Cuba. On what had become his annual vacation to the Keys, the poet might have also witnessed the spoil island's namesake, the quarantine ship *Wistaria*, catch fire and burn in February 1933. The following year, Stevens took up residence at Casa Marina Hotel, again to limn the country's borders and explore Key West's "summer without end." Stevens admired Key West as a primitive paradise. In a letter to Harriet Monroe, Stevens contrasted Key West with the "pure pastiche" of places like Coral Gables: "Key West, however is the real thing . . . of the sweetest doing nothing contrived."[2] This environment would provide a stirring context for Stevens to explore the dynamic relations of the imagined and the real. During his first trip to the Keys in 1922, the poet witnessed the area's magical realities at a fish camp on Long Key: "The ground is white coral broken up, as white as this paper, dazzling in the sunshine. . . . The place is a paradise—midsummer weather, the sky brilliantly clear and intensely blue, the sea blue and green beyond what you have ever seen."[3] By 1935, Key West had lost some of its appeal to Stevens. No longer the wonderfully primitive and "most old-fashioned place in the world," it was "becoming rather literary and artistic."[4] But the poet had already found an incomparable paradise.

Stevens returned north to Hartford later in the Spring and completed the Key West poem in October 1934. "The Idea of Order in Key West" is not a poem about spoil islands, but it is a poem about a particular place where sea and land, nature and human, imagination and reality converge. It is a marginal, sometimes disordered, place where order is discovered rather than imposed. It is a meditation on the type of space that occurs between Key West and the newly forming, human-made islands to its immediate western edge. The process of discovering order and its concomitant topography, recounted by Stevens in the Key West poem, parallels the idea of order—and disorder—found in spoil islands. Stevens' poem is a nautical map for understanding Key West's unique permutation of dredged islands and for reconsidering the broader implications of the east coast's spoil island archipelago.

TWO MORE KEYS

At a nation's terminus, Key West shares its sunsets with two spoil islands. It also now concedes its westernmost development and its penchant for nonconformism to the islands, one developed for high-end residence and the other resisting annexation in spite of its long histories of utility. To those looking west from Mallory docks, the two islands present a stereographic view that oscillates between order and disorder: on the left, Sunset Key, previously called Tank Island, is densely developed with houses offering a formal plan of development but patently disrupting the natural sunset, and on the right, Wisteria Island is the background for transient sailors and squatters and waits for decisions about its fate. Wisteria is also known as Christmas Tree Island for its stand of Australian pines that were stripped during Hurricane Wilma in 2005 and now sway with the trade winds like the masts of the sailboats that fill its popular anchorage. This latter site began as a shoal where the disinfecting barge *Wistaria* ran aground, served as a quarantine facility, and then burned, to be replaced by a shark camp preparing the hides for a lucrative market. Both islands grew out of the Navy's dredging program for its strategic Key West base, and both are linked by the developer's gaze of Pritam Singh, who squatted on Wisteria while dreaming how to develop Sunset. The two islands are topographical emblems for the order and disorder that characterize spoil islands.

As the final installment in the sequence of spoil island topographies, this chapter demonstrates how spoil islands resist political and social order and at the same time confirm an unlikely logic of manufactured land. These islands offer an informative and sometimes imaginative proximity between order and disorder, marrying resistance with stability and matching openness with restraint. The specific landscape of Key West is a unique place of reckoning, a vulnerable site for storied residents seeking an antidote to continental pretensions. These islands are not just a mainland edge but also carry the weight of a nation pressed into a narrow frontier not much wider than the overseas highway's right of way. It is the end of the road at the limits of a continent. This outpost quite literally tests the waters of extended infrastructure, expanded naval defense, and an outsized culture of independence and play. And the spoil islands play at being keys. Naturally-formed keys are built on the surface of coral reefs where currents have deposited residual sediment and skeletal remnants from other reef areas. Tank Island was renamed "Sunset Key" to align itself with this biogenesis, and both spoil islands engage an artificially ordered version of this constructive process with their own deposits of sand and marl placed atop bank and reef. Such island-building simulates key formation.

Wallace Stevens' 1934 poem "The Idea of Order in Key West" frames a principled, if sometimes ambiguous, topos where the poet attempts to reconcile artifice and nature. The poem does not directly reference the spoil islands, but it touches on similar ideas and moves the setting of human experience from mainland's solid foundations to the instability of places where water meets land. These are places where the "dark voice of the sea" rises and "water-walled" coral sinks. Stevens shifts questions about what orders human activity from earth to water. Similarly, the cover image of J. Hillis Miller's *Topographies*, an extended commentary on Stevens' work, presents a map of Key West that centers not on the main island but on the two spoil islands.[5] In his own reframing of Key West's aqueous topography, Stevens captures the place's "blessed rage for order" and, in the process, outlines what can serve as an ethics of making and occupying islands. This is "the maker's rage to order words of the sea." In the seeming disorder of spoil islands, we might find a provisional—sometimes private, at other times shared—order in waste, nature, and culture. It is fitting that the first of the two spoil islands was named for a shipwreck.

WISTARIA THE SHIP AND WISTERIA THE ISLAND

In the early morning hours of October 11, 1909, the lull in major hurricanes to strike Key West ended. Carrying Stevens' "heaving speech of air," winds in excess of one hundred miles per hour and barometric pressure of 28.42 inches pushed flotsam, boats, and sea water against Curry's Docks.[6] Among the fragmented vessels and splintered masts was the quarantine steamer *McAdam*, named for Assistant Surgeon William McAdam, who had died ten years earlier—almost to the day—of yellow fever.[7] Key West was particularly vulnerable to disease as well as hurricanes, and the U.S. Marine Hospital Service had recently transferred the steamer to the island from Biscayne Bay. Key West was growing rapidly in the first decade of the twentieth century, and it soon outpaced Miami's marine-hospital in the number of inspections, processing more than forty thousand passengers and crew members. With the loss of the *McAdam*, the Marine Hospital Service sought a replacement vessel with room for both inspection and detention. Another storm in October 1910 slowed the response, but on February 24, 1911, the service finally commissioned the 150-foot *Wistaria*.[8] A seasoned vessel built in 1882 by

Pusey and Jones Company, *Wistaria* had previously tended lighthouses for twenty-nine years along the east coast from Charleston, South Carolina to Portland, Maine.[9] The retrofitted ship provided spaces for inspection, isolation, and treatment; and it was equipped with a steam disinfecting chamber, a thirty-six-foot boarding launch, and room for about twenty detainees and patients.[10]

The most powerful hurricane to hit Key West disrupted quarantine services again in 1919. The September 10th storm unmoored the *Wistaria* and drove most of the vessel's ten-foot draft into the spoil area at the southern edge of the Frankford Bank. The iron-sided steamer first launched into the Delaware River's dredged channel had come to rest on a human-made shoal at the nation's terminus. In its fixed position, the *Wistaria*'s decks continued to host inspections that were approaching one hundred thousand each year, and the vessel remained in service until the mid-1920s. *Wistaria*'s decks soon became the site of shark-skinning operations. By the late 1920s, leather production from shark skins was well-established in Key West, and an article published in *Popular Mechanics* noted two shark-skinning stations in the Keys, one at Big Pine Key and one in the vicinity of Key West along Frankford and Pearl Banks.[11] A Keystone View Company stereograph shows the May 1926 activities at this latter location.[12] In 1928, Lowe Fish Company signed a lease agreement with the Secretary of Navy for the official use of Wisteria Island as a site for preparing shark hides. Representing the company, Benjamin Lowe had requested permission to use part of "an exposed spoil bank of coral lying to the westward" of Key West. Citing the Bank's isolated location and its lack of use at the time and deeming "mutual benefits" of the proposal, the Navy authorized a revocable license that lasted for one year at thirty dollars per month and permitted the construction of temporary structures.[13] Although not well-documented, Ray Knopp, a building inspector for the city of Key West, also set up a shark camp on the spoil island itself before World War II.[14]

Frankford Bank's shoals hosted precedents for human-made constructions, and processing sea catches offshore in the Key West harbor was not new. As early as 1896, "slaughter pens" are shown on the U.S. Coast Survey's nautical maps of the area. U.S. Navy Rear Admiral James Forsyth had earlier ordered these pens to be removed from the coastal waters, amidst the rising tensions that would lead to the Spanish-American War. By 1907, according to the Coast Survey's "Key West Harbor" map, the pens had returned. These offshore areas had also served as dock extensions. In 1885, Manatee Dry Dock Company applied for and received permission to construct dry docks—"not exceeding ten acres in area"—on the tidal lands of Frankford Bank west of Man of War Harbor. The company planned to build warehouses, wharves, and platforms for the dockage and repair of vessels.[15]

Wistaria's decks were reportedly also the scene of "pleasure parties." In the 1920s and 1930s, shark-skinners shared the platform with partiers and picnickers. Boaters found that the stranded ship, as well as its island, afforded a place of escape, an offshore site that would later harbor more than traditional picnics and "happy gatherings" with its reputation for hedonism and lawlessness. It might have been one of these earlier, less riotous, events that led to the spectacular burning of the vessel on February 10, 1933: "Flaming skies and flying sparks in the northern heavens 2:15 o'clock this morning apparently portended a raging conflagration."[16] Fearing from a distance that mainland property was being destroyed, hundreds came to watch the flames from Key West's waterfront. By early morning, the side-wheeled ship had burned "from stem to stern." The iron hull remained, but a "goodly number of shark hides were destroyed by the flames." Among the crowds, we might imagine the poet Wallace Stevens watching the wind-born sparks and "flying brands" that threatened to ignite the main island's docks. Looking to the northwest through the "theatrical distance" between island and ship, Stevens would have seen the mound of spoil, backlit by flame-soaked air. Transferring outmoded spelling to common name, the remains of *Wistaria* the ship would come to identify Wisteria the spoil island. By a coincidence that is quixotic but wholly fitting, this name identifies a non-native ornamental plant, bringing to mind fragrant lavender flowers as well as its vine's vigorous invasiveness. First marginalized by quarantine and waste deposit, ship and island share space on the rising shoals of Frankford Bank. As the vessel receded, dredging advanced its namesake.[17]

GATE TO THE GULF

> Key West is practically the gate to the Gulf of Mexico, and war vessels necessary to protect the commerce passing to and from Gulf ports should be able to take station at this place or to enter the same freely.[18]

> Key West has been called the Gate to the Gulf of Mexico; she undoubtedly holds a commanding position both as a military and naval base for strategic reason, and is a commercial center through her proximity to Cuba, proximity to Panama, and quick rail connection with the American mainland.[19]

Pressing out into the lower Gulf of Mexico and upper Caribbean, the island of Key West anchors a tenuous filament of coral and sand that stretches from southeast Florida to the nation's southernmost point. Though long considered a distant outpost, the island and its harbor are inextricably linked to commercial aspiration and military strategy. Key West's naval base dates back to 1821, when John Simonton bought Key West from the Spaniard Juan Pablo Salas and—fearing the effects of the town's well-known lawlessness on its future development—lobbied the U.S. government to take advantage of its strategic location, closer to Havana than Miami. Mallory Square was the site for the first military base in 1823 when piracy threatened U.S. trade routes. With Florida's statehood in 1845, the Keys were reserved for military activities; and when the rest of the Keys reverted to the state in 1855, part of Key West and the series of keys to its north and west remained under the control of the federal government for naval operations.[20] In the decades to follow, the base's naval squadron battled illegal slave trade, remained a Union stronghold during the Civil War, and backed up calls to apply the Monroe Doctrine and achieve greater Caribbean influence.

In 1908, Commodore W. H. Beehler began his report to the Naval Institute in Annapolis with the following summary of Key West's strategic importance:

> The principles of strategy are the same for all ages, independent of the changes in tactics that are introduced by the changes in weapons, and a commanding site on the highways of commerce must always be of great strategic value. No change in armament or methods of warfare can ever affect the strategic importance of such a site as that of Key West. . . . With a radius of one thousand miles, Key West (the gate of the gulf) is the center of a circle that passes through Washington, San Juan, Port Rico [sic], and Panama.[21]

Accompanying the Commodore's presentation, a diagram presented the "lines of force" that radiate from Key West's location and demonstrated the circular range of "reliable wireless communication" that encompassed the Gulf and the Caribbean and extended up to New York. Beehler's expansive comments coincided with the first of two Executive Orders to affirm the reservation of Key West's surrounding islands for naval use. Theodore Roosevelt's order setting aside a strategic archipelago—"the islands situated in Key West Harbor and its approaches"— also affected future public and private uses of Wisteria and Tank Islands, where civilian strategies for development and aspirations for preservation would meet.[22]

Echoing naval commanders and military engineers, Key West's early boosters and businessmen promoted the island's strategic importance as a commercial gateway. Writing as the railroad made its way to Key West and publishing his work in 1912, the same year Flagler boarded the inaugural train ride, Jefferson Browne effusively described his native city's potential for commerce in spite of its historic isolation: "That there would be a city of 22,000 population on an island in the gulf, without a railroad or a wagon road connecting it with the county of which it politically forms a part, is the best evidence of the commercial importance of Key West."[23] The Florida senator, who later served as the state's chief justice, went on to extol its gateway status and noted that ships putting into Key West only added "about ten miles to their voyage—an advantage possessed by no other port in the United

States." Eagerly promoting his island, Browne frequently elided geographic constraint and engineering costs. He also served as chairman of the Florida Railroad Commission in 1904, and eight years earlier he had written a *National Geographic* article that made a case for extending rail service across the Keys. In the article, he also argued for the ease of construction and characterized the keys as natural formations that could be perfected with modest changes: "When cleared of a few inches of vegetable mold and loose stones, the surface of the islands is as level and smooth as a ballroom floor."[24] Browne may have underestimated construction logistics, but he did not miscalculate the commercial value of the topographical changes around his native island. He would later seek to buy Wisteria Island, and his application to the IIF would trigger the second Executive Order from the President's office.

Key West's significance derives from its naturally deep harbor and its location within a network of channels, which plot its "gateway" position. Brown hyperbolizes but effectively introduces this position: "No other city in the United States occupies or has occupied such a unique position. Its harbor, landlocked by keys and reefs, in which the largest ships can float, has four entrances. . . ."[25] Coral heads, sand banks, and shoals surrounded Key West but made room for a network of no fewer than eight channels that have historically provided access to and from the main island. Skirted to the west by the Northwest Channel and to the east by Key West's main harbor, Frankford Bank forms a crescent tilted slightly west of a true north. Its southerly arc ends at Wisteria and Tank Islands, where the Northwest Channel branches from the Main Channel. Further to the west, the Middle Ground rises up between two branches of the Northwest Channel on its way past east and west jetties that mark entry to the Gulf of Mexico's open waters and a "shortcut" to Tampa. The West Channel sweeps south of an irregular line of keys—Mule, Crawfish, Man, Ballast, and Woman. Discontinuous, undersized, and smaller than the artificial island Wisteria, these keys do not challenge Key West as the chain's terminus, but they do mark the channel's northern boundary and the western passage to Marquesas Keys and the Dry Tortugas. Back east, the Main Ship Channel connects Key West's harbor with the Florida Straits along a north-south axis. Its approach passes close to the island's western shoreline and just east of the natural shoals, which became Wisteria and Tank Islands, before opening into Man of War Harbor.

Before its current name, the harbor reflected not battle among humans but refuge from conflicts with weather and nature. Hurricane Harbor became Man of War Harbor in 1855 when the Navy officially reserved the harbor. Defending against storms as well as northeasterly trade winds, its naturally protective attributes fell into military service to maintain a nation's security. This natural harbor forms the center of what is now a system of filled land and artificial shorelines. Extending northward from the city, the harbor's natural depths historically ranged from twenty-four to thirty feet. Its secure calmness would inspire poetic language in naval reports—"magnificent sheet of water"—and superlatives in early piloting guides—"large and commodious . . . one of the best anchorages for large vessels south of Chesapeake Bay."[26] But this splendid water and its channels required maintenance and deepening for modern vessels. At the conclusion of his Naval Institute presentation, Commodore Beehler alluded to an isolation and eccentricity, which would historically characterize Key West's military and civilian operations alike, when he decried the lack of funds for dredging operations: "With the exception of some dredging in the northwest channel, for the convenience of . . . mail steamers, there has never been any money expended for the improvement of this harbor."[27]

HAPHAZARD

> Key West has not been systematically developed but grew up in a haphazard way as emergencies arose in former wars and as conditions presented temporarily in peace time.[28]

Eight years later, Key West's dredging operations had reached a larger scale but remained disorganized. A 1916 report noted the area's unsystematic and "haphazard" operations in which commercial and military interests combined to maintain and deepen this network of channels. This work afforded opportunities to expand Key West's land area with dredged fill. Commercial interests helped build the terminus of the Overseas Railroad at Trumbo Point, created Key West Bight, and augmented the island's western edge near Mallory Docks. These activities contributed to the newly made land, but it was the Navy's presence that sculpted Key West's profile and populated its harbor with artificial islands. Dredged spoil filled the area around Fort Taylor, the 1845 fort previously accessible only by causeway; further expanded Trumbo Point as a military installation; and generated the Truman Annex, with its outer mole and submarine base. By 1971, the island of Key West had nearly doubled in size to reach an area of three thousand acres.[29]

Making land was linked to the art of war and defense. Cycles of dredging and filling around Key West alternately previewed and served as postscripts to regional and global conflicts. As hurricanes threatened its ships and made *Wistaria* an artificial reef, war's looming tempests spawned Key West's artificial islands and eventually made Wisteria and Tank Islands critical features of its defense works. Dating the origins and the fluctuating deposits of the two spoil islands is difficult, but wartime exigencies and peacetime development record these displacements of coral, marl, shell, stone and sand.[30] Wisteria first emerged on Frankford Bank when the main ship channel was cleared of coral heads, reefs, and other obstructions in 1908. During World

War I, naval officers planned a dock with six berthing slips on the harbor side of this spoil area. The May 1917 project called for a "secure and safe wharfage on Frankford Bank, with ideal Aviation Site and unlimited room for dry-dock and other naval expansion . . . taking advantage to the fullest extent of all the deepest channels."[31] After the war, in January 1919, permits were issued for additional dredging in the harbor for a submarine basin and protective seawalls, breakwaters, and piers.

By the 1920s, Wisteria had become a significant topographic feature of almost three acres. A drawing produced by the Department of the Navy on January 21, 1927 documents the island's location, volume, and features in great detail. Precisely established by longitude and latitude coordinates, the center of the spoil "bank" is a half mile from the Navy Yard, it is 2.95 acres, and the majority of the land is three feet above mean high water (MHW), having been filled by a hydraulic dredge. With its crescent-shaped length running east to west, the island curiously takes the form of a sprouting bean. A radicle sprout at its western end supports a shack and a timber wharf, which rests half over water and half rooted on its landward side. A hatched area along the southeastern sweep of its curve is noted as a pile of six thousand cubic yards of dredged coral rock. Deposited by a dipper dredge, this embryonic core of the island rises 25 feet, an elevation making it the highest land feature in the Florida Keys at the time. Despite this prominence, a thin lagoon curves between this elevated area and the rest of the three-foot spoil deposit, indicative of the island's vulnerability to tidal change. Charted to the southeast of its eponymous wreck, Wisteria's deposit remained in this basic form with a more pronounced lagoon up to 1940, when the early stages of World War II necessitated deeper channels and more island infrastructure. By 1943, Navy and Army engineers were collaborating on dredging operations that deepened the main channel and extended the spoil island northward to its full extent of thirty-nine acres.[32]

At this time, Tank Island—Wisteria's spoil sibling—emerged as two small islands almost due west of Mallory Docks. In the first years of 1940, dredging expanded the submarine basin, extended Fort Taylor's land area by twenty-one acres, and deepened the main shipping channel to thirty feet. In July 1943, the War Department recommended a "proposed spoil area" at the southern limit of Frankford Bank, where Tank would begin to materialize. After the war, dredging for the twenty-six-foot turning basin near Fort Taylor added slightly to the island's accumulated spoil in late 1951 and early 1952, and aerial views from 1959 show a three-acre island—evocatively similar to Wisteria's embryonic shape—resting a quarter mile offshore from Key West and also supporting a small building on its western tip. In 1964, a project to enlarge the naval air station's turning basin and to deepen the channels to thirty and thirty-four feet further augmented the spoil area, and a year later the Navy oversaw continued dredging of its basin on Key West's southwestern edge to accommodate diesel submarines that required a depth of thirty-six feet. The U.S. Army Corps of Engineers began dumping dredged material on Wisteria but soon shifted the deposits to the sandbars closer to the basin. The project's ten million cubic feet of spoil created a twenty-two-foot-high mound, surpassing by six feet the Keys' highest elevation until it was spread out and leveled to nine feet. By 1966, Tank Island was fully formed, its two tanks added a few years later along with cables and pipelines to the main island.[33]

Clear indications of Key West's "haphazard" growth, the spoil islands carry part of the burden of its strategic location. In addition to their original function as waste deposit sites, Wisteria and Tank have offered a combination of efficient propinquity and necessary distance. They are close to the naval air station's center of operations, but far enough away for accessory purposes. In the early decades of the twentieth century, the islands served as material reserves for naval construction projects to further extend Key West's filled lands. But the spoil areas also amplified the natural protection afforded by the shoals of Frankford Bank, adding a layer of defense from storms and enemies and protecting Man of War Harbor, which had been designated as a "hurricane harbor" of safety from storms. Commodore Beehler recognized the importance of this protective barrier and in April 1908 asked the Secretary of the Navy for permission to supervise ongoing dredging operations that threatened to remove material from the Bank's southern extent. In spite of the Commodore's assurances in his letter, dredging activities had become a source of friction between the Navy and the Army, who contracted the dredge project to expand Fort Taylor. Beehler feared the Army's contractors would use fill that entailed the least cost "regardless of its effect upon the channels," and he suggested widening and deepening the channels rather than the "indiscriminate removal of Frankford Bank."[34] In 1918, a Navy report also contended that breakwaters would be unnecessary for naval station development because protection could be achieved around Wisteria by "filling in on Frankford Bank and the planting of mangroves in the fill to insure permanency."[35]

Commander Warren Terhune, Beehler's successor at Key West, also saw Frankford Bank's partially submerged keys, shoals, and spoils as substrates for the Navy's expansion and as sites "capable of development." The spoil islands offered safely distant but still proximate sites for hazardous and potentially explosive infrastructure such as fuel tanks. Installation of fuel storage on Tank Island reduced the dangers of explosion within the main base. In 1916, naval commanding officers recommended the expansion of Key West's station as a base for warships, submarine facilities, an aerial base and aeronautic station, and a mine-layer depot. This latter program, with the dangers of "mines and magazines," was planned for upland spoil areas on

Frankford Bank, "separated from the proposed aeronautic station and fuel storage plant on Fleming Cay." [36]

In 1908, the Navy identified the shoals of Fleming Key as sites for the expansion of its station. After examining Key West's waters for two years with a "wire-drag" system, the Coast and Geodetic Survey determined that the removal of only a "few obstructions" could yield channels with widths of one thousand feet and with depths of thirty-five feet for Man of War Harbor's battleship anchorage. At this time, the Department of the Navy outlined its proposal for Fleming Key:

> The bureau approves the proposition, which is strongly recommended by the commandant of the station that a torpedo depot be established on Fleming Key in Key West Harbor. It is proposed to build a retaining wall on the key to raise the greater part of the surface (about 60 acres) to a level about 8½ feet above low water. The Florida East Coast Railway Company has agreed to deposit 165,000 cubic yards of material (taken from the adjacent shoals last year) on the key when called upon, and the remaining material necessary for the fill can be obtained by dredging. [37]

Three decades would pass before the area around Fleming Key was proposed as suitable site for ammunition storage. Before the artificial island was finally constructed after the start of World War II, Wisteria was viewed as a reserve of land that could be developed if necessary, but the Navy's 1917 plans for the aviation facility and large docking area for military vessels were eventually superseded by Fleming Key's development. From July 1941 through the end of 1942, the Navy dredged channels for its seaplanes, and spoil was deposited across more than two hundred acres of partially submerged land to the north of Trumbo Point. The two-mile-long island included thirty-two fenced and camouflaged magazine areas, a boathouse, a watch tower, and service buildings. [38] With the completion of Fleming Key, Wisteria remained a strategic component of the military's broader field of operations that extended outward to include Woman and Man Keys, which had been reserved by executive order for naval practice grounds and war games. Wisteria's proximity to Fleming Key's secure zone also meant that security could best be preserved by leaving the island empty. Wisteria's significance within the military reservation would soon reach the U.S. President's desk for a second time, signaling a complicated future that mixed layers of administration with civilian aspirations.

DEVELOPER'S GAZE

Despite continued military presence, Wisteria Island almost went to the highest bidder in 1924. Though not even three acres of spoil material, the island was beginning to draw attention beyond its strategic reserve for waste:

> Notice is hereby given that the Trustees of the Internal Improvement Fund of the State of Florida, will hold a meeting at 11 o'clock A. M., Tuesday, May 20th, 1924, for the purpose of considering sale of a submerged tract of land in Monroe County, described as follows, to-wit:
>
> An island in the vicinity of Key West Island, caused by the deposit of excavated material from the Ship Channel. The deposit lies 1800 feet, more or less in a Northwesterly direction from the Porter Docks, which are at the end of Fitzpatrick Street, City of Key West. It extends 400 feet, more or less, in a Northeasterly direction, 300 feet, more or less, in a Northwesterly direction, and contains approximately 2.8 acres. All in Section 6, Township 68 South, Range 25 East. Exact description to be furnished with deed. Certified check for $500.00 must accompany all bids.[39]

On February 26, 1924, two Florida residents, Judge Jefferson B. Browne of Tallahassee and Daniel L. Navarro of Key West, applied to purchase the "shallow bank" and island in tidal waters northwest of Key West. They agreed to bid a minimum of five hundred dollars for the land. Complying with Florida statutes, the Internal Improvement Fund (IIF) placed an advertisement in the *Key West Citizen* newspaper. This ad provided the next meeting time, the island's origins in excavated spoil material from the Ship Channel, and a brief description of its location and size. Notified of the pending sale, the Navy filed objections to the transaction. At its May meeting, the IIF Board postponed action until the following month, when the Trustees decided to "withdraw the island from sale and reject all bids."[40]

Undoubtedly aware of increasing development pressures at the height of Florida's land boom, Secretary of Navy Curtis Wilbur promptly drafted an executive order that would reserve "all islands" in the vicinity of Key West. In his cover letter, he wrote to President Calvin Coolidge that the "Navy Department has for over 30 years held undisputed possession of Fleming Key and the adjacent shoals, including Frankford Bank and Man of War Harbor. In view of their strategic location for naval purposes and in order to avoid confusion in the future, it is desired to have them formally reserved for naval purposes."[41] Indicative of the request's urgency, two days later on August 11, 1924, President Coolidge issued Executive Order 4060 that reiterated the reservation, for naval purposes, of this area of Key West's waters that included the spoil island Wisteria.[42]

In his letter, the Navy secretary specifically cited his concern about private ownership of the islands. Noting the location of Key West's defensive armaments, he pointed out that Frankford Bank and its shoals were "directly under the guns" of Fort Taylor. He then summarized the hazards of errant shots and the economics of friendly fire: "If privately owned and developed, this location might become a constant expense on account of the claims by private parties for damages incidental to gun fire." The islands were in range. They could also later serve as outer defense works if Key West's military installation was modernized. As an attachment to his request to President Coolidge, the Secretary included a chart that delineated the proposed military reservation area. The map is an extraordinary document for its combination of rectilinear boundaries with contoured topographic edges on its southwestern limit that forms a prong stretching the reserve to the extreme southern end of Frankford Bank and capturing the spoil banks, less than one thousand feet across the shipping channel from Key West. The majority of President Coolidge's Executive Order is taken up with the metes and bounds narrative necessary to describe this bounding line as it rambles southward to capture the inchoate spoil islands.[43] By "rounding the southern point of Frankford Bank for a distance of 1,500 feet more or less," the reserve reaches out to its topographic quarry.[44]

With wartime memories receding and pressures mounting from local politicians, the IIF Board finally accepted the sale of Wisteria's spoils. On July 24, 1951, Paul Sawyer, on behalf of state house representative Bernie C. Papy, applied to purchase the thirty-nine-acre area. The proposal offered fifty dollars per acre for the property described as "a parcel of bay bottom land and spoil area northwest of the island of Key West."[45] Urgent correspondence followed within the Navy's offices. On August 15, Key West's commandant sent a Naval Speed-Letter to Washington to outline the naval air station's concerns. Signed by Captain W. T. Eckberg, who served as the Naval District Engineer, the letter cites Coolidge's previous executive order from 1924, notes that Navy funds for dredging created the majority of the spoil island in 1943, and calls attention to the island's immediacy and its strategic value: "Due to the proximity of this spoil area to highly classified Naval activities, it is considered a dangerous security risk to allow this property to fall in the hands of private developers . . . the strategic location of this spoil area makes its use for military purposes highly possible, and its use for a fuel storage area is now under consideration."[46]

At the IIF Board's subsequent September 4th meeting, competitive bidding resulted in the sale to Sawyer and Papy for seventy-one dollars per acre. Although the sale was finalized, the deed had not yet been delivered when the Navy filed its official objection to the transaction on September 27th. At the January 8, 1952 meeting, Land Clerk Sinclair Wells recommended that IIF Trustees confirm the sale and transfer the deed if Papy was "willing to take the risk of condemnation by the Government." The deed was assigned the following day.[47] In office for thirteen terms as the state's longest-serving legislator at that time and unopposed for his fourteenth term when he died, Papy now owned this controversial real estate for $2,769.[48] The Navy's objections went unfulfilled, and Papy went on to make a handsome profit. He sold the spoil area in March 1956 to Amaryk Aldo for $115,000. In the interim, Navy correspondence illustrated the continued disagreement—and friction—over the island's ownership. In 1953, the Bureau of Yards and Docks wrote the Bureau of Land Management's Director that its office had for many years considered Frankford Bank of "great strategic importance" and then—almost ironically—"but apparently the State of Florida does not agree."[49] In 1956, after Papy sold the island, the Navy considered acquiring Wisteria but decided to review pre-existing property claims and to determine if it had been "built for federal use"—a condition that would seem to be evident from the island's dredging history.[50] This exploration of the "reason of the deposit of the spoil" was not completed amidst a three-decade lull in official federal discussions of the island. In the meantime, it was the developer's gaze—rather than artillery sight-line—that was leveled on the spoil island. The property was transferred to Aldo's development company Wisteria, Inc. in 1962 and then sold to F.E.B. Corporation in January 1967 for $317, 750.[51]

> Wisteria Island—that 39-acre piece of choice real estate covered with whispering Australian pines and inhibited [sic] by a few renegades living on a deserted barge—will be transformed into a multi-million dollar hideaway for wealthy boat owners cruising the Caribbean.[52]

Wisteria Island lines up with one of Key West's major axes. David Wolkowsky knew this when he set out to build the Pier House Resort Motel in 1967 at the northern terminus of Duval Street, where it meets the two thousand feet of water between the dock and Wisteria Island. A year earlier, he proposed a multimillion-dollar resort on the island he preferred to call "Christmas," noting "how much more intriguing" this name sounds.[53] One story of the island's name recounts that when wives of naval officers planned a picnic to the spoil area, they advertised their excursion to "Christmas Island." Having seen the swaying Australian pines offshore from the naval base, their audience would know exactly where they meant to go. Other residents of Key West also saw the allusion. For them, whether originating from the Navy wives or not, it was "Christmas Tree Island."

South on Duval Street, soon after returning home to southern Florida from Philadelphia, Wolkowsky had already set up Pirate's Alley, which transformed an area of lower Duval Street into a tourist shopping area that included the Key West Cigar Factory, employing Talmaege Culmer, the last of the island's cigar makers. With Pirate's Alley at one end and the Pier House at the other, Wisteria Island extended Duval Street's axis as well as the developer's vision for a private archipelago of luxury. While Pier House was under construction, Wolkowsky also purchased Ballast Key. Considered the southernmost land in the contiguous United States, the key is set within the Key West National Wildlife Refuge, and its ownership projected private development into public land holdings. Eight miles from Key West, Ballast forms a background for the developer's near-shore aspirations.

Wolkowsky knew Key West and he wanted Wisteria. His plan for Wisteria approximated a lagoon island. The rendering of the multimillion-dollar development "for the wealthy who sail the Caribbean and seek privacy and comfort" shows a yacht club, a yacht basin carved out of Wisteria Island's center surrounded by homes on one hundred foot lots, and a three-hole golf course on the northeastern quadrant. With the yacht basin as a lagoon, the project exploits the island's artificiality and simulates not the coral foundations of a Florida key but the coral-rimmed volcanic islands of the Pacific along with their associations of escape, abundance, and paradise. Wolkowsky, with his partner Clyde Partin, a businessman from Orlando's citrus industry, planned to engage architect Edward Stone's firm in the formulation of plans for the island. It would be ten years before Stone would complete the new State Capitol building in Tallahassee; but the architect was already an international figure known for the Kennedy Center in Washington and the St. Louis Cardinals' Busch Stadium, and his name added weight to a project that for some may have seemed impossible or at least mystifying. Veiled in showman-like mystery, the developers' broader plan included a "secret project" for one hundred acres of submerged land north and northeast of Wisteria.

Seven years later, another developer's gaze originated from Wisteria Island's community of squatters.[54] When Paul LaBombard, Jr. camped on Wisteria in the winter season of 1973, Tank Island had only recently emerged out of the Navy's dredging program and the twenty-year-old transient had not yet embarked on his career as a developer. But LaBombard would have seen the newly formed island—its ground still a bare bank of sand and its two fuel tanks still empty—and perhaps dreamed of its potential. In conventional expectations of how developers begin their careers, LaBombard was a nonconformist, a characteristic that some Key Westers would be wary of but many others would embrace as appropriate for the free spirit of the "Conch Republic."

The idiosyncrasy of his early life is not unlike the eccentricity of the island where he camped, and camping is certainly an atypical mode of real estate speculation. LaBombard first surveyed Key West's real estate two years earlier from the Old Post Office Inn's porch, a similarly marginal but mainland location. Diving for shells to sell to tourists, he lived on the porch from January until April 1971, after hitchhiking to Key West at the age of eighteen. He left in late April to attend the May Day protests in Washington, D.C., where he was arrested for civil disobedience. Soon after these trips, LaBombard converted to Sikhism, changed his name to Pritam Singh, and lived in a Massachusetts ashram for five years. In 1979, his transformation of a Maine house into ten apartments began a real estate career that less than seven years later would allow him to purchase Key West's most valued property. At this time, the environmental impact statement for Tank Island and Truman Annex was passing through draft form as a part of the "disposal of surplus military properties in Florida."[55]

The year after Singh left Key West, the city faced the closure of its Naval base. In March 1974, military personnel were transferred and civilian workers were laid off.[56] Key West formed a Redevelopment Agency (RDA) to lease the facilities, and its tax-generating goals soon conflicted with many community members who wanted the project to include open space and public waterfront access. During the Reagan administration, amidst increased military activity and production, the Navy reclaimed part of the base from the General Services Administration (GSA) and the RDA, but it kept aside the forty-two-acre parcel of the Truman Annex and Tank Island. Reflecting increased public concerns, the City Commission in 1983 passed resolutions requiring that any new housing must be designated affordable and that an eleven-acre park must be included in plans for Truman Annex. With continued discord and the failed coordination of a development plan, Key West's RDA broke up two years later, and the Navy—frustrated by the city's lack of consensus about the property—made plans to auction the property the following year.

When Pritam Singh read the *New York Times* article about the imminent auction of Truman Annex property, which included Tank Island, he had three days to travel to Key West. On September 10, 1986, Singh arrived in time for the auction, posted the requisite certified check for $250,000, and in less than an hour outbid eleven other developer groups for the property.[57] The developer paid $17.25 million for 103 acres, including the forty-two-acre parcel of the Truman Annex and the twenty-seven-acre island. It had taken more than twelve years for the Navy and GSA to negotiate this transfer, and much of the media's focus centered on the Annex property, which included numerous historic structures, most notably the "Little White House" where President Harry S. Truman spent his working vacations in Key West. But it would be the island—by some reports included in the sale parcel as an afterthought—that was more emblematic of Singh's distinctive approach and the development pressures influencing dialogues about Key West's future.

Only days after the auction, confusion about jurisdiction and ownership of Tank Island emerged. It was not clear whether the island fell within city or county boundaries, and only in 1987 would the city's annexation of the island—at Singh's request—settle the issue. At one point, the Navy had given the island to the city as the site for a sewage treatment facility, which was never built, and the island reverted to the Navy.[58] The island's ambiguous status also affected the final sale price. Of the fifty-five acres associated with the island parcel, slightly more than half were submerged lands, some of which belonged to the state rather than the federal government. In the auction, it turned out that the government sold $250,000 of state land that it did not own, reducing the sale price to $17 million. One of these anomalous tracts was less than a tenth of an acre of upland on Tank Island. Its genealogy can be traced to a curious—and unlikely—collaboration between Bernie Papy and Admiral R. Y. McElroy of Key West's naval base. On May 2, 1961, they made a request for the transfer of submerged land and two small spoil islands that would seem to have been already reserved for the Navy by the precedent set with Wisteria. Their application, which was accepted, called for the deposit of spoil material from the dredging of the turning basin onto the fifty-five acres that was "desired for construction of additional Naval facilities." Papy's motives were not clear, but in the meeting he did point out that the "expansion and development would benefit the City of Key West and the people of Monroe County."[59] An investigation determined that the Navy exceeded contractual boundaries when it built the island, and the Federal government paid the state of Florida $24,534 after the auction.

Soon after the auction and well before the tanks were removed in 1988 to become artificial reefs in two hundred feet of water south of Key West, Pritam Singh renamed his island. The translation of Tank Island to Sunset Key is as much the extension of Key West's geography as it is also an indication of Singh's early philosophy for development of the property. "Sunset" denotes the natural—even cosmological—event already celebrated from Mallory Square, which at the time was Key West's only public waterfront. "Key"—not island—connotes another naturalized connection to the archipelago's geographic chain. Islands can be built, but keys must be grown over long periods of time with the accumulation of coral. Sunset Key pushes the 157-mile chain of developed keys one-quarter mile farther.

Early on, Singh expressed a sense of environmental and social responsibility for his newly purchased property. Reflecting his long-held interest in adapting older structures in the northeast, Singh noted that it "would be foolish not to" restore the Annex's historic buildings, even though many of the development restrictions could not be enforced.[60] Whether based on a desire for sustained control or an interest in design—he had attended architecture school at night for a short time in Boston, Singh developed a twenty-seven-page set of design guidelines drawn from the site's Victorian and Bahamian architectural context. For the Annex, he also proposed a meditation center, an aviary, and much-needed public space along Key West's waterfront. Though more conservation by degrees than full-fledged resource protection, Singh's eccentric approach toward Tank Island appealed to environmental groups. The Florida Audubon Society presented him with an award for waiting to clear Tank Island's vegetation until a colony of terns hatched their eggs.

It was the spoil island where Singh's aspirations and his project's exigencies and paradoxes were most evident. Rather than historic structures, the island had Australian pines and fuel storage tanks. For Singh, it lacked the natural attributes that would prompt conservation: "I think that all land that is not already touched by human beings should be left alone. I think a piece of land with an oil storage site is all right for our use."[61] With the extent of human intervention, this definition keeps very little, if any, land from the scope development and recalibrates Singh's position as an "enviro-developer." Reports at the time noted his interest in reconciling ethics and profits, and his Sikhism resonated with the goals of an "enlightened capitalist."[62] Singh hoped to translate his faith's "spirituality, democracy, and egalitarian economic approach" into his development practices. The clearing of the island's vegetation returned it to the tree-less state Singh would have observed from his Wisteria campsite, and the tabula rasa of this human-made land opened the possibility for applying Sikh values but also marked the start of a project that would permanently alter the view from the public spaces he praised.

His island formed a highly visible part of Key West's public panorama, and its auspicious position in the foreground of each day's sunset could not be resolved with the site's development opportunities. In 1989, Singh presented plans for a 285-room Ritz Carlton hotel on Sunset Key. The project also included villas, houses, retail space, and four restaurants. Singh wanted its architecture to reflect the architectural styles he had seen in Old Town. Underscoring these goals to infuse the island's identity with far-flung places, Singh reported that he would import eleven thousand tons of Bahamian beach sand to soften the rocky, rip-rap shore of the island. Singh presented his hyperbolic plans with equally theatrical speeches at public meetings, which were themselves associated with Key West's celebrated twilight phenomena: "To hear him speak is an event. It's like going to a marvelous opera or a really beautiful sunset."[63] In the three years following his purchase, Singh gained numerous concessions from the city that allowed him to redefine affordable housing with a growth management ordinance that did not impose income restrictions on buyers and that allowed yards and medians to count toward open space and parkland requirements.[64]

In spite of these allowances, Singh declared bankruptcy in 1990, and the island's future followed changes in the real estate market and vagaries of development. Two more years passed before Singh continued formulating the project with new partners from the Truman Annex Real Estate Company. The start of construction on the island stalled; and in 1994, Ocean Properties bought Sunset Key from Singh's creditors and constructed forty-five houses on the island by 1996.[65] Today, the island is exclusive and private. The public components of the island's plans had already begun eroding by the time of its fourth revision in September 1989 when island perimeter available to the public had been reduced from a half mile to 650 feet.[66] Managed by Westin Resorts, the island's waterfront is now reserved for the owners of its forty-eight single-family homes and the tenants of its forty guest cottages.[67] Rates range from $595 to $2,225 per night, Oprah Winfrey rented all forty cottages for her forty-fourth birthday, and its twenty-one vacant lots are each valued at more than $1.5 million.[68] Singh's environmental vision for development, tempered by the human-made island's artificiality, was also lost. Although its promotional literature cites ecological consciousness and certification by Florida DEP's Green Lodging program, the resort's "environmental commitments" are vague—"we participate in environmental projects to improve our community and educate our neighbors"—and ordinary—"we provide all guests with the option to re-use linen and bath towels."[69] Singh's "enlightened capitalist" approach had fallen victim to market demands, and Sunset superseded much of Key West's real estate value.

The spoil island's waste had allowed Singh to play the part of environmentalist and real estate developer. As "enviro-developer," Singh cleaned up Tank Island of its eponymous, though never-used, industrial residue.[70] As astute businessman, Singh sought to maximize profit from the real estate venture dubbed "Sunset." But Singh's vision also included a self-described "hippie outlook."[71] This characteristic—and his need for a place to live—had led him to the margins of an already liminal town.[72] Wisteria became a squatter's home and allowed for a comfortable distance to look back on Key West and also to look over to Tank Island. Wisteria's indeterminate status provided a place to live, and its pairing with Tank provided a place to dream. The squatter's view became the developer's vision.

MESS

> There was just a whole lot of misinformation flying around. It is a mess. You know, we've been stewards of that thing for 30-odd years and it's never been as dirty as it is now. It's gotten significantly worse after the hurricanes.
>
> People talk about the island like it's some paradise; it's a mess. We're trying to do something that's an asset.[73]

To some, Wisteria anchors a disordered landscape. Unruly Australian pines, stripped bare by the 2005 hurricane, wave in the winds that continue to erode the island. A motley collection of live-aboard sailboats bobs in the waters that surround it. There is garbage, but that problem is not unique to the spoil island. Wisteria's messiness reveals something else. In one sense, it recalls the island's origins in waste. Without bulkheads or rip-rap, the deposit of dredged material is inexact and subject to the whims of natural forces. Piled up by dredgers, pushed around by hurricanes, and overgrown with uncultivated and untended vegetation, Wisteria is naturally untidy. In another sense, the island's material was unwanted until it served as a barrier to hurricanes or a repository for fill in other parts of Key West. Augmenting Frankford Bank, the island established a first line of westerly defense for Fleming Key's naval operations and also protected the area's unsanctioned anchorage. But hurricanes come and go, and projects to extend Key West's island with fill have all but ended.

No longer a deposit area for dredged waste or a critical site for naval defense, Wisteria is now just a spoil island. For some, it lacks a clear purpose, and from that viewpoint, dirt is considered to be dirty. For others, it should remain undeveloped and resolutely public as a conserved park area. For most, its layered histories and administrative complexity make the island a conundrum—beautifully composed but unrelentingly disordered, like its namesake's flowering vine. It has harbored a makeshift array of programs—from shark camps to homeless camps, it has sparked debate between military divisions about its best use, its presence has directly influenced two executive orders, and surveying errors and controversies of jurisdiction have alternately erased and exposed it.

In 2007, the developer Roger Bernstein used the term "mess" to describe Wisteria. In one sense, he was citing the complexity of the issues and viewpoints—the messy situation that is confusing and offers difficulties for resolution. In a second sense, messiness is the dirt that generates disorder. Contrasted with paradise, mess is troubling and detrimental to social norms. Euphemistically, mess can also be excrement, particularly that which is out of place. But it is not an aesthetic vision or a sense of impropriety that troubles those who believe that Wisteria is out of order. The problem for this view is that the island has not met its potential as real estate paradise. Whether underutilized paradise, much-needed public green space, or just dirt, the messiness of the island has redefined planning methods and frameworks. Spoil island joins non-conventional typologies like mooring fields and offshore islands to alter discussions about the ordering of land use in general and the conserving of public land in particular.

Forty years after his family purchased Wisteria, Roger Bernstein partnered his corporation with the developers of Sunset Key. F.E.B. Corporation and Ocean Properties began the process of the island's development. Despite its location two thousand feet from the city, Wisteria is in Monroe County, not in Key West, so the first step in this process was annexation. On May 1, 2007, a first reading of an ordinance to annex the island came before the Key West City Commission.[74] Initiated by the property's owners, this annexation was voluntary, and some commissioners saw an opportunity to increase the city's tax revenue. Based on Sunset Key's density, the city planning department estimated that Wisteria's twenty-one upland acres could support 168 units, significantly more than Sunset's development.[75] City commissioners initially voted to approve the resolution five to two, but a second reading was required, and two weeks later a final decision was narrowly postponed. A local group—the Wisteria Island Committee—collected petition signatures for a charter amendment requiring all annexation decisions to go to public vote. At the same time, Key West's mayor sponsored a non-binding referendum to gauge whether voters wanted the city to purchase the island. Bernstein withdrew his application for annexation, no trespassing signs were posted around Wisteria, and voters passed the annexation amendment and turned down the mayor's referendum for the island's purchase. The "messy" debate about Wisteria—one commentator noted how much its name sounds like "hysteria"—had revealed public interest in stewardship and a spoil island's vulnerability.[76]

OFFSHORE ISLAND

Three years after seeking annexation by Key West, Bernstein's F.E.B. Corporation presented its Wisteria development plans to Monroe County. On June 25, 2010, despite opposition from the state's Department of Community Affairs (DCA), the county's planning commission approved the proposal for thirty-five market-rate housing units, thirty-five transient units with eighty-five bedrooms, a 116-ball mooring field, a restaurant, and a bar. The proposal went to the county commission, which, before casting its votes, asked Key West's city commission to provide its opinion. Key West's commissioners unanimously voted against the proposal, and DCA's concerns about increased density requiring extended service infrastructure and creating strains on hurricane evacuation centered debate around the island's land-use classification.[77] But this also proved to be messy.

Wisteria might have been charted, but it had not been mapped. The spoil island was left out of the future land use map (FLUM) when planners drafted Monroe County's comprehensive land use plan in 1992. In planning jargon, it had not been "FLUM'd." Drawing boundaries, establishing shorelines, and differentiating islands is not easy in the Keys, where everything is an island and land's edges are determined by tidal flux. Defining "offshore island" might be clear from a mainland perspective, but from island to island this designation becomes ambiguous. And in practice as a planning term, "offshore island," or "OS," eludes definition, as Monroe County's growth management staff noted: "There is no definition of offshore island in the Comprehensive Plan or the Land Development Code."[78] In Monroe County's land development code, "Offshore Island" (OS) serves as a catch-all term for areas "without a specific land use designation."[79] But the land development code's description of OS's purpose does make clear what is not an offshore island—a differentiation specific to the Florida Keys. If an island is not connected U.S. 1, it is an offshore island. The land code goes on to define the purpose of OS districts as the establishment of protected areas with minimal habitation—such as camping—that can be served by self-contained facilities.[80]

In the Fall of 2010, this purpose was a main point of contention among developers, planners, and citizens. Monroe County's comprehensive plan categorizes OS districts as Tier 1, natural areas of environmental sensitivity and conservation.[81] With this background and in their deliberation over an appropriate FLUM for Wisteria, the county's planning officials recommended Wisteria's designation as Residential Conservation (RC). Wisteria's developers sought zoning for mixed-use commercial. They argued that Wisteria was not an offshore island because it was a spoil island and its origins in waste and its unnatural state made it unsuitable for the RC designation. Their case focused on the characterization of the "predominance of undisturbed native vegetation" in areas designated RC.[82] At commission meetings, developers indicated that the island's habitats were disturbed and included only patches of native vegetation. But local groups—including Last Stand—countered with evidence that the island supported native vegetation and wildlife, including a bald eagle, white-crowned pigeons, and numerous native plant species. In the discussions, planners noted that the comprehensive plan does not distinguish between spoil islands and offshore islands. On July 27, 2011, the county's planning commission approved the FLUM designation of Residential Conservation, limiting development to a maximum of five houses.[83]

Before the decision, Bernstein began the process to amend the land use plan with a new category, the "maritime harbor island" designation. The revised proposal included less than two acres providing public access to the island and tying into a managed mooring field.[84] Wisteria provides refuge for two anchorages, one north of the island and the other to its southeast, where many ships have historically anchored on the western edge of Man of War Harbor.[85] These informal mooring fields, particularly the latter, have come to be identified with Wisteria. In 1972, Bernstein's F.E.B. Corporation purchased the submerged land underneath the northern anchorage.[86] Even though it was clear from the outset that it could not be developed, the land became part of discussions about a mooring field, which coincided with a study recommending the establishment of a managed field in this location and a pump-out program throughout Key West harbor to protect its waters from the dumping of polluted effluent from vessels.[87]

In one of his frequent contributions as a guest columnist for *Keys News*, Bernstein argued that the proposed mooring area would improve the current field of "derelict" vessels and provide an "affordable cruising destination" with its proximity to Key West.[88] In his commentary, the informality of the existing mooring field mirrors the marginal status of the "man-made spoil (dredged) island," which harbors "squatters and criminal activity" and is covered "with invasive exotic species." The developers emphasized the public benefit and environmental improvement of a managed mooring field for Wisteria, but critics noted that the proposal entailed patron public use—by those who subscribed to the development's marina—rather than an open public use.[89] Nonetheless, after the planning commission's July decision, Bernstein reiterated F.E.B.'s project goals, effortlessly deploying each issue: "how best to transform an abused spoil island into an accessible environmental playground."[90] But Wisteria's past history as another kind of offshore island—for military purposes—would further obscure these plans.

When President James K. Polk reserved offshore shoals and islands for military objectives, this broadly sweeping executive order set the context for Wisteria's formation and its preservation as federally held land. And after Florida's first governor reiterated these goals for the Keys in 1845, groundwork was also laid for the offshore island's public status.[91] Six days after the planning commission's vote to designate Wisteria as an RC district, the naval air station's Public Affairs Office issued a news release on August 2, 2011, reporting that the Navy had no claim to Wisteria Island.[92] At first glance, this was good news for the island's developers, but the news release included a detailed genealogy of the island's ownership, concluding with a reference to Public Land Order No. 6214 of 1982.[93] This document revoked the previous executive orders and reservation for the Navy, but federal claims over the offshore lands remained and reverted to the Bureau of Land Management (BLM).

On November 7, 2011, a letter from BLM provided a significant turning point in discussions about the ownership and development of Wisteria Island. Its reply to an inquiry about the island's federal ownership concluded that legal title appeared to be "still vested in the United States."[94] The letter also noted other layers of federal jurisdiction—management by the Department of Interior, use by the Department of Treasury, and administration by the United States Fish and Wildlife Service's Key West National Wildlife Refuge. The letter transformed the framework for discussions of Wisteria's future, affecting F.E.B. Corporation's application for the land use amendment, which was to be reviewed by the county planning commission. Federal ownership also widened the gap between F.E.B.'s proposed private use and the island's status as a public place of conservation. BLM's mandate, under Public Land Orders, limits activities, maintains public values, and reserves areas for public purpose. With the BLM's letter, Wisteria moved further from the developers' aspirations and closer to the Refuge lands in which it had tenuously rested for more than a century.

Wisteria remained resolutely offshore as a green backyard for its anchorage's live-aboards and as an undeveloped backdrop for those watching from Key West's "mainland." Amplified by a real estate boom in the 1980s and increasing costs of living in Key West, urban growth has caught up with its human-made island counterpart to make both island and anchorage alternative places for affordable housing. Those same pressures have also made Wisteria the continued focus of development plans. Arguments for development propose closer connection to city regulations and ordering principles, while residents of the mooring field rely on this closeness but also favor the isolation and escape the island's position—with its anchorage—affords. Though sometimes made circuitous by multimillion-dollar yachts, a short dinghy ride—less than a half mile—brings full-time residents to Key West Bight.

The layout of boats in Wisteria's eastern anchorage is a stochastic exercise in social distance and natural forces. Personal space and shoaling determine placement. In aerial views, there is a marked contrast between Wisteria's unpopulated green terrain surrounded by its confetti of moored boats and the glistening surfaces of Sunset Key's roofs set within its largely unoccupied aquamarine field. The former island is a background for public living set within an extended field of floating dwellings, while the latter is an island where private development has been maximized, public mooring is prohibited, and its waters are too close to channels and too exposed to winds for a boater's refuge. Eight hundred feet away, moored like its live-aboards atop Frankford Bank, Wisteria has defied development. While ruderal vegetation has stabilized physical placement of its layered spoil, layers of federal jurisdiction have now affirmed its public status. It is a mess, but its marching orders were there all along.

STEREOSCOPES AND RECKONING: "IDEAS OF ORDER"

> Oh! Blessed rage for order, pale Ramon,
> The maker's rage to order words of the sea,
> Words of the fragrant portals, dimly-starred,
> And of ourselves and of our origins,
> In ghostlier demarcations, keener sounds.[95]

A woman walks next to the sea at twilight in Wallace Stevens' poem "The Idea of Order in Key West." She sings, and the sea's "dark voice" rises. But the two sounds are not "medleyed." They remain distinct so that her listeners—we assume Stevens is among them—can parse her song. The sea remains distinct—"merely a place by which she walked to sing"—but is also a part of the process that will make for "keener sounds." Her voice makes a world, one with a horizon "acutest at its vanishing" and with a "measured" solitude between voice and world. When she stops singing, her listeners turn to see the town's ships. They see them anew, as human-made constructs that have "portioned out the sea" and "mastered the night." Masts of the boats measure vertical distance, and their bows register the fluctuations of tide and current. The vessels' lights flicker, pinpointing location and indicating atmospheric change. They reflect on the sea as they also mirror the fallen sun. These vessels augment, contrast and sometimes order the sea's apparent formlessness. The boats, like the song, are ways of "fixing, . . . arranging, deepening, enchanting." The singer begins "beyond the genius of the sea," but ends with the sea as the context for a process of ordering. As such, the singer enlists the "spirit" of the sea. In the middle of the poem, Stevens asks "Whose spirit is this?" The answer to this query is more than the sea's voice and more than the singer's voice.

Making land also entails more than nature and more than human enterprise. The mix of land and water itself requires new frameworks. Making land accepts the sea as the context for ordering such human endeavors. Like moored ships, spoil islands will measure and record their changes at sea. They will rise and fall with dredging deposition and water-driven erosion. Their morphology and vegetation will change with tidal currents and microclimates. They will also register choices about development and conservation. In one respect, the pairing of Sunset-Tank and Christmas-Wisteria posits such conditions as discrete alternatives. The former has been developed as real estate, while the latter remains undeveloped. But we must recall that both have their origins in human activities, and this polarity is more complex, a function of layered histories and administration. This is not a simple choice between order and disorder, between asset and mess. The island pairing provides an approach to the idea of order itself. This idea can be approached, geographically, in two ways: from the water and from land.

From land, the islands can be viewed stereoscopically. Such a view from the docks assumes two perspectives on islands of similar origins, just as a stereoscope employs two images of the same object taken from a different angle. This subtle shift allows the stereoscope's viewer to see a two-dimensional object in relief, providing each eye with its uniquely projected vision. This view begins with the islands' shared origins in human agency and waste deposit and then defines differences to understand the "depth" of their meaning. Stereoscopically, the two islands might shift one over the other for a new dimensionality to elicit Stevens' imagination that addresses the world's messy realities. In his poetry, such human-made things can be "surrogates" for imagination.[96] These paired spoil islands, west of Key West, function similarly. The islands are stand-ins for natural islands. Confusion about their origins opens space for speculation about their future. Offering alternatives, they take on the role of a new, more westerly Key West. For some, this is the sanitized streets of Sunset Key; and for others, this is the open reserve of Wisteria Island.

From the sea, the islands offer a means of reckoning, a way of finding an indefinite position between land and water, between order and disorder. This speculative view should be balanced by the histories of these islands. At sea, from the northwest approach, you join shrimpers returning with their catch. Look back southeastward to Key West and let the spoil island pairing frame your approach. At starboard is the glint of galvanized roofing, and off your port bow is the calm repose of "green relief." Right and left, to set up a reckoning with the past. Reckoning seeks to settle difference between two separate parties. It is also a means of navigation that estimates position, working between computations of the course steered and observations of phenomenon observed. Eschewing astronomy, dead reckoning makes compromises between what is seen and what has been logged. It involves contemplation and calculation. This view works initially from the differences of these islands, arriving then at their shared heritage as places cast off from land. Attempts at order in the words of the sea.

NOTES

1. Wallace Stevens, "The Idea of Order at Key West," *The Collected Poems of Wallace Stevens* (New York: Vintage, 1990), 128–130.
2. Wallace Stevens, Letter to Harriet Monroe, March 2, 1934, *Letters of Wallace Stevens*, ed. Holly Stevens (Berkeley: University of California Press, 1996), 269.
3. Wallace Stevens, Letter to Elsie Stevens, January 10, 1922, *Letters of Wallace Stevens*, 225.
4. Wallace Stevens, Letter to Emma Stevens Jobbins, December 22, 1953, *Letters*, 806; Letter to Harriet Monroe, March 13, 1935, *Letters*, 278. Stevens had not been back to Key West since the early 1940s.
5. Fitting the book's title and Stevens' interest in collisions of earth and water, the cover image uses a topographical map to depict a place that is primarily defined by marine features and nautical activities. See J. Hillis Miller, *Topographies* (Stanford, CA: Stanford University Press, 1995).
6. Jefferson B. Browne, *Key West: The Old and the New* (St. Augustine: The Record Company, 1912), 159.
7. Assistant Surgeon McAdam died on October 12, 1899 after reporting the yellow fever outbreak in September. See Treasury Department, U.S. Marine-Hospital Service, *Public Health Reports* XIV (October 13, 1899). *McAdam* the vessel was commissioned in 1901 in response to the continued threat of the arrival of disease in Biscayne Bay. A photograph dated October 11, 1909, in the Florida State Archives shows the *McAdam* lodged in the wreckage at William Curry's docks (image number N041472).
8. *Annual report of the Surgeon General of the Public Health Service: 1910 (*Washington, D.C.: Government Printing Office, 1911), 96. The flat-bottom river steamer had previously served as Biscayne Bay's floating quarantine facility since 1901.
9. See Paul H. Silverstone, *The New Navy: 1883–1922* (London: Routledge, 2007), 211–212. See also Department of Commerce, *Fifty-fourth Annual List of Merchant Vessels of the United States* (Washington, D.C.: Government Printing Office, 1922), 562.
10. See the *Annual report of the Surgeon General of the Public Health Service: 1912* (Washington, D.C.: Government Printing Office, 1913), 73. For a full description of *Wistaria*,

see the *U.S. Congress Hearings of the Subcommittee of House Committee on Appropriations* (Washington, D.C.: Government Printing Office, 1922).

11. Hamilton M. Wright, "Leather from the Sea," *Popular Mechanics* (February 1928): 275–278. New York journalist Wright's travels through the Keys suggest that this area had become one of the centers for shark leather. Wright's article was also included in the Summer 2002 issue of the *Florida Keys Sea Heritage Journal*. The journal article includes a photograph of the "shark skinning station in Key West Harbor built over the remains of the Lighthouse Tender *Wistaria*," held in the archives of the Monroe County Library.

12. See Library of Congress, "Skinning the Sharks—a phase of Florida's fish industry, Key West," May 27, 1926, item 2002711258.

13. "Revocable License," 17 August 1928, signed by the Acting Secretary of the Navy, Archie D. Lowe and Benjamin F. Lowe (Archives of the Naval Air Station, Jacksonville, Florida).

14. The February 24, 1966 article "Plan for Posh Island Development Revealed" in the *Key West Citizen* cites Knopp's claim and notes that Thompson Enterprises also ran a "shark factory" on Wisteria Island (4).

15. *Legislature of Florida at its Thirteenth Session, 1885* (Tallahassee, FL: Charles F. Dyke, 1885), 115. This Act, included in Chapter 3670, became law on February 22, 1885. See Manatee Dry Dock's advertisement published in the *Key West Democrat*, 1 September 1883, 4.

16. *Key West Citizen*, 10 February 1933. See also Carole Heinlein, "Lighthouse Tender *Wistaria* Becomes Island after Fire," *Florida Keys Sea Heritage Journal* (Fall 1992): 7.

17. *Wistaria*'s wreckage was still clearly visible on the western edge of the island in aerial photographs from 1950. At this time, above-water sand deposits half the size of Wisteria Island can also be seen to the island's northwest, and the sand bar that today arcs prominently north and west is visible only as a submerged sandy area. At this time, patches of exposed sand are also visible where Tank Island would soon be formed. (See photograph by U.S. Naval Air Station, Key West, Florida, archived as item number 2208 at the Monroe County Library.) Today, preservationists seek the protection of *Wistaria*'s sunken hull, and snorkelers explore the fish and other sea life attracted to its artificial reef.

18. Brigadier General A. Mackenzie, Chief of Engineers and Lansing H. Beach, Major, Corps of Engineers, "Preliminary Examination of Key West Harbor, Florida," included in "Letter from the Secretary of War," May 27, 1908, referred to the Committee on Rivers and Harbors," *House Documents of 60th Congress, 1st Session*, volume 20 (Washington, D.C.: Government Printing Office, 1908), 3.

19. "New Trade," *Pan-American Magazine* 15 (1912): 42.

20. The letter from Key West NAS Commandant Commodore William H. Beehler to the Secretary of the Navy, dated April 14, 1908 and received on April 17, 1908, includes a reference to a letter from the Federal Land Office dated March 9, 1855. Addressed to the Surveyor General of Florida, the letter shows that the "shoals and Fleming Key are reserved for military and naval purposes" and that "Fleming Key, its adjacent shoals, and Man of War Harbor are reserved for the Naval Depot." Two years earlier, Fort Taylor construction work had resumed in May 1853. See House of Representatives Documents, during the First Session of the 33rd Congress (Washington, D.C.: A.O.P. Nicholson, 1854), 172. In the Report, dated November 30, 1853, the Chief Engineer on fortifications made the case for Key West's strategic location: "I would say that even now too great liberality cannot be extended to certain defences [sic] in progress, especially those at Key West and the Tortugas, and, as before stated, those at San Francisco." (157) The area of Fleming Key, immediately to the north of Key West, was reserved by the President for military purposes on September 17, 1845.

21. Commodore W. H. Beehler, "The Strategic Importance of the Naval Base at Key West, and Its Strategic Lines of Force," *U.S. Naval Institute Proceedings* 34 (Annapolis, Maryland: U.S. Naval Institute, 1908), 605.

22. Theodore Roosevelt, Executive Order No.808, June 8, 1908 (National Archives).

23. Browne, 99.

24. Jefferson B. Browne, "Across the Gulf by Rail to Key West," *The National Geographic* 7 (January 1896): 205.

25. Browne, *Key West*, 99.

26. Beehler, 615; and U.S. Coast and Geodetic Survey, *United States Coast Pilot, Atlantic Coast* (April 11, 1896), 108.

27. Dredging of the Northwest Channel, also called the "Northwest Passage," was authorized on September 19, 1890. An "experimental cut" had been dredged by the federal government in 1881 to a depth of fifteen feet, but it filled up soon after. See William Endicott, "Northwest Passage, Key West Harbor, Florida," *Letter from the Secretary of War*, transmitted to the Committee on Rivers and Harbors as a part of "A Preliminary Report of the Board of Engineers upon the further improvement of the entrance to Key West Harbor, Florida," 2nd Session of the 50th Congress, House of Representatives, Executive Document (February 20, 1889), 2.

28. Entry dated October 9, 1916, *Report of Navy Yard Commission* (U.S. Naval Station, Key West, Florida, 1916), 279. This report corroborated the earlier 1908 Naval Institute report: "The advantages of Key West arise from its strategic location at the southernmost continental limits of the Nation with direct railway communication; at the turning point for entry and departure of all Gulf water-borne commerce via Straits of Florida and Yucatan and Bahama Channels; its proximity to Cuba and the Panama States; . . . with Fleming Cay, Frankford Bank, and other partially submerged cays and shoals capable of development; and the existence of fortifications which should be improved." (279)

29. C. Sherrill and R. Aiello, *Key West: The Last Resort* (Key West: Key West Book and Card Company, 1978), cited in T. Murray, J. London, and M. Shivlani, and C. Kruer, *The Impact of the Cruise Ship Industry on the Quality of Life in Key West, Florida* (Key West Naval Properties Local Redevelopment Authority, 2005), 167. Sherrill and Aiello estimated the size of Key West Island as 1,575 acres in 1829 and 3,000 acres in 1971.

30. Early records of dredging and filling can be found in the following: Navy Department, "Specification No.1453 for Filling and Grading at the U.S. Naval Station, Key West, Florida, Acts approved July 7, 1898 and March 3, 1905" (Bureau of Yards and Docks, 1905); "Specification No. 1579, for Diking, Dredging, and Filling at U.S. Naval Station, Key West, Florida, under appropriation 'Dredging and Filling In,' Acts approved July 7, 1898, March 3, 1905, June 29, 1906, and March 2, 1907" (Bureau of Docks, 1907); and J. R. Mickler and Charles E. Roordan, "Key West in World War II: A History of the Naval Station and Naval Operating Base" (Naval Station and Naval Operating Base, 1945), manuscript in Monroe County Public Library archives. Other information has been drawn from a table of projects titled "Dredge History of Key West Ship Channel and Truman Harbor," provided by the Department of Environmental Protection (Bureau of Beaches and Coastal Systems) to Curtis Kruer for his book *The Impact of the Cruise Ship Industry on the Quality of Life in Key West, Florida* (2005).

31. Navy Department, "Proposed Development for Submarine Base in U.S. Naval Station, Key West," blueprint dated May 11, 1917, Bureau of Yards and Docks, 1917. The document is also stamped with the date June 26, 1924. (Archives of the Naval Air Station, Jacksonville, Florida)

32. U.S. Coast Guard, "Key West Harbor," 1940. Scaled 1: 10,000. See Letter dated September 15, 1953, from the Real Estate Division of the Bureau of Yards and Docks to the Director of the Bureau of Land Management. (Archives of the Naval Air Station, Jacksonville, Florida)

33. The Navy later planned a fuel depot with twelve tanks for the island, but only two tanks—eighty feet in diameter and forty feet high—were built. Finally removed in 1988, the tanks were never used and were only filled partially with water to protect them during hurricanes. Pipes connected the island to Piers A and B and the Westin Marina.

34. Letter from Key West NAS Commandant Commodore William H. Beehler to the Secretary of the Navy, dated April 14, 1908 and received on April 17, 1908. (Archives of the Naval Air Station, Jacksonville, Florida, and National Archives)

35. Entry dated November 3, 1917, House Documents, vol. 32, part 4, report no. 6 of Commission on Navy Yards and Naval Stations, *Report of Navy Yard Commission* (U.S. Naval Station, Key West, Florida, 1918), 49. The report notes that the Florida East Coast Railroad had found this method "efficacious" in its construction work across the Keys.

36. *Report of Navy Yard Commission*, 279. The report notes that enlarging Fleming Key with fill would provide grounds for "aeronautics and fuel storage," while filled areas of Frankford Bank could be used for "explosive stowage." (280, 283)

37. Department of the Navy, Bureau of Ordnance, *Annual Report of the Chief of the Bureau of Ordnance to the Secretary of the Navy* (Washington, D.C.: Government Printing Office, November 19, 1908), 36.

38. August 24, 1944, *Minutes of the Trustees of the Internal Improvement Fund* 24 (Tallahassee, FL: 1945): 826–827. In the description of the "Fleming Key Spoil Area" in the September 5, 1944 advertisement approved by the IIF Board, the total area was 221 acres; but the text explicitly excludes the ten acres of Fleming Key. The advertisement notes that this natural island was originally withdrawn from state jurisdiction with Executive Order 4060 in 1924. For an exhaustive description of Fleming Key's construction and its naval magazine, see Winfred R. Hoey's 1945 report on the Naval Air Station Trumbo Point (Seaplane Base). An excerpt, titled "History of the U.S. Naval Magazine" and lavishly illustrated, was included in the *Florida Keys Sea Heritage Journal* 21 (Spring 2011): 3–8.

39. April 4, 1924, *Minutes of the Trustees of the Internal Improvement Fund* 15 (Tallahassee, FL: 1925): 191. Dated April 5, 1924, this notice appeared in the *Key West Citizen* newspaper.

40. June 11, 1924, *Minutes of the Trustees of the Internal Improvement Fund* 15 (Tallahassee, FL: 1925): 215.

41. Letter from Curtis D. Wilbur to Calvin Coolidge, August 9, 1924, Department of the Navy, Washington, D.C. (Archives of the Naval Air Station, Jacksonville, Florida, and National Archives). Wilbur notes that the Secretary of the Interior supported the proposed Executive Order in a letter dated July 25, 1924. In September 1944, the Navy reiterated its title to the offshore lands and islands.

42. Calvin Coolidge, "Executive Order," no. 4060, August 11, 1924.

43. The U.S. Coast and Geodetic Survey's Chart No. 584 serves as the background for the delineation of the reserve's boundary.

44. See the metes and bounds description in Executive Order 4060.

45. September 4, 1951, *Minutes of the Trustees of the Internal Improvement Fund* 28 (Tallahassee, FL: 1952): 486. This page includes typed marginalia that notes the September 27, 1951 objection filed by the Navy and refers to the minutes of January 8, 1952 for the sale's confirmation.

46. Letter from Commandant, Sixth Naval District, to Chief, Bureau of Yards and Docks, Navy Department, August 15, 1951, Code 60 (DE-200), Ser. 1499 (Archives of the Naval Air Station, Jacksonville, Florida).

47. January 8, 1952, *Minutes of the Trustees of the Internal Improvement Fund* 28 (Tallahassee, FL: 1952): 627. See also Deed 119/4, filed January 9, 1952, to Paul E. Sawyer (document in Bureau of State Lands Management's archives, State of Florida, Department of Natural Resources). These documents include a map derived from the U.S. Coast and Geodetic Survey's Chart No. 584. Dated July 11, 1951 and titled "Plat showing proposed bay bottom land and spoil area to be acquired from state I. I. Board by Mr. Paul Sawyer," the map depicts Simonton and Duval Streets and from Simonton measures 2,250 feet northwest and then marks out a rectangular area 1,000 feet by 1,700 feet. The Tank Island spoil area is shown as a cluster of two small islets.

48. Document stamps indicate a sale price of $6,100 when Paul Sawyer transferred the property to Papy and his wife Pauline on January 28, 1952. See also Timothy O'Hara and Cheryl Smith, "Feds: We Own Wisteria," *Key West Citizen*, 19 November 2011, 2.

49. Letter from Real Estate Division, Bureau of Yards and Docks, Department of Navy, to Director, Bureau of Land Management, Department of the Interior, September 15, 1953. After its introduction to the problem, the letter summarizes the history of executive orders and land transactions. (Archives of the Naval Air Station, Jacksonville, Florida, and National Archives)

50. Letter from Chief, Bureau of Yards and Docks, to District Public Works Officer, Sixth Naval District, with subject line "Information concerning island formed by dredge deposit," October 15, 1956. The letter includes the following opinion from the Bureau's Office of Counsel: "(1) If the island was originally built up for Federal use, title to the land remains in the

United States, or (2) if it cannot be established that the island was built up for Federal use, the Navy is without a strong argument to claim it." The letter also proposes researching the issue: "It is requested that a study be made for the reason for the deposit of the spoil which created this island and a report made to this Bureau for use in the event the item is submitted for inclusion in a subsequent program."

51. O'Hara and Smith, 3.

52. "Plan for Posh Island Development Revealed," *Key West Citizen*, 24 February 1966, 1.

53. "Plan for Posh Island Development Revealed," 1.

54. For an exhaustive, if sometimes anecdotal, account of Wisteria Island's squatter community beginning in the 1960s, see Tamara Lush, "Key West Wrangle: Sides square off over Christmas Tree Island," *Miami New Times*, 1 November 2007, 1–5.

55. *EIS: Digests of Environmental Impact Statements* (Arlington, VA: Information Resources Press, 1981), 54, 86. The extended document was published as: *Disposal of Surplus Federal Properties in Florida: Portions of Harry S. Truman Annex and Trumbo Point Annex of Key West Naval Air Station and Former Coast Guard Station, Key West* (Atlanta: General Services Administration, 1979), and the draft was finalized in 1980.

56. Catherine Hill, "Measuring Success in the Redevelopment of Former Military Bases: Evidence from a Case Study of the Truman Annex in Key West, Florida," *Economic Development Quarterly* 14 (2000): 267-276.

57. "Sold! Truman's 'Little White House' Brings $17.25 Million Bid," *New York Times*, 11 September 1986, A16.

58. "Complicated History Clouds Island Issue," *Miami Herald*, 14 September 1986, 1B.

59. May 2, 1961, *Minutes of the Trustees of the Internal Improvement Fund* 33 (Tallahassee, FL: 1952): 269. The application requested the "conveyance of two small spoil islands originally created by harbor and channel dredging, plus surrounding submerged lands lying west of the Naval Station." See also Susan Ornstein, "Mistakes Nibble at Annex Price Deal to Close at $250,000 Below Bid," *Miami Herald*, 10 March 1987, 1B.

60. "Sold! Truman's 'Little White House' Brings $17.25 Million Bid," A16.

61. Joel Achenbach and Warren Getler, "Heaven's Gate: Who is this man who says he's building paradise in Key West?" *Miami Herald*, 1 July 1990, 8.

62. Achenbach and Getler, 8.

63. Susan Ornstein, "Key West Warms Up to Singh," *Miami Herald*, March 8, 1987. Ornstein is quoting Jane O'Reilly, a writer for *Time*, who was in residency at the Navy base in 1987.

64. Hill, 269.

65. Hill, 270.

66. Achenbach and Getler, 8. Requirements for the provision of "free" access to the Annex were also changed to "convenient" access at this time.

67. Public access to the island is limited because there is nowhere for boaters to moor their vessels.

68. These figures were derived from the Monroe County Property Appraiser's Office, 2006 Tax Assessment.

69. "Luxury Key West Resort: Sunset Key West Cottages," accessed January 22, 2012, http://www.westinsunsetkeycottages.com.

70. See Achenbach and Getler, 8.

71. Mandy Brown, "Hippie . . . Buddhist . . . Builder," *Key West Citizen*, 30 January 1999.

72. Richard Conniff, "Pritam Singh's Strange Career: An Eccentric developer brings good taste to a tacky island," *Time*, 11 December 1989, 21–22, 27. Evident in its sub-title, the article presents Key West as a "tacky" context in which Singh plans to convey an order of "good taste."

73. Roger Bernstein, quoted in Kyle Teal, "Annexation deal is dead: Verge might still seek referendum on Wisteria future," *Florida Keys Keynoter*, 11 July 2007.

74. May 1, 2007, Minutes of Key West City Commission Meeting.

75. According to City Planning Director Gail Kenson, the Monroe County Property Appraiser's Office calculated the annual tax revenue for Sunset Key's 48 homes and 21 vacant lots as $308, 412.68. See Kyle Teal, "City moves to annex Christmas Tree Island," *Florida Keys*

Keynoter, 18 April 2007; and Teal, "Annexation deal is dead," *Florida Keys Keynoter*, 11 July 2007.

76. It was determined that the island is on FEMA's "red list," generated by a 2005 federal injunction that prohibits the writing of new flood insurance policies for coastal property that might serve as habitat for endangered species. A similar change to the state constitution was proposed in 2010, requiring land-use amendments to go before the applicable voting public.

77. In its opposition, the Department of Community Affairs cited the Keys' designation as an Area of Critical State Concern in 1972 when the Florida Legislature's enacted Section 380.05 of the Statutes. They argued that development of islands such as Wisteria increased population without increasing infrastructure for hurricane evacuations.

78. See Monroe County Staff's response to questions concerning the definition of Offshore I.slands in the comprehensive plan, in Monroe County Growth Management Division, "Memorandum," from Mitchell Harvey to Townsley Schwab, October 26, 2010.

79. Section 101-2(13)(e) of *Monroe County Land Development Code* (Tallahassee, FL: Municipal Code Corporation, 2008).

80. See Section 130-40, "Purpose of offshore island district (OS)." Its purpose is explicitly provided: "to establish areas that are not connected to U.S. 1 as protected areas, while permitting low-intensity residential uses and campground spaces in upland areas that can be served by cisterns, generators, and other self-contained facilities."

81. Policy 102.7.3 of the Monroe County 2010 Comprehensive Plan. Wisteria was also not included on the Tier Overlay Map Series, but planners finally included it in the 2011 revisions. See http://keyscompplan.com/project-overview.

82. In its amendments, section 28-20.100, the Comprehensive Plan's policy document (101.4.1) includes the following: "The principal purpose of the Residential Conservation land use category is to encourage preservation of open space and natural resources while providing for very low-density residential development in areas characterized by a predominance of undisturbed native vegetation. Very low-intensity public and utilities uses are also allowed. Maximum permitted densities shall be based upon the results of the habitat analysis required by Division 8 of the Monroe County Land Development Regulations, as amended."

83. Offshore Island designations restrict development to one house per ten acres.

84. Administratively defined and geographically bounded, mooring fields are designated areas where mooring anchors offer fixed points for tying up boats. Anchors for mooring are marked by floating buoys normally held in place by dead-weight anchors or screw-down augurs embedded within the sea floor. Moored boats are typically sailing vessels, which have deep keels that make dry-docking and out-of-water transportation difficult.

85. Wisteria's anchorage has a long history of occupation and also contrasts with Key West's official, city-owned mooring field. The area around Man of War Harbor has long been an anchorage. In these waters, nineteenth-century charts show a network of mooring buoys, including the yellow buoy signaling the quarantine boarding station north of Fort Taylor and the quarantine anchorage to the west of Fleming Key. See early documents such as U.S. Coast and Geodetic Survey, "Chesapeake Bay Entrance to Key West," *United States Coast Pilot: Atlantic Coast*, part VII (Washington, D.C.: Government Printing Office, 1896), 108–109. Birds-eye views of Key West from 1838 and 1884 show ships and small keys to the northwest of the main island of Key West. In particular, W. A. Whitehead's sketch, titled "The Business Part of Key West" and based on the view from Tift and Company's warehouse cupola, shows the activities of the anchorage northwest of the island.

86. In September 1967, F.E.B. Corporation purchased 125 acres of submerged land, immediately to the north of Wisteria Island, for $300 per acre. Reducing the total acreage from 170 acres, the tract's configuration had been changed to conform to the Board of Conservation biologist's recommendations for the preservation of particular grassy areas. See September 26, 1967, *Minutes of the Trustees of the Internal Improvement Fund* 36 (Tallahassee, FL: 1968): 310–311. Despite the approval, environmental studies slowed the transfer of the deed, and three years later the IIF Board revisited the issue at its December 12, 1972 meeting, in which IIF's staff advised the applicants that the environmental concerns made it "extremely doubtful" the submerged land could be developed and offered to facilitate a refund for the property; but F.E.B. asked for the issuance of the deed, which was recorded on December 15, 1972, with a

sale price of $41,901.60. See December 12, 1972, *Minutes of the Trustees of the Internal Improvement Fund* 39 (Tallahassee, FL: 1975): 194–195.

87. SWC Consulting Services, "Key West Mooring Field Pilot Study, Final Report," 8 April 2011.

88. Roger M. Bernstein, "Mooring Field Offers Many Benefits," *Keys News*, 16 July 2010. He argued that it would differ from "the current city mooring field by its location in Key West Harbor, the close support facilities on Wisteria Island, and its stunning views."

89. Sean Kinney, "Meetings address Wisteria Island plans," *Florida Keys Keynoter*, 26 May 2010.

90. Sean Kinney, "Wisteria Island owner vows to push development," *Florida Keys Keynoter*, 30 July 2011.

91. See U.S. General Land Office, *Annual Report of the Commissioner of the General Land Office* (Washington, D.C., 1845). The document cites the designation of "Key West Shoals, S.W. point of Key West" under Executive Order, September 17, 1845.

92. Public Affairs Office, U.S. Naval Air Station, "Research concludes Navy has no claim to Wisteria Island," August 2, 2011.

93. U.S. Department of the Interior, "Revocation of Executive Orders Nos. 808 and 4060 and Public Land Order No. 30," *Code of Federal Regulation*, Table of Public Land Orders, 1981–1990, no.53 (March 18, 1982): 11668. Public Land Order No. 6214 was recorded on March 11, 1982.

94. Letter from Frankie Morgan United States Department of the Interior, Bureau of Land Management, to Naja Girard, November 7, 2011 (Archives of the Naval Air Station, Jacksonville, Florida). On August 9, a week after the Navy's news release, Girard, a member of Last Stand, drafted a letter to BLM to follow up on the Navy's implication that BLM claimed the offshore area. Last Stand is a membership-based organization that seeks to preserve quality of life in the Florida Keys with an emphasis on environmentally sustainable development practices.

95. Wallace Stevens, "The Idea of Order at Key West," 130. Wallace Stevens used "Ideas of Order" as the title for the original poetry collection that included "The Idea of Order in Key West."

96. Miller, 257. Discussing the power of imagination over reality in Stevens' poetry, Miller makes the case that the jar in the poem "Anecdote of the Jar," as a man-made object, is a "surrogate for imagination."

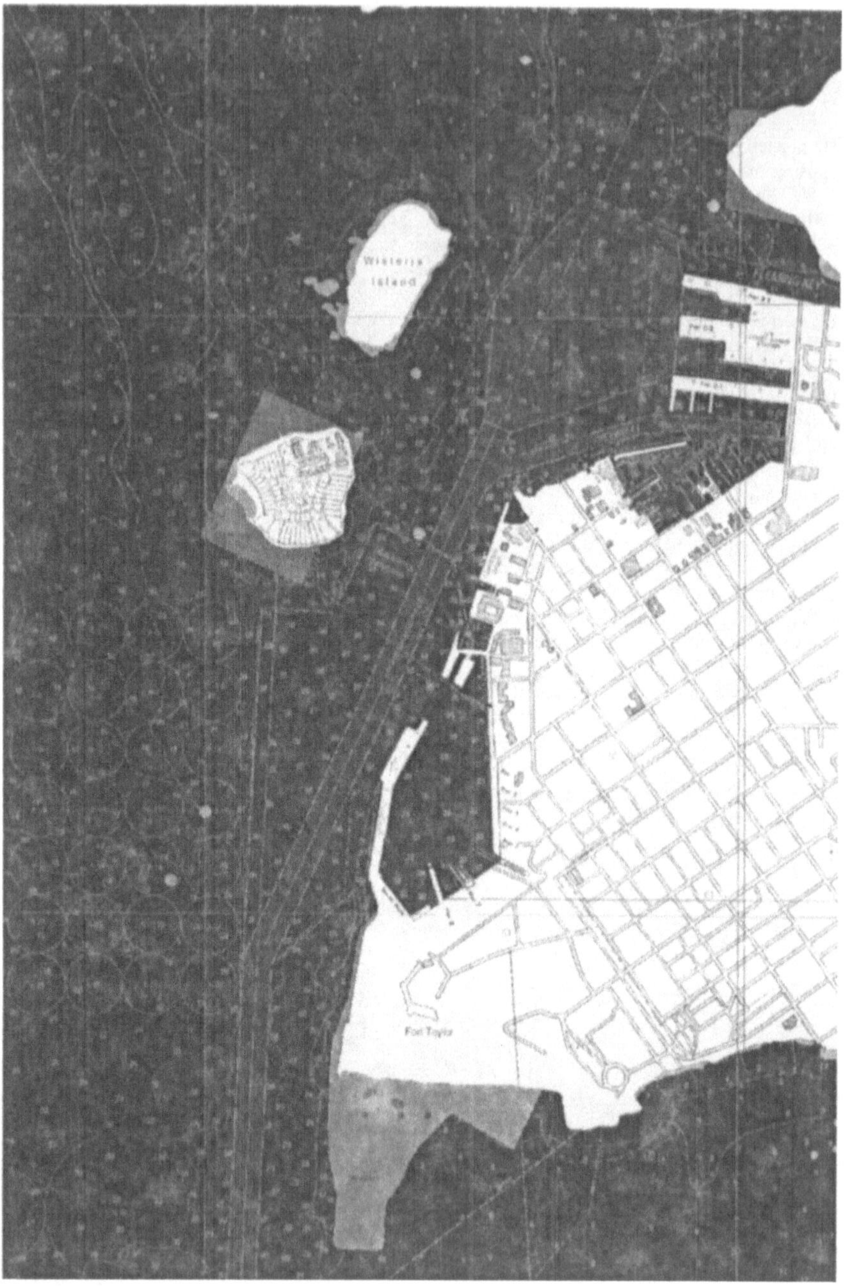
Wisteria Island
Fort Taylor

Barge

A little after 1:30 in the afternoon on September 23rd, 2005—as Wisteria Island residents watched Hurricane Rita move westward—a floating island passed a fixed island at the other end of the intracoastal waterway. Both islands originated with human intervention and neither allowed human occupation. The one that moved was Robert Smithson's *Floating Island*, proposed thirty-five years earlier as a barge to hold earth, trees, bushes, rocks, and moss. The one that stayed in position was U Thant Island, the outcropping built from subway tunnel spoil a century earlier. Both were also relatively small, as far as islands and barges go. The barge's dimensions—thirty feet by ninety feet—only approximated Smithson's drawing of a stubbier vessel but typified the twenty-first century's garbage barges. Four of these barges equal the low-tide land area of U Thant Island.

The 2005 barge project worked from Smithson's single 1970 diagram. Its sketched components were now quantified. Pulled by the tug *Rachel Marie*, rented from Island Towing company, the barge was filled with fifty tons of dirt from a composting heap in Fairfield, New Jersey. Placed under the soil, eighteen tons of hay bales reduced the overall weight. Nine trees were planted: two weeping willows, two river birches, one bur oak, two red maples, one sugar maple, and one beech tree, which was later replanted in the southwest corner of Central Park's East Meadow. The barge also made room for shrubs and bushes—witch hazel, chokeberry, hydrangea, blueberry, and sumac. Borrowed from Central Park, three Manhattan schist rocks and moss were also included.

The *Floating Island*'s mobility was now tangible. The barge contracted Central Park's expansiveness and put it in motion. Freed from Manhattan's landmass and street grid, the barge also revealed the itinerancy of infrastructure. In spite of these surreal displacements, the barge's movement was not beyond rationalization. Schedules published for the public to track the barge's circuit between September 17th and 25th read like Metrorail timetables. The barge may have reminded some New Yorkers of Central Park, but many other residents associated the barge and tow with waste. In the end, it was not one or the other, neither park nor barge, neither soil nor waste, neither nature nor infrastructure. It quite simply became a part of the city. Even with its arboreal cargo, the vessel was hard to distinguish among other barges, tows, and small islands like U Thant.

In spite of blending with its context, *Floating Island* was meant to be viewed rather than occupied. It reversed the Circle Line's relationship to the city. The barge was an object of perception—an addition to the harbor's watery spectacle—rather than a sight-seeing vessel for touring landmarks and other attractions. Some responded to this visual distance with attempts to occupy the barge. Artist Duke Riley tried to board it, but he was unable to inhabit the floating island as he had briefly colonized the fixed location of U Thant Island the previous year. And another group approached the barge with a replica of one of the fabric gates deployed by Christo in Central Park earlier in the year. They described their proposal as the "anti-*Gates* project," seeking to mobilize the fixed position of the gates and to demystify the experience of the barge.

Smithson wanted us to watch the project the way we watch airplanes move across the sky, trains cross the countryside, and ships move in and out of port. He also wanted us to think about nature, not as trees, bushes, and birds, but as the context for infrastructure itself and as a place for thought. Experience of this project then comes about not by physically moving through the artificially constructed nature of the barge, but by assessing our location now, the location from which we view it. When Smithson constructs park and barge as a permutation of site and nonsite, he exchanges the opposition of nature and infrastructure for the terms of mind and matter. Like the barge project, spoil islands help to close these distances.

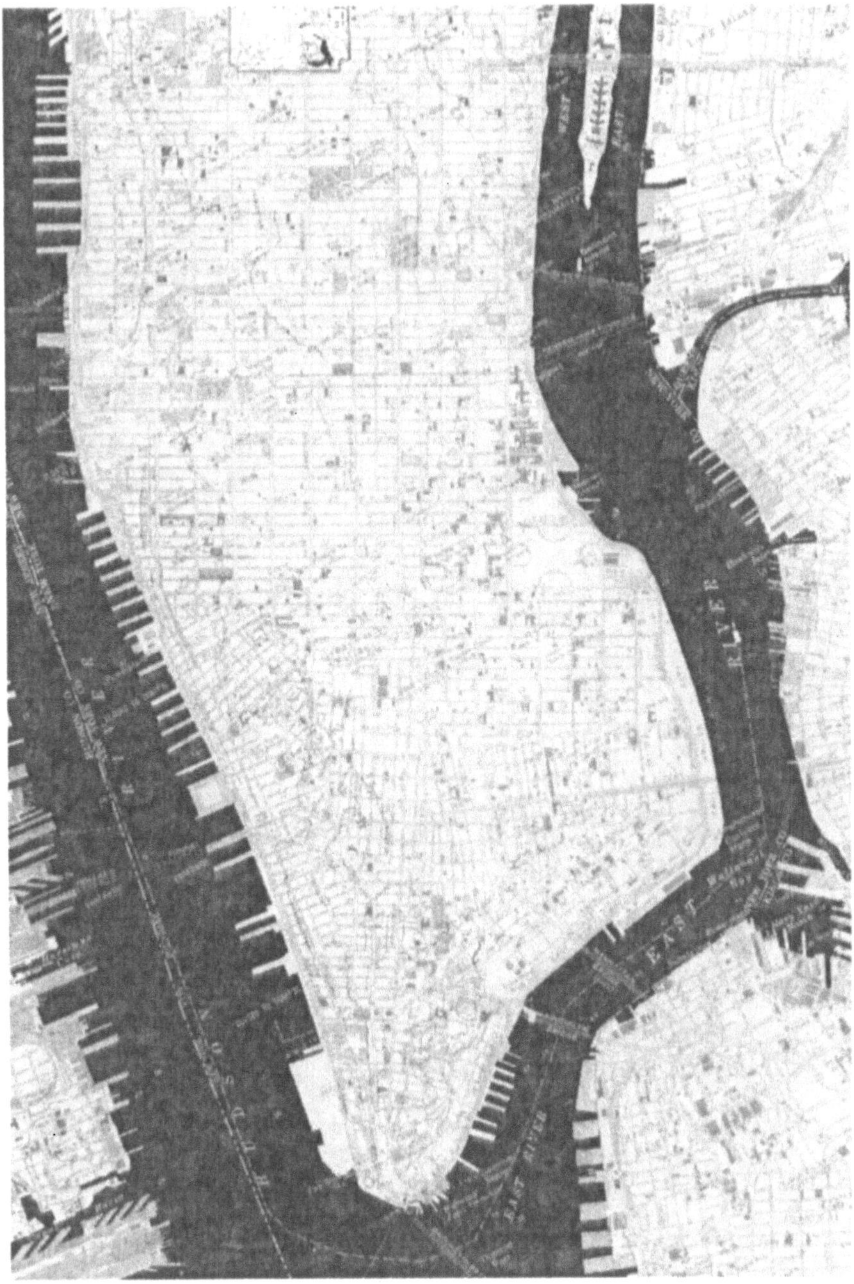
EAST
RIVER

Archipelago

> The island is not meant to save anything or anybody, but to reveal things as they are.[1]

SEA GLASS

On one of those pleasant late mornings in June, the kind that lures Miami's visitors past their wintering season, our kayaks drifted over the calm, flat waters of Biscayne Bay. North of the 79th Street causeway, an early summer camp disembarked from Pelican Island to practice sailing, and the mid-morning sun backlit the apartment towers of Harbor Island. When we arrived at Pelican we had the island to ourselves. In the shade of two gumbo limbos, their branches good for climbing, lay the commemorative block for Christo's and Jeanne-Claude's project. Directing vision downward, its funerary profile memorializes the ground itself. Nothing and everything are buried here.

But it is a sign among signs, and along the concrete paths we are asked to report suspicious people, to refrain from camping overnight and from bringing our pets, to call 911 in emergencies, and to enjoy fishing, grills, and volleyball but not ground fires. We are also reminded of island and waste: "Help maintain the beauty of this island. *Please* dispose of your trash on the mainland. Por favor no botar basura aqui." But there is very little misplaced garbage here. The barbecue grills, gazebos, decks, volleyball court, and picnic tables rest comfortably within a well-kept landscape. Pigeons on one of the tables complete the city park atmosphere.

To the north, a pelican splashes down on the piano sandbar, and another pair of spoil islands continue the bay's archipelago. South from Pelican, construction on the 79th Street bridge hums against the wind and lapping water. Workers raise and lower the drawbridge, testing its swing, and a light chop slows our approach to Bird Key. Its residents call from their sanctuary, which is one of the bay's few islands that preceded the many spoil deposits. The low-slung key's mangrove fringe offers a clue to these origins, but Pelican spoil island has also been colonized by the land-makers and not all of the bay's human-made islands have uplands, their profiles resting low in the water. A more discernable difference is the garbage that covers this island. And on the spit connecting the two parts of the key, we discover sea glass.

The bay's spoil islands are made of limestone and coral. Oolitic chunks of Miami limestone pile up on the shorelines, water-carved fragments grind together underfoot, some of the stones hold shells and fossils, and hardened coral branches look like bones. Among this lithic debris, there is glass. We call it sea glass, but technically it is probably bay glass. Even more precisely, since the northern bay has been wrapped and nearly closed by artificial land, it is lagoon glass. Some of it has been recently broken, but polishing has already begun. The bay serves as lapidary of both stone and glass.

White, brown, gin-bottle green, and sapphire blue, like shallow water over sand. Sea glass is waste that has been turned over to nature. Some is accidental and inadvertent, and much comes to this place out of indifference—not all island visitors obey another of Pelican Island's signs: "No glass containers." Water-borne sand works on this glass as wind and water also modify the island, and sea glass joins the dredged material as visible island substrate. Edges are smoothed, more fragments break apart, pieces are displaced and then accumulate. We like it because it is out of our control. It registers a process that we also inhabit and that also changes us. Spoil islands and sea glass are like us.

Visits to spoil islands inspire penchants for amateur archeology. They also tug at our inclination, which is also perhaps an obsession, to collect. The glass in particular did not lure us to these islands. What did attract us was the belief that we were sure to find something new, at the very least something unexpected. For Robert Smithson, parks should also carry this unexpected quality. His floating barge worked in this way as both site and nonsite. It indexed the park's elements, but also mobilized their stability. It extended the park's boundaries beyond Manhattan, but it resisted habitation. In its waste was a public resource.

Spoil islands emerge from a similar relation of site and nonsite to inform each island's potential use and meaning. These islands are hypothetical and actual at the same time. Sea glass transmits and reflects light. It is smooth, faceted, and hard. It also inspires speculation, indexes materiality, and embodies multiplicity. And that is then the sense of discovery that each test case—whether glass or island—presents to us as kayakers and mainlanders alike. The context for this unearthing is a public space, unprecedented in its archipelagic extent and unparalleled in its paradoxes. Spoil islands are wrecks of technological precision. Their waste becomes resource, their disorder is a kind of ordering, and their formlessness shapes our nature. What the islands mean is what they reveal.

ISLAND OF BROKEN GLASS

> You have a string of hypothetical islands that finally find their result in an actual island.[2]

In April 1969, Smithson stopped in Florida on his way to the Yucatan. On Captiva Island, he visited Robert Rauschenberg. During his stay, he set up the mirrors, which he had packed for his Mexico trip, on the beaches of neighboring Sanibel Island. This experiment began the mirror travel that continued in the Yucatan and prefigured a similar setup on an island in the Usumacinta River. Smithson described the site for his eighth Yucatan mirror displacement in terms of mind, matter, and entropy: "The island annihilates itself in the presence of the river, both in fact and mind. Small bits of sediment dropped away from the sand flats into the river. Small bits of perception dropped away from the edges of eyesight. Where is the island? The unknowable zero island?"[3] What follows is a remarkable passage, even for Smithson's essays in general and this "Mirror-Travel" essay in particular. With vivid directness, Smithson registers the "unconsolidated views" created by the mix of island and mirror. The essay's language, rhythm, and syntax take on the context's particulate and erosive nature. Mirror, island, and one's own vision are "dropping, draining, eroding, trickling, spilling." Piles of language—Smithson had earlier called such accumulation the "heap of language"—are like so much dredged material. He asks how visibility might be "put together again." It cannot, but the act of viewing remains. And the continued dialogue between perception and matter can gauge transformations when "floors of bleached detritus [have] replaced the flatness of rooms."[4]

Smithson's sole footnote to this essay comes here, near the end of this pour of language. In the main body of the text, he introduces the series of island maps and projects to follow: "Outside this island are other islands of incommensurable dimension."[5] And in the footnote, he continues: "This is just one of thousands of hypothetical arguments in favor of Atlantis. Conjectural maps that point to this non-existent site fill many unread atlases . . . the documents of the lost island proliferate."[6] Smithson's subsequent projects will seek out these lost islands, and he mentions here the upcoming installation in Loveladies. With this project, Smithson brought Atlantis to the New Jersey coast, and *Map of Broken Glass (Atlantis)* remained in place for 20 days in July 1969 as one of his first "hypothetical continents." Produced at about the same time and similarly titled, a pencil drawing relates to the Loveladies installation and points toward future hypothetical islands.[7] This *Map of Broken Clear Glass* also indexes the project's attributes. It cites its formal and historical origins—"after map of Lewis Spence . . . see History of Atlantis," quantifies requisite materials—"several tons of broken clear glass needed," directs setup—"trace limits (approx.) on floor lightly, then fill it in," and scales the shards and offers suggestions for assembly—"get few big pieces . . . stick them upright" and "balance big pieces against each other."[8] Despite these directives, photographs taken during the installation show Smithson breaking the sheets of glass with a spade shovel.[9]

But this action is part of the process. It is the "wreck of precision" that allows Smithson to bring disorder and order next to each other. Breaking glass is not unlike cutting earth. Shattering is dredging. Like these early broken glass maps, spoil islands are products of what Smithson called a "jeopardized map making."[10] The islands verge on disaster—disrupting environments, effecting shipwrecks, and generally creating "spoil areas." As waste, they occur at the edge of engineered responses to environmental constraint. Channels and harbors must be deepened, and they are the result. Like spoil islands and the mirror displacements that inspired them, the glass maps are also precarious in their construction and arrangement, providing materiality to the "mental disasters" of mirrors and their "convergences that couldn't converge." Its density like another heap of language, the footnote ends with a description of the island map that also neatly summarizes the material paradox of spoil islands: "The map is a series of 'upheavals' and 'collapses'—a strata of unstable fragments is arrested by the friction of stability."[11] Spoil deposits become islands when they hold together, but they have originated in an "upheaval" and they remain susceptible to an accompanying "collapse" caused by the "extra forces" of erosion or submersion. The wreck of precision means that there are forces that remain beyond control. When Smithson compares the security of the artist's studio to the isolation of the scientist's laboratory, he points out that this stability is "constantly challenged by the instability of these extra forces, like the storm over the island."[12]

Late in 1969, Smithson continued his travels, which were also unequivocal movements away from museum spaces. And in Vancouver, he proposed what would have been his first permanent earthwork. Smithson saw a direct lineage from Sanibel, to the Yucatan, and through the map of glass installations and drawings, here to Canada's west coast. In December, he told an interviewer in Rome about his proposed project: "I'm very interested right now only in land for the work to be on, in other words the work of art needs actual land. When I go back to the United States, I have planned to cover a whole island with broken glass."[13] Smithson had visited Vancouver the previous month in search of a site for the glass island. On behalf of the artist, Douglas Christmas sent a letter to the Minister of Lands, Forests, and Water Resources to inquire about the "purchase or acquisition of a small island with an area no larger than 1/4 of a small block." The island should be "wind swept, preferably flatish [sic] and barren, in short an island that would have no commercial value."[14] Of three potential sites in Stuart Channel, Smithson preferred an island located west of Vancouver and known as Miami Islet. Small, isolated, and visible mainly to shipping traffic, the rock outcropping met the characteristics Smithson sought. Like the mirror displacement sites in the Yucatan, it could be an "allusive" and "ungraspable" place.[15]

Its elusiveness was also its danger, and its name came from the ship that sank on its rocks. In its sea-going history as a collier steamship, the *Miami* had in fact wrecked twice, on both sides of the continent, and its final grounding at 7 a.m. on January 25, 1900, marked the end of a career that had taken it up and down the U.S. east coast, through its namesake Florida city, and to a berthing in Key West. Although Stuart Channel's rocky area was known as Danger Reef, it is not clear if Smithson knew this history, but the wreck resonates with the role of accident and event-space in the glass projects. This eccentric history and position underscores Smithson's interest in center and periphery. The islet is at the edge of navigational and geographical order. It rests at the margins of what is mapped and what is not mapped. Such locations are referenced defensively as places to avoid rather than visit. Similarly, spoil islands are not always mapped, or at least not located precisely. Dredge deposits are "spoil areas" marked out as rectilinear zones that allow for margins of error on nautical maps. Such areas might also be excluded altogether—planning guidelines left out Key West's Wisteria Island, and the USGS did not include the Barge Canal's chain of spoil islands in its topographic series. It is hard to map places that are out of order.

To make a case for the "ordering" of Miami Islet's remote site to government officials, Christmas cited the "cultural value" that Smithson's project would confer on the island. By this argument, the islet lacked use value; and a cultural program would fill a void and make use of previously wasted space. Proponents of Christo's project made a similar case that art could improve the disused—and misused—spoil islands in Biscayne Bay. Like the status of the U.S. east coast's spoil islands, Christmas also noted that the island should remain in the public domain and allow public access.[16] Christmas' overtures paid off, the Ministry presumably supported the project, and an order for one hundred tons of green-tinted glass was placed. Focused on island acquisition, the gallerist's description of the project did not convey the full depth of the artist's own goals. Smithson was working toward the "physical equivalence of a mental image" in order to present a dialogue between perception and matter.[17] His project would achieve what spoil islands do inadvertently. It would mix an actual place with displaced material. As a work of art, it would extend this interaction of place and material to site and nonsite and matter and mind. It would come to exist as a dialogue between order and disorder.

Discussions of waste played a part in the island project's concept, its critical acclaim, and its public resistance. Like spoil islands, Smithson's projects approach wasteland. Interviewing Smithson, Dennis Wheeler noted: "you're always moving toward the wasteland, toward the reduced landscape."[18] But initial support for the project deployed references to waste in a different way. In an article titled "Working for Nature," one critic compared the island to a "wasteland area" that would be reclaimed by art. Some critics of the project took issue with this idea that a previously unoccupied—by all accounts "natural"—setting could be improved with art, much less by human intrusion. As Grant Arnold points out, opposition used not only environmental but also nationalistic frameworks. Some saw the project as a case of U.S. intervention within Canadian affairs and lands. Such criticism drew from the politics of waste, equating broken glass with "U.S. garbage" that would be dumped across the border. One critic even suggested that the glass should instead be placed in Miami—the American city rather than the Canadian island.[19] Shifting the project to Miami would have preceded Christo by a decade and a half.

For these critics, matter out of place, or "MOOP," was disordered—and politically charged—garbage, but for Smithson it was a way of ordering.[20] For him, such displaced material was no longer waste, but a "sacrifice of matter."[21] Smithson frames aesthetically what spoil islands are in technical and economic terms—necessary offshoots, sometimes even tithes to efficiency, that arise from engaging the environment in a productive process. After its "sacrifice," this matter out of place becomes part of the "arrested movement" of disintegration.[22] This process is not just erosion. It is also an opportunity to investigate the porous boundaries between elements of land, water, sky, and fire—this latter element characterized by the island's glass shards. Signaling shifted location, material displacement plays a part in the movement from center to periphery. Smithson installed his island projects along beaches, in lagoons and river channels, and within straits. Like spoil islands, they are far enough away to be marginal but near enough to sustain relations with the mainland. These locational and phenomenal displacements define MOOP. Sites shift, experiences of elements overlap, and understanding such matter requires work and approaches ineffability. Modeling the dynamic—and entropic—landscape of spoil islands is as much art as science, just as Smithson's entropic landscapes adhere to an idea that "matter is allusive, fugitive as the mind; both matter and mind are in a sense inexplicable, unknowable."[23]

TABOO TERRITORY

Matter out of place also has social and cultural implications. In January 1970, just two days after the 70th anniversary of the wreck of the *Miami*, the Minister of Lands sent a telegram to stop *Island of Broken Glass* from going forward. Smithson's project had become taboo. For critics, it encroached on forbidden areas of nature and sovereignty, and it was ecologically impure, geographically untidy, and politically dangerous. As garbage, it upset a broader ecological system; but for Smithson it helped make sense of that same system. Anthropologist Mary Douglas' discussion of dirt helps frame this perspective: "If we can abstract pathogenicity and hygiene from our notion of dirt, we are left with the old definition of dirt as matter out of place. This is a very suggestive approach. It implies two conditions: a set of ordered relations and a contravention of that order. Dirt, then, is never a unique, isolated event. Where there is dirt there is system."[24] While the project's critics did not make this abstraction, Smithson engaged the broken orders of glass and island. Disorder offered potential, and the broken glass helped make another kind of taboo, a territory that confronted ambiguities of matter, nature, and ecology without imposing order.[25] Smithson's project sought a new order.

In an interview with Dennis Wheeler, Smithson talked about his explorations of Vancouver's islands, some of which were restricted. Their exchange frames another view of taboo spaces: "Once again you can even conceive of this like a taboo territory, this is very primitive in a sense. . . . If we get into primitive structural set-ups, nonsite is . . . [Wheeler interjects] . . . Totemic . . . [and Smithson continues] . . . In a real conscious sense, not in any fantasy sense."[26] For Smithson, taboo territories are the actual places where his hypothetical island maps might be realized. Douglas defines taboo as "a spontaneous coding practice which sets up a vocabulary of spatial limits and physical and verbal signs to hedge around vulnerable relations."[27] Smithson extends this definition, but does not "hedge" and instead confronts the indefinite relations of mind and matter evoked by both island and glass. For Smithson and Douglas, dirt is a matter of perspective, and Smithson's taboo territories work from multiple viewpoints, elide codifications, and seek out multivalent experiences. Collision of glass and island make a space where "disparate elements are related and disparate experience is given meaning."[28]

Spoil and island also constitute a taboo territory. With spoil islands, taboo takes on an extended meaning transformed by diverse perspectives and varied geographical contexts. For dredgers, spoil deposits were dirt placed outside the clearly defined spatial limits of the excavated site, whether it was channel, port, or waterway. The material was rejected as waste, and mostly ignored. For others, spoil created spaces of opportunity at the margins of law and jurisdiction. Those who inhabited spoil islands crossed a spatial limit set up by a taboo framework that differentiated the islands as waste. Before this appropriation of spoil islands, calling them "foul areas" or "spoil areas" protected "distinctive categories of the universe."[29] This perspective affirms dirt and takes the polluted territory—the rejected spoil—as an out-of-the-ordinary place of retreat. In the public domain, such spaces were open to those who spied them, following a logic that it was permissible to recycle what had been discarded. After environmental regulations in the 1960s and 1970s and with increasing urban growth along the coast, many spoil islands were conserved in another version of taboo territory—as areas off-limits because of environmental sensitivity. Conservation required the islands to be purged of foreign attributes—exotic plants, garbage, human settlement. This perspective often ignored the islands' past genealogies as waste from the dredging process. And this process consequently revisited past rituals of taboo and spoil that sought to purify contamination and to return to the unspoiled.[30] Conservation of spoil islands installed one taboo space over another, and it ran the danger of oversimplifying the taboo islands as abnormalities.[31] This protected zone must first reckon with the realities of its otherness.

PUBLIC COMMONS

Even though spoil islands have become increasingly codified, the taboo remains. Possibilities for its engagement also remain. In much the same way that spoil islands—as forgotten spaces—became informal vacation places and squatters' abodes, they still offer a break from the everyday that might be loosely managed but is also sometimes not yet explicitly forbidden. Multiple experiences of spoil islands—many of them individualized, temporary, and ephemeral events—engage their multivalent histories of taboo and often still defy classification. The adaptation of such areas affords what has been described as a "new genre of public space." West 8's Adriaan Geuze found this type of space adjacent to the Maasvlakte, the deep-sea port facility constructed offshore from Rotterdam. Here, adjacent to what he calls the "city of taboos," leftover sandbanks have accommodated a "planning orphanage" with the additional infrastructure of power plants, wind turbines, trout farms, and disaster training centers.[32] They also provide the more populist, more informal grounds for mass recreation. Here, in a collective experiment of surrealist urbanism, city dwellers become explorers, spectators become actors, and anonymity yields to exhibitionism. Users have defined the programs of use, and as Geuze notes: "The sandy plain looks like a practice area for sledge dogs or autocrossing, the dredging depot like a hang glider slope, the block dam like a fossil hunter's cliff face, the salt-water sand quarry like a deep-sea diving site."[33]

Residents have imposed their own fantasies on the spoil islands of the U.S. east coast. These islands are still the "few patches of white, sans code, sans allocation" that Geuze found in Rotterdam's planning maps.[34] Based on infrastructure and artificial nature, spoil islands engage multiple users, blur conventional boundaries of use—such as recreation and heavy industry, and require new modes of planning and conserving. Though lacking the informality that it once had and that Maasvlakte retains, Peanut Island mixes continued dredging with picnicking as well as a diverse range of other recreational activities. Wisteria remains a contentious public reserve that anchors Key West's imagination as well as its mooring field. Miami's picnic islands provide surreal moments of green relief within its Central Park void of Biscayne Bay. And the Indian River's linear archipelago stretches a canvas of public green space for more than 150 miles. Continued planning of these areas requires alternative approaches. Formulating the guidelines for the second Maasvlakte, Rotterdam's planners became phenomenologists when they resisted determinate use for the port's dredged spoil areas. Programs such as a beach esplanade with restaurants did not "fit in with the function of the land reclamation (as a port and industrial area)." But sensual experience did: "it is a place where people can enjoy sun, sea, beach, wind and peace."[35]

With language that sometimes sounds as if it were written by Geuze himself, these planning documents provide a first step to answer his challenge "to create space and textures for city dwellers to colonize in their turn."[36] Here, industry and nature collide but appear to coexist. Geuze describes this situation as one in which nature plays a role, but it is neither the starting point nor the dominant factor. Smithson hypothesized this kind of space: "But you will be able to perhaps confer a different value through a different kind of cultivation. I think it is possible to cultivate waste, spoil banks, or the cuts in the strip-mining areas. It's a matter of developing a different value structure, and also to have some kind of relationship between the industrial process and the ecological controls."[37] Like Maasvlakte, spoil islands proceed where Smithson's island experiments left off. Their cultivation arises out of informal use as much as managed oversight. When Rotterdam's recreational guidelines claim that "everyone will find a place on the Maasvlakte 2 beach," they are not merely expressing idealism or utopianism. Such prospects are necessities for understanding—and planning—places like spoil islands, where what is human-made and what is natural mix like earth and water, and a new public commons is possible.

MONUMENT TO ECOLOGY

After the rejection of *Island of Broken Glass*, Smithson proposed a new permutation of his map of glass. The *Island of the Dismantled Building* would place materials from a demolition site in Vancouver on the islet. Smithson posited, and Christmas promoted, this project as an alternative that would benefit the water channel's population of seals and birds, noting that the displaced concrete material would encourage cormorants to nest and would provide resting platforms for seals because it would "rise above sea level."[38] A drawing from the time shows the origin of the proposal's material and a sketch of its placement on the island. On the left, a four-story apartment block is partially demolished with a ragged cut. On the right, birds fly above an island with broken slabs of concrete, vertically protruding rebar (where a few birds have perched), and windowed walls falling into the sea. This part's subtitle alludes to the second project for Miami Islet, but it also suggests Smithson's ironic approach to this revision: "Concrete" has replaced "Atlantis" as the parenthetical word after the project's main title. Matter has been substituted for imagination. Smithson had already alluded to the hazards of a perspective that privileged nature over human intervention, ignored entropy for stability, and denied that islands are as much idea as physical sites. When he wrote "the memory of what is not may be better than the amnesia of what is," he anticipated this situation where concrete replaces memory, and waste becomes waste.[39] In this new proposal, Smithson presented what he saw as the problem of ecology.

In February 1970, as the Corps of Engineers struggled to deal with NEPA, and as Craney Island took on extra spoil, Smithson called his revised Vancouver project a "monument to ecology."[40] This designation reveals multiple aspects of ecology. In one sense, the island would be a monument to the dialogues about ecology and environmentalism that emerged out of projects like *Island of Broken Glass* in places like the Stuart Channel. For Christo, the dynamics of public debate and preparations leading up to the installation were a significant part of the overall project. With the controversy of *Surrounded Islands*, Miami's chain of spoil islands became a "temporary monument to the environment." But Smithson's monument was less didactic and more polemical. The *Island of the Dismantled Building* was a monument to an ecological approach that did not oppose human to nature and did not place nature somewhere else. It engaged wasteland as well as wilderness. Smithson was foreshadowing arguments that a Romantic view of nature can get in the way of ecology's balanced interrelationships between human systems and their environment.[41] Smithson critiqued what he saw as ecology's reliance on economic systems, and he viewed some environmentalists' calls for rigidly ecological approaches as "nostalgia . . . for a view of landscape that at one time existed."[42] Truly unspoiled places do not exist, and restoring sites to unspoiled conditions is untenable.[43]

Smithson's proposal also revealed the need to reconsider nature's dialectics. In an edited interview that reads much like a manifesto on earthworks and land art, Smithson indexes the mutable definitions of nature:

> There is no going back to Paradise or 19th century landscape which is basically what the conservationist attitude is. People have always had different views of what nature is . . . a nostalgia for the enclosed garden . . . the notion that the earth was corrupted by the fall . . . the attitude that the mountains weren't

> spoiled . . . aestheticized mountains . . . landscape as "beauty spot" . . . war between the formal and anti-formal . . . political implications . . . withdrawal from nature . . . fear of nature . . . dominating the land. . . . A lot of the land that people work on is usually desolate. I'm interested in all different types of places, not just the outdoors; that can become a stance that doesn't have any resolution, or become overly precious. I'm more interested in the way things are.[44]

As this inventory suggests, nature is a variable term. And in cases where nature is feared or corrupted by humans, this distanced nature would require ecology's withdrawal. With his "Monument to Ecology," Smithson instead confronts ecological concerns and invokes nature as a dialogue between absence and presence. Like the islands he explores, it might at any moment disappear. The coastal zone that surrounds an island and defines its only boundary is also a zone of vulnerability.[45] What he calls the "islanding situation" accommodates the physical site but acknowledges its transformation, even its potential negation.[46] This nature also does not subscribe to being "natural." It is neither artificial nor natural, and it does not vary by degree. Its dialectics instead work through changes of state. This alteration is how islands of waste might become landscapes of recreation, habitation, and infrastructure. Spoil islands make up a denatured landscape.[47]

This is the nature of spoil islands. They cannot return to a Romantic idea of nature, nor can they be parsed in terms of whether they are natural or unnatural. Each island offers a different kind of nature, while it also remains in an indefinite state with its transformations of solid earth and flowing water.[48] Spoil islands resist naïve responses of conservation, while they also defy explanations of scientific analysis. Reading this archipelago entails experience. As a part of these changes of state, spoils accommodate both-and conditions. Nature becomes infrastructure, which elicits meditations on the nature of being. Individual habitation of the islands affirms public experience within common wildness. Use precedes meaning, and the useful and the cultural share a common ground. Island realities maintain the irrational truths of their formation as well as their continued surreal presence. And their disorder is a kind of ordering. Reckoning does not try to put the islands back together but closely watches how these transformations take shape. It recognizes that the islands will continue to change and so will we.

After the final rejection of his Vancouver island project, Smithson investigated other possible sites in Chesapeake Bay and along the Gulf coast of Texas.[49] He also explored floating islands as a way to mobilize his project—avoiding the practical constraints of a fixed site, but also continuing to discover the dialectics of nature as well as mind and matter. After proposing the *Floating Island* project to circumnavigate Manhattan, Smithson wrote: "One's mind and the earth are in a constant state of erosion. . . . The entire body is pulled into a cerebral sediment, where particles and fragments make themselves known as solid consciousness. A bleached and fractured world surrounds the artist. To organize this mess of corrosion into patterns, grids, and subdivisions is an aesthetic process that has scarcely been touched."[50] Sri Chinmoy touched on this process with his meditations on U Thant Island, and Wallace Stevens sounded the complex poetics of Key West's island and sea chaos.

Smithson's project offers one tool for making sense of the disordered, degraded, and displaced environment we have inherited. Spoil islands offer a similar means only when they are understood dialogically, in all of their paradoxes and inconsistencies. Smithson noted the pitfalls of conservation, and it is interesting that planners of the 2005 *Floating Island* project interpreted his call for trees "common to the region" as those indigenous to Manhattan. Weeds are just as common; and a prominent tree in Smithson's sketch, the weeping willow, is not native to New York. The "greening" of Smithson's barge proposal parallels current tendencies to call for sustainable and "green" approaches without querying the different views of nature and nativeness that exist. In that respect, Smithson's barge is more of a spoil island than many of the recent floating islands that idealize their insular place. Contemporary conceptions of these engineered ecologies run a risk of ignoring the lessons of spoil islands as human-made landscapes when they seek to construct isolated natural or urban settings. Tourist brochures tout human-made eco-islands where environmentally sensitive landscapes provide the setting for best practices of sustainability. Artificial floating habitats help avian nesting and operate at ecological intersections, but other floating islands remain limited as private investigations of dwelling or adventure and as isolated urban theme parks.[51]

In broad terms, public activities might instead set out to tend and preserve nature in a similar way that domestic life harbors aspects of wildness. Managing spoil islands might find ways to cultivate the indefinite—a sometimes unanticipated backcountry—within the islands' common wildness. Such models would include setting responsible limits so that respect for the otherness of natural systems could be maintained while drawing close—and providing access to—the values we associate with wildness in general and spoil islands in particular.[52] Working with spoil islands requires an openness to the unexpected—a quality that Smithson saw in Olmsted's projects and something that he followed with his own idea of park. Such parks and such islands reveal our relation to natural systems. Spoil islands are lodged between the relation of man and nature, in the way they are made and in what they mean.

Spoil islands demonstrate a new understanding of the relation between nature and culture. These islands are gardens where what we have made collides with what we might be. Spoil islands are "contrived" in the sense that they have been built but remain "largely composed of natural things."[53] When I first reached the Cross Florida Barge Canal's spoil islands, I did not know what to make of their landscape. I did not have the tools to read this place. They required that I dispense with Romantic ideas of what remote, relatively untouched islands should be. They also did not fit with my view of cultural landscapes and my preconceptions about waste-sites. At the same time, I found these islands striking in their immediacy and their intermediacy, and I came to associate them with gardens. I have since recalled Henry Beston's extraordinary—yet quite ordinarily presented—insights about our relation to gardens: "In making a garden there is something to be sought, and something to be found."[54] In spoil island gardens, there is meaningful order that we can pursue. And they offer a more ephemeral and sometimes disordered, though no less significant, set of experiences that we can discover.

I also came to realize that a significant part of these islands' character is the improbability of our being there. Entirely fabricated, the islands themselves should not really be there, but our presence on them is relatively unrestricted and open. Logistically, their offshore location requires a boat, but technically they are accessible to everyone. They are truly public gardens. Zooming out to see the entire archipelago along the east coast takes this idea further. These "grotesque gardens" constitute an unprecedented and extraordinary linear park system made up of thousands of islands, thousands of what Smithson called "carriers of the unexpected and of contradiction on all levels of human activity, be it societal, political, or natural."[55] They form a natural and cultural landscape that is wholly built, truly wild, latently common, and necessarily public.

HYPOTHETICAL ISLANDS

Robert Smithson drew an archipelago of projects. During a series of conversations with Dennis Wheeler, the artist's works emerged to float across a sheet of polar coordinate paper, like that commonly used for navigation. One drawing is titled "A Surd View for an Afternoon," and it is dated 1970, although this particular dialogue took place late in 1969 in Vancouver during the planning of *Island of Broken Glass*. Two prominent lines tether diagrams to the graph paper. A horizontal line marked "Perception at the zero degree" crosses the middle of the page through the central coordinate, while two other less prominent but equally significant horizon lines anchor the perceptual fields of specific projects. Smithson also inscribed a thirty-degree line that is the "entropy slope." Angling from the upper right where it is labeled "abstract concepts," it falls toward the lower left where it extends "towards the low levels of perception." These two lines cross at the graph's zero coordinate, marked as the air terminal project for the Dallas-Fort Worth Regional Airport (1967). But, for Smithson the center is defined by the edge, and around this crossing circulate his islands. An "Unlocatable Island," an

"Island or Map of Matter," the Loveladies site, and a *Map of Broken Glass* float among many other projects. Adjacent to the air terminal, the *Island of Broken Glass*, ongoing at the time of the drawings, rests at the origin of an Archimedean spiral.[56]

Drawing maps during interviews and conversations was common for Smithson.[57] Mapping combined placing matter and imagining new places and interrelations within a conceptual field. Each drawing became a hypothetical atlas of materials and the questions they invoked—what they revealed. Drawing was dredging, and the artist unearthed concepts and relations to link disparate places and ideas.[58] The graphite was itself so much displaced material. With a particular site, the islands become material maps, and Smithson's nonsites are "like three dimensional maps that point to an area." Each spoil island is likewise a map that points toward its origins—its foundation in displaced material and human intervention. These areas also point back in time to ancient accumulations of earth and land masses. A similar recognition is at work when Smithson finds prehistory in waste areas and reduced landscapes, particularly during his investigation of *Spiral Jetty*'s site: "The mere sight of the trapped fragments of junk and waste transported one into a world of modern prehistory."[59] Considering islands as material maps reveals the process of their making, the context for that construction, and the displacement of the project. Their materials refer to a remote site, but they also engage memory—not as a nostalgic exercise but as a continuing exploration of both presence and absence. The excavation of material—whether raw materials or phenomenal matter such as light, air, and color—is as important as its deposition. That is why, in Smithson's work, there is a correspondence between mirror, glass, and island.

These maps also ask questions of their hypothetical archipelago. One afternoon late in 1969, Robert Smithson sketched his first map with Dennis Wheeler. On wide-ruled paper, the drawing includes words, diagrams, notations, and line work. In the top half of the page, Smithson transcribed a question: "where are you revealing orders as opposed to inventing new orders?" Heavy black graphite emphasizes the last two words, and "new" is underlined twice. To the left, he has listed his island projects under a heading "Map" and below it the underlined term "Surd." An arrow points up to "mental" and "Abst," and a line works in the other direction to connect a series of tiny island-like markings that also approximate bulleted tags for each island project. This axis culminates in another arrow that points to the center of a spiral. Along its edge, he has written "island of broken glass."

This is a traveler's quixotic map of future destinations and places already traversed. It is also an index of Smithson's artistic work. His islands took on the map's query, and they revealed existing orders and proposed new orders. And spoil islands might do the same for us. Each island is a working hypothesis because it has already been made and the question has already been asked, even if implicitly. These islands continually recast their questions, asking us who have made them: Where is nature? How will we engage it? What will we make next? Key West asks what orders are revealed by Wisteria's presence? With Peanut Island, what can we learn from telling the history of a place that is also the history of waste? In New York's archipelago, what is the nature of infrastructure? The Indian River Lagoon asks how to manage the fragments of a linear park, and what if it is extended up and down the coast—not just of Florida but of the eastern United States?[60]

I suspect that Robert Smithson liked spoil islands. Many of the islands he proposed and some that he made combine island with waste—as displaced or discarded material. Spoil islands fall in with the landscapes that inspired his work: "Pavements, holes, trenches, mounds, heaps, paths, ditches, roads, terraces, etc., all have esthetic [sic] potential."[61] So, in April 1971, when he reached into the tide-turned ocean flats of the Florida Keys to continue experiments in island-making, he was consciously adding to the east coast's archipelago of dredged islands. He was doing by hand what the U.S. Army Corps of Engineers refined mechanically throughout the last century. Smithson's islands are hand-made and very personal. They are changes of state from water to land, from private to public, from disorder to moments of fleeting order.

The shallow water on the bay side of the Keys is milky white with suspended sand, oolitic fines, and filtered sunlight. North of U.S. 1's route across Summerland Key, Smithson found a shallow lagoon where he composed *Oolite Island* with coralline material dredged by hand and barged to the site on a plastic raft. For the second island, Smithson gathered larger coral and limestone fragments that he found on the bottom of the lagoon. In his interview with Gregoire Müller, he noted that the "rocks were encrusted with small slimy things . . . called 'dead man's fingers.'"[62] In contrast to *Oolite*'s mounded form, most of *Sunken Island* remained underwater, and these chalina sponges added a living aspect to the composition. Both islands were five feet in diameter, but their documentation remained without scale. Smithson photographed them at the center of the square-format, chromogenic-development slides, with a horizontal sliver of sand and mangroves and equally thin band of sky at the top. The photographs index the four elements' "interconnected phenomena."[63] Light reflecting off the surface behind *Sunken Island* complements displaced land, humid air, and skinny water.

With his islands, Smithson set these phenomena in motion. Smithson identified with the island experience, and the artist's and the maker's presence is undeniable. Part of the allure of spoil islands is what they tell us about ourselves. Like sea glass, they remind us of our lack of control over natural systems—the same forces that patiently mold islands and glass. Like islands of broken glass, they remind us that we do affect change along the porous boundaries of our environment. They remind us that the places we make must account for both. The islands judge us.

> It is not for us to judge the island but for the island to judge us.[64]

NOTES

1. Robert Smithson, "Rejoinder to Environmental Critics," *Collapse*, no.2: 124 (n.d.), cited in Grant Arnold, ed., *Robert Smithson in Vancouver: A Fragment of a Greater Fragmentation* (Vancouver: Vancouver Art Gallery, 2004), 25.
2. Robert Smithson, *Robert Smithson: The Collected Writings*, ed. Jack Flam (Berkeley: University of California Press, 1996), 200.
3. Smithson, 129. This essay titled "Incidents of Mirror-Travel in the Yucatan" was originally published in *Artforum* VII (September 1969): 32.
4. Smithson, 130. For Smithson's work with the "heap of language," see his drawing titled *A Heap of Language*, 1966, and the accompanying essay "Language to be looked at and/or things to be read" (Smithson, 61). In the drawing, cursive writing along the horizontal lines of grid paper forms a gradually sloped mound. The word "Language" stands alone at the top of this tumulus of words.
5. Smithson, 131.
6. Smithson, 133.
7. Numerous other island projects swirl around this prospective map of Atlantis in 1969. Smithson will install the "earth map of tumbled rocks to be placed on quicksand" with *Hypothetical Continent in Stone, Cathaysia* in Alfred, New York. And his *Island of Broken Shells* for Sanibel Island's Gulf coast beach follows the drawing titled *Hypothetical Continent of Lemuria*, in which he proposed a map of seashells on "swampy ground." These two projects approach the midden typology that parallels spoil island formation along the Indian River and is also common in the southwestern region of Florida, where Smithson's project was installed.
8. There are at least two versions of this drawing. The *Map of Broken Glass (Atlantis)*, included in *Robert Smithson in Vancouver: A Fragmént of a Greater Fragmentation*, contains fewer notes and calls for "6 tons of broken glass" (31). In the text, I am drawing from the *Map of Broken Clear Glass (Atlantis)*, archived and made available on the Estate of Robert Smithson's website, accessed April 11, 2012, http://www.robertsmithson.com.
9. Dennis Wheeler, in his interview with Smithson, alludes to the mirror displacements—and by extension, the shards of the glass maps—as karate chops. Expanding this idea to Smithson's agreement, he notes: "It's like the gyrostasis made material in its energy. Like something moving out, as you said, in the cyclonic action, but capturing the shattering as if it was frozen . . . and the shatterings were caught in the earth in the spinoff and inserted. And that's why they're so energized . . . it's like a karate chop." Wheeler also compares this process to punching holes in the earth; and by these formulae, spoil islands also originate with an earthmoving karate chop. (230)
10. Smithson, *Collected Writings*, 212.
11. Smithson, 133.
12. Smithson, 229.
13. Robert Smithson, "Robert Smithson has been interviewed in Rome by Achille Bonito Oliva," *Domus* 481 (December 1969): 42. Grant Arnold also cites this passage from the interview in his essay for *Robert Smithson in Vancouver: A Fragment of a Greater Fragmentation* (2004).
14. Excerpt from Douglas Christmas' letter to Roy Williston, 3 December 1969, cited in Grant Arnold's essay in *Robert Smithson in Vancouver: A Fragment of a Greater Fragmentation*, 15. The letter is located in a file titled "Robert Smithson Glass Island Project" in the Vancouver Art Gallery Archives.
15. Smithson, 64.
16. Smithson, 15, 17.

17. Smithson, "Robert Smithson has been interviewed . . .", 42.
18. Smithson, *Collected Writings*, 230.
19. See Grant Arnold's summary of these exchanges in *Robert Smithson in Vancouver: A Fragment of a Greater Fragmentation*, 22–23.
20. Matter out of place is known popularly as MOOP. For example, at Burning Man, the collection of MOOP left by the annual event's participants has become part of its leave-no-trace practices.
21. Smithson, 229.
22. Smithson, 256–257.
23. Smithson, 42. Hydrodynamics and the complex systems where water meets land pose difficulties to scientific computation. See Y. K. Vyes and J. B. Herbich, "Erosion of Dredge-Material Islands Due to Waves and Currents," *Proceedings of the 2nd Annual Symposium on Dredging Technology*, vol.1 (Cranfield, UK: Center for Dredging Studies and BHRA Fluid Engineering, 1977), H2-14. In 1977, researchers at Texas A&M noted the difficulties of predicting the erosion of spoil islands: "no analytical or empirical method for predicting erosion rates of dredged-material islands is known to exist." So, they built a model, scaled horizontally 1 to 100 and vertically 1 to 4. The latter scaling allowed them to exaggerate the island's height for a closer study of erosion effects. Their case study was a circular spoil area known as Crane Island. In its horizontal dimension, the island's full-scale radius translated to a six-foot diameter spoil deposit of sand and gravel. A thirty-two foot wide and eighty-six-foot long "moveable-bed model" provided a context for the island and for the surrounding wave action. Concrete blocks formed the wave basin, and removal of one row of blocks simulated the dredged channel. Given its dynamic complexity, the research required consideration not only of space but also time—both hydraulic and sedimentation time—as well as the consideration of small magnitude currents with waves created by boat traffic and daily tidal fluctuations, both of which increased the dispersion rate of island sediments. Their research concluded that the proper location of dredge islands can result in "more stable island geometry and lower infilling rates of navigational channels." It is interesting that this experiment comes late in the history of twentieth-century dredging practice, after the presumptive end of unconfined spoil deposition, amidst continued discussions about the environmental impact of dredging and its placement, and six years after Smithson's own full-scale experiments. Researchers have also used dimensional analysis and experimental evaluations to try to understand the complex interactions of solid and fluid at the cutter heads of hydraulic dredges. See an early attempt in L. N. Gladigau, "Interactions Between Sand and Water," *Proceedings of WODCON World Dredging Conference* (1975), 261–276.
24. Mary Douglas, *Purity and Danger* (New York: Routledge Classics, 2002), 44.
25. Smithson follows Douglas' contention that dirt is a matter of perception, but he reverses the idea that matter out of place must be approached through order. MOOP can be approached on its own terms, and ordering emerges from the site of displacement itself. Patterning, for Smithson, requires neither the application of an external order nor the exclusion of disorder. Working from a perspective of instability rather than stability, Smithson also engages the idea that dirt is a kind of creative formlessness (Douglas, 198–199), but he is careful not to oppose form to anti-form. Douglas summarizes the complexity of dirt's potential as an ordering device and as an element in creation: "Granted that disorder spoils pattern; it also provides the materials of pattern. Order implies restriction; from all possible materials, a limited selection has been made and from all possible relations a limited set has been used. So, disorder by implication is unlimited, no pattern has been realized in it, but its potential for patterning is indefinite. This is why, though we seek to create order, we do not simply condemn disorder. We recognize that it is destructive to existing patterns; also that it has potentiality." (117)
26. Smithson, 206. In the ensuing conversation with Wheeler, Smithson posits an investigation that parallels this book's study, framed by Smithson's particular project and carried out through the spoil islands more generally: "So that might be an interesting thing to pursue in terms of the whole notion of taboo territory, like the Line of Wreckage could certainly be taboo, like pollution, foul area." (206)
27. Douglas, xii.

28. Douglas, 3. William James had earlier observed the relation of rational and irrational elements in human experience: "Here we have the interesting notion . . . of there being elements of the universe which may make no rational whole in conjunction with the other elements, and which, from the point of view of any system which those elements make up, can only be considered so much irrelevance and accident—so much 'dirt' as it were, and matter out of place." See *The Varieties of Religious Experience* (Rockville, MD: Arc Manor, 2008), 104. Douglas cites James in *Purity and Danger* (2002), 203.

29. Douglas, xi.

30. James Frazer noted how African tribes, victorious in battle, had to cleanse spoils of war before making them their own: "the spoil taken from the enemy had also to be purified, according to its nature, either by fire or water" See *The Golden Bough* (Cambridge University Press, 1911), 177. Smithson cites Frazer in his discussion with Dennis Wheeler (206), and T. S. Eliot acknowledged Frazer's influence on *The Waste Land*.

31. By defining "nativeness," conservation principles have the potential to oversimplify the very process of conservation, much like Frazer oversimplified taboo as abnormality. See Douglas' critique of the generalizations present in his nineteenth-century anthropological views (*Purity and Danger*, 27–30).

32. Adriaan Geuze, "Accelerating Darwin," *The Artificial Landscape*, ed. Hans Ibelings (Rotterdam: NAi Publishers, 2000), 256–257.

33. Geuze, 256.

34. Geuze, 255.

35. See Port of Rotterdam, "Maasvlakte 2: Recreation," accessed June 2, 2012, www.maasvlakte2.com.en/index/show/id/662/Recreation.

36. Geuze, 256. Extending westward into San Francisco Bay and offering views of the Golden Gate Bridge, the Albany Bulb presents a similar exercise in the urban colonization of waste areas. Formed from construction debris in the 1960s, the extension of Fleming Point over time accumulated dirt, supported vegetation, hosted homeless encampments, provided a setting for local artists, and became a popular place for dog-walking. The tidelands site remains a contentious zone amidst sweeps to clear homeless residents, plans to make it a park, and concerns raised by environmental advocates. See "A Brief (and approximate) History of the Albany Bulb," accessed June 3, 2011, http://environmentalsciences.berkeley.edu/es100/downloads.

37. Robert Smithson, *Robert Smithson*, ed. Eugenie Tsai (Berkeley: University of California Press, 2004), 94.

38. Smithson, 25. It was noted that the slabs of concrete would be "placed artistically." Grant Arnold includes this drawing in his essay (27). Similar to Smithson's proposal for the slabs of concrete on Miami Islet, the ruins of the buildings on Swinburne Island offer nesting sites for cormorants and other birds, particularly along the sloped hip of their roofs.

39. Smithson, 131. Smithson drew other islands similar to the *Dismantled Building* and pressed further into the ironies of his proposal. In *Island Project* (1970), he constructed a landscape that combines mining operations with industrial structures. Block walls spiral, fire burns ritualistically atop cylindrical steps, smoke billows, conveyors and paths cross mounds of earth, ladders connect levels, and a gantry supports what looks be a massive fish hook. He also samples some of the monuments from his Passaic photowork (1967), and pipes along the island's edge protrude like the "Fountain Monument" he had documented earlier in New Jersey. Smithson has effectively translated Piranesi's prisons (*carceri*) into ecological monuments. A second drawing from the same time period further fractures the mined landscape, and the island has itself been fragmented into islets. Dwarfed by these impossible structures, a small barge-like vessel floats in a channel between two islets and references the barges Smithson proposed after the dissolution of his island map projects. See *Robert Smithson*, ed. Eugenie Tsai (Berkeley: University of California Press, 2004), 170.

40. In his essay for *Robert Smithson in Vancouver: A Fragment of a Greater Fragmentation*, Grant Arnold excerpts the 13 February 1970 letter from Douglas Christmas to Roy Williston, in land file 0294633, Land and Water British Columbia, Inc. (25).

41. See Timothy Morton, *Ecology without Nature: Rethinking Environmental Aesthetics* (Cambridge: Harvard University Press, 2007), particularly his discussion of waste, ecology,

and anti-aesthetics (51–52). Waste in the public eye forces a revaluation not just of how a setting looks (what might be considered aesthetically pleasing) but also how an industrial process enters into the discussion of shared landscape (what might be deemed productive). When we recognize waste and its subsequent location, then we are facing our role as caretakers of the world. Aesthetics and politics meet in places like spoil islands. See also Paul Wapner, *Living Through the End of Nature* (Cambridge: MIT Press, 2010).

42. Smithson, *Robert Smithson*, 93–4.

43. What has been called "Humpty-Dumpty ecology" looks at how environmental restoration addresses immediate questions of need and use, posing questions like "should a particular ecosystem be restored to its past conditions, or should it meet current socio-cultural goals." David Lowenthal discusses these issues in "Reflections on Humpty-Dumpty Ecology" in *Restoration and History: The Search for a Usable Environmental Past*, ed. Marcus Hall (New York: Routledge, 2010). Earlier, but still after Smithson's writings, Ashby W. Clark wrote an article titled "Another Look at Humpty-Dumpty" for *Ecology* 60: 445–446. Smithson recites the nursery rhyme in his discussion of the problems of mining, and he notes Humpty Dumpty is a "nice succinct" definition of entropy (*Collected Writings*, 301).

44. Smithson, *Collected Writings*, 237. Paul Toner and Robert Smithson edited the interview in 1970.

45. Although she would not share Smithson's position on ecology, Rachel Carson's interest in land at the waterline does share his explorations of primitiveness and change. In *The Edge of the Sea* (New York: Houghton Mifflin, 1955), Carson writes: "Like the sea itself, the shore fascinates us who return to it, the place of our dim ancestral beginnings. . . . In the recurrent rhythms . . . and varied life of the tide lines there is the obvious attraction of movement and change and beauty. There is also, I am convinced, a deeper fascination born of inner meaning and significance. When we go down to the low-tide line, we enter a world that is as old as the earth itself—the primeval meeting place of the elements of earth and water, a place of compromise and conflict and eternal change." (vii) As a marine biologist studying the delicate balance of flora and fauna on oceanic islands, Carson saw the "problem of the islands" as an urgent ecological question. See "Lost Worlds: The Challenge of the Islands," *The Wood Thrush* (May – June 1949): 4. Islands, like humans, are particularly vulnerable to change. Despite their different perspectives, Smithson and Carson both wanted to point out the hazards of becoming immersed in what Smithson called the "technological miracles." Carson wrote: "Mankind has gone very far into an artificial world of his own creation. He has sought to insulate himself, with steel and concrete, from the realities of earth and water." See her 1954 essay "The Real World Around Us" in *Environment: An Interdisciplinary Anthology*, ed. Glenn Adelson (New Haven: Yale University Press, 2008), 559.

46. Smithson, 236.

47. Smithson used the term "denatured" in parallel with "dedifferentiation" and "decreation" in his conversation with Wheeler (132).

48. Writing about *Spiral Jetty*, Smithson described a dialectics of site and nonsite that "whirled into an indeterminate state, where solid and liquid lost themselves in each other." (146)

49. In February 1970, Smithson's made inquiries about purchasing islands in Chesapeake Bay and along the Texas Gulf coast. See Ann Reynolds, *Robert Smithson: Learning from New Jersey and Elsewhere* (Cambridge: MIT Press, 2004), 289.

50. See my earlier discussions of the 1970 and 2005 barge projects. Rendered similarly to the box-like barge of Manhattan's *Floating Island* and *Barge of Sulphur Passing Through Panama Canal From East To West* (1970), a small floating vessel appears in the *Island Project* drawing of 1970. Smithson's project for the *Floating Island* can also be compared with Stanley Crawford's 1972 novel *The Log of the S.S. the Mrs. Unguentine*, in which a couple repurpose a barge to begin a life on the sea (Chicago: Dalkey Archive Press, 2008). Just as Smithson's monument resituates a liberated, but now unoccupiable, Central Park, Crawford's fictional vessel displaces home for his Unguentine couple and in the process translates the wasteland to a Garden of Eden, which has now become a space of incarceration rather than expulsion. In these appropriations of the barge space, the industrial site of waste becomes an adapted paradise, one

that is mobilized by desires for the return to nature. Spoil islands affix these floating conditions of waste, nature, and culture.

51. Artist Lynne Hull has designed floating wildlife sanctuaries such as the Bird Barge and the Polar Platform. See her website http://eco-art.org and Floating Island International's more commercial proposals http://www.floatingislandinternational.com. But many other floating island projects operate in a seclusion much like the isolated consciousness modeled by artist Sally Moore's *Glass Island* (2004). British artist Richard Sowa's two Spiral Island projects raise ecological questions about island-making, waste-making, and recycling. Before it was destroyed in 2005 by Hurricane Emily, his first project floated on more than a quarter of a million plastic bottles held together by netting. It benefited from a location south of Cancún, Mexico, near a machine-dredged canal where dispersed sand helped provide anchorage. His second Spiral Island has become a tourist attraction and has been classified by the Mexican government as an "eco-boat." In a more transitory project, David de Rothschild and a crew of six sailed across the Pacific Ocean on *Plastiki*, a catamaran made of recycled plastic bottles, in order to draw attention to marine pollution like the gyres of plastic refuse known as "garbage patches." Still other floating islands thematize urban experience. The Gulf Coast gambling boats of Biloxi or New Orleans seek offshore separation from mainland rules and regulations; and on the Han River in Seoul, Korea, three floating islands contain a cultural center with more than two hundred thousand square feet of space, where events related to performance, water sports, and aquatic events will serve as a tourist attraction.

52. See William Cronon, "The Trouble with Wilderness; or, Getting Back to the Wrong Nature," ed. William Cronon, *Uncommon Ground: Rethinking the Human Place in Nature* (New York: W.W. Norton, 1995), 86.

53. Edward Casey, *Getting Back into Place* (Bloomington: Indiana University Press, 1993), 154. Casey writes: "Gardens are forms of building that answer to this circumstance. They are built, or perhaps we should say contrived, places and yet one largely composed of natural things." Spoil islands, like gardens, are "kinds of building" that correspond to the experience of determining how and what to build when immersed in what is ostensibly the natural world. Casey continues: "When I stand in a garden, I find myself in a scene intermediate between the completely constructed and the frankly wild." (154)

54. Henry Beston, *Herbs and the Earth* (Boston: David R. Godine, 1990), 6–7.

55. Smithson, 160.

56. The Archimedean Spiral is a curve that can be readily expressed as a formula with polar coordinates.

57. Eva Schmidt, who edited and annotated the section "Four conversations between Dennis Wheeler and Robert Smithson" in *Robert Smithson: The Collected Writings*, notes the centrality of *Island of Broken Glass* in their dialogue (196–233) and in the drawing "A Surd View for an Afternoon" (198). In his essay for *Robert Smithson in Vancouver: A Fragment of a Greater Fragmentation* (Vancouver: Vancouver Art Gallery, 2004), Arnold includes another drawing with notes and diagrams from the first conversation between Smithson and Wheeler (13). In 1970, Smithson generated a similar map on lined paper during his interview with Paul Toner in 1970. See Smithson, *Collected Writings*, 237.

58. Tools of dredging and excavating also a played a part in Smithson's construction of his projects. For *Spiral Jetty*, he used draglines to build dikes in Salt Lake, and his drawings for *Spiral Hill* and *Broken Circle* illustrate how drawing scripts the dredging exercise.

59. Smithson, 145. I found similar resonances in the Indian River, where spoil islands recall the earlier histories of the middens.

60. Other proposals for human-made archipelagoes have sought purpose-built solutions to natural events and territorial expansion. After the Deepwater Horizon oil spill in 2010, Louisiana governor Bobby Jindal called for the construction of what he called "sand berms" off the coast of his state. The governor lobbied for these artificial barrier islands as "nature's best defense" against the effects of hurricanes and erosion on low-lying lands. After the spill, Jindal further politicized the artificial barrier islands (and their spoil materials) when he put them forward as more effective, though more costly, alternatives to the floating containment booms that slow the movement of oil. "Fake island" became an operative term for Jindal's berms and perhaps unwittingly referenced the human-made islands created by oil companies in offshore

territories where fewer regulations apply for oil extraction and processing. Another purpose-built archipelago, outside the scope of this book but significant to the broader meaning of artificial islands, is Dubai's Palm Islands and World Islands. In one sense, these projects are the "technological miracles" Smithson ironically identifies, but they have proven hazardous to navigational systems and vulnerable to economic downturns. It was reported in 2011 that the World Islands were sinking and, in March 2012, that the project would not proceed. Whether as a part of land-making projects or channel-modification, dredging will remain a critical factor in landscapes influenced by economic and political transformations; and it will be equally critical to question every aspect of dredging and its potential land production. Increasingly larger amounts of dredged material will result from the deeper channels necessary to accommodate the deeper-draft vessels for the global transport of goods and tourism. With its final permits approved late in May 2012, the Port of Miami's expansion project known as "Deep Dredge" will increase depths to fifty feet with an estimated cost of two billion dollars.

61. Smithson, *Collected Writings*, 56.

62. Robert Smithson, interview with Gregoire Müller, ". . . The Earth, Subject to Cataclysms, Is a Cruel Master," *Collected Writings*, 261.

63. For a third island, he anchored drifting mangrove seedlings in a circle with a one-hundred-foot diameter. Smithson's May 1971 drawing of *Mangrove Ring* shows the perfect circle's expected growth—in its first, second, and third years. About this project, he notes that there is "something about water that quickens my motivation" (261).The serial and circular aspects of this project parallel the drawing *9 Islands in Circular Pond* (1971).

64. Robert Smithson, "Rejoinder to Environmental Critics," *Collapse*, no.2: 124 (n.d.), cited in Grant Arnold, ed., *Robert Smithson in Vancouver: A Fragment of a Greater Fragment ation* (Vancouver: Vancouver Art Gallery, 2004), 25.

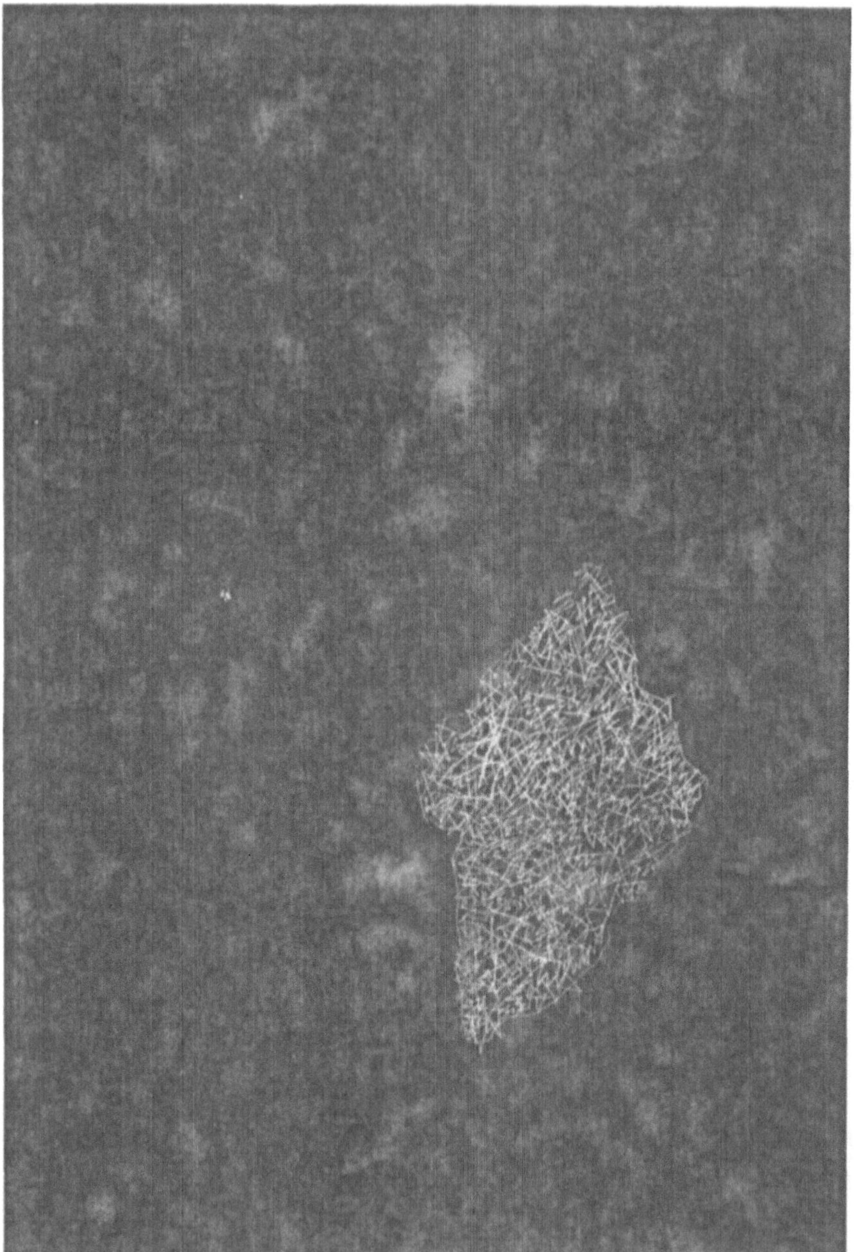

Select Bibliography

Ackroyd, Peter. *Venice: Pure City*. New York: Doubleday, 2010.

Arnold, Grant, ed., *Robert Smithson in Vancouver: A Fragment of a Greater Fragmentation*. Vancouver: Vancouver Art Gallery, 2004.

Baedeker, Karl. *The United States: Handbook for Travelers*. New York: Charles Scribner's Sons, 1904.

Bataille, Georges, ed., "Critical Dictionary," *Encyclopaedia Acéphalica: Comprising the Critical Dictionary and Related Texts*. Edited by Robert Lebel and Isabelle Waldberg. Trans. Iain White. London: Atlas Press, 1995.

Beston, Henry. *Herbs and the Earth*. Boston: David R. Godine, 1990.

Bingham, June. *U Thant: The Search for Peace*. New York: Alfred A. Knopf, 1966.

Brinker, Helmut, and Hiroshi Kanazawa, and Andreas Leisinger. *Zen Masters of Meditation in Images and Writings*. Honolulu: University of Hawai'i Press, 1996.

Browne, Jefferson B. *Key West: The Old and the New*. St. Augustine: The Record Company, 1912.

Brown-Peterson, Nancy and Ross W. Eames. *Indian River Lagoon Spoil Island Management Plan*. Bureau of Aquatic Preserves, Division of State Lands, Florida Department of Natural Resources, 1990.

Caro, Robert. *The Power Broker*. New York: Vintage, 1975.

Carson, Rachel. *The Edge of the Sea*. New York: Houghton Mifflin, 1955.

Casey, Edward. *Getting Back into Place*. Bloomington: Indiana University Press, 1993.

Chernow, Burt and Wolfgang Volz. *Christo and Jeanne-Claude: A Biography*. New York: St. Martin's Press, 2002.

Chinmoy, Sri. *My Meditation-Service at the United Nations for 25 Years*. Jamaica, NY: Agni Press, 1995.

Christo. *Christo: Surrounded Islands*. Trans. Stephen Reader. New York: Harry Abrams, 1985.

Cornish, Vaughan. *Waves of Sand and Snow and the Eddies Which Make Them*. London: T.F. Unwin, 1914.

Crawford, Stanley. *The Log of the S.S. the Mrs. Unguentine*. Chicago: Dalkey Archive Press, 2008.

Crawford, William G. *Florida's Big Dig: The Atlantic Intracoastal Waterway*. Cocoa, FL: Florida Historical Society Press, 2007.

Cronon, William. *Uncommon Ground: Rethinking the Human Place in Nature*. New York: W.W. Norton, 1995.

Deleuze, Gilles. *Desert Islands and Other Texts*. Los Angeles: Semiotext(e), 2004.

Douglas, Mary. *Purity and Danger*. New York: Routledge Classics, 2002.

Eliot, T. S. "The Waste Land." *The Complete Poems and Plays*. New York: Harcourt, Brace, and World, Inc., 1980.
Fisher, Jane. *Fabulous Hoosier: A Story of American Achievement*. New York: Robert M. McBride and Co., 1947.
Fuller, R. Buckminster. *Your Private Sky*. Edited by Joachim Krausse and Claude Lichtenstein. Baden: Lars Müller Publishers, 1999.
Gandy, Matthew. *Concrete and Clay: Reworking Nature in New York City*. Cambridge: MIT Press, 2003.
Gee, Sophie. *Making Waste: Leftovers and the Eighteenth-Century Imagination*. Princeton: Princeton University Press, 2010.
Gerozissis, Christoph. *The Florida Landscape: Revisited*. Lakeland, FL: Polk Museum of Art, 1992.
Gilpin, William. *Three Essays on Picturesque Beauty, on Picturesque Travel, and on Sketching Landscape*. London: R. Blamire, 1792.
Hailey, Charlie. *Camps: A Guide to 21st-Century Space*. Cambridge, MIT Press, 2009.
Hailey, Charlie. *Campsite: Architectures of Duration and Place*. Baton Rouge: Louisiana State University Press, 2008.
Hayden, Dolores, ed. *Preserving Cultural Landscapes in America*. Baltimore: Johns Hopkins University Press, 2000.
Heaney, Seamus. *Sweeney Astray*. London: Faber and Faber, 2001.
Hood, Clifton. *722 Miles: The Building of the Subways and How They Transformed New York*. Baltimore: Johns Hopkins University Press, 2004.
Huston, John. *Hydraulic Dredging*. Cambridge, MD: Cornell Maritime Press, 1970.
Jackson, John Brinckerhoff. *Discovering the Vernacular Landscape*. New Haven: Yale University Press, 1984.
Kleinberg, Howard. *Miami Beach: A History*. Miami: Centennial Press, 1994.
Kuitert, Wybe. *Themes in the History of Japanese Garden Art*. Honolulu: University of Hawai'i Press, 2002.
Lawrence, D. H. "The Man Who Loved Islands." *Dial* 83, no.1 (1927): 1–25.
Mabey, Richard. *Weeds: In Defense of Nature's Most Unloved Plants*. New York: Harper Collins, 2010.
Macfarlane, Robert. *The Wild Places*. London: Penguin, 2008.
Miller, J. Hillis. *Topographies*. Stanford, CA: Stanford University Press, 1995.
Morton, Timothy. *Ecology without Nature: Rethinking Environmental Aesthetics*. Cambridge: Harvard University Press, 2007.
Moses, Robert. *Public Works: A Dangerous Trade*. New York: McGraw Hill, 1970.
Muir, John. *A Thousand-Mile Walk to the Gulf*. New York: Houghton Mifflin Co., 1916.
Munari, Bruno. *From Afar It Is an Island*. Cleveland, OH: World Publishing Company, 1971.
National Park Service. *General Management Plan: Gateway National Recreation Area, New York / New Jersey*. U.S. Department of the Interior, August 1979.
Neville, Brian and Johanne Villeneuve, eds. *Waste-Site Stories*. Albany: State University of New York Press, 2002.
Nietzsche, Friedrich. *The Will To Power*. Trans. Walter Kaufman and R. J. Hollingdale. New York: Vintage, 1968.
Noll, Steven and David Tegeder. *Ditch of Dreams*. Gainesville, FL: University Press of Florida, 2009.
Office of Coastal and Aquatic Managed Areas. *Indian River Lagoon State Buffer Preserve Land Management Plan*. Tallahassee, FL: Florida Department of Environmental Protection, 2004.
Olson, Charles. *Call Me Ishmael*. San Francisco: City Lights Books, Inc., 1947.
Plato. *Statesman*. Edited by J. Annas and R. Waterfield. Trans. Robin Waterfield. Cambridge: Cambridge University Press, 1995.
Powell, Richard. *Pioneer, Go Home!* New York: Charles Scribner's Sons, 1959.
Redford, Polly. *Billion-Dollar Sandbar: A Biography of Henry Flagler*. New York: Dutton, 1970.

Sanderson, Eric. *Manhatta: A Natural History of New York* City. New York: Harry Abrams, 2009.

Schama, Simon. *Landscape and Memory*. New York: Alfred A. Knopf, 1995.

Seitz, Sharon and Stuart Miller. *The Other Islands of New York City: A History and Guide*. Woodstock, VT: Countryman Press, 2011.

Serres, Michel. *The Natural Contract*. Trans. Elizabeth MacArthur and William Paulson. Ann Arbor: University of Michigan Press, 1995.

Shakespeare, William. *The Tempest*. New York: Simon and Schuster, 2004.

Sloterdijk, Peter. *Critique of Cynical Reason*. Trans. Michael Eldred. Minneapolis: University of Minnesota Press, 1987.

Smithson, Robert. *Robert Smithson*. Edited by Eugenie Tsai. Berkeley: University of California Press, 2004.

Smithson, Robert. *Robert Smithson: The Collected Writings*. Berkeley: University of California Press, 1996.

Snyder, Gary. *The Practice of the Wild*. San Francisco: North Point Press, 1990.

Stevens, Wallace. "The Idea of Order at Key West." *The Collected Poems of Wallace Stevens*. New York: Vintage, 1990.

Tafuri, Manfredo. *Venice and the Renaissance*. Trans. Jessica Levine. Cambridge: MIT Press, 1995.

U Thant. *View from the UN*. New York: Doubleday, 1978.

Thompson, Michael. *Rubbish Theory: The Creation and Destruction of Value*. Oxford: Oxford University Press, 1979.

Thoreau, Henry David. *Walking*. New York: HarperCollins, 1994.

U.S. Army Corps of Engineers. *Peanut Island, Palm Beach County, Florida: Final Environmental Restoration Report and Final Environmental Assessment*, Environmental Restoration, Section 1135, Jacksonville District, March 2000.

Van Dyke, John C. *The New New York*. New York: Macmillan Company, 1909.

Wapner, Paul. *Living Through the End of Nature*. Cambridge: MIT Press, 2010.

Ward, Colin and Dennis Hardy. *Arcadia for All: The Legacy of a Makeshift Landscape*. Nottingham: Five Leaves Publications, 2003.

Young, Claiborne S. *Cruising Guide to Eastern Florida*. Gretna, LA: Pelican Publishing Company, 1996.

Index

About the Author

Charlie Hailey teaches design, theory, and history at the University of Florida's School of Architecture, where he received his doctorate. A licensed architect, he also studied at Princeton University and the University of Texas and has worked with the designer-builders Jersey Devil in Florida and California. Previous work examines camping as placemaking (*Campsite: Architectures of Duration and Place*) and investigates how the contemporary spaces of camps register cultural and political forces (*Camps: A Guide to 21st-Century Space*).

www.ingramcontent.com/pod-product-compliance
Lightning Source LLC
LaVergne TN
LVHW091334280926
848997LV00053B/9
* 9 7 8 0 7 3 9 1 7 3 0 6 0 *